Topics in Mining, Metallurgy and Materials Engineering

“Topics in Mining, Metallurgy and Materials Engineering” welcomes manuscripts in these three main focus areas: Extractive Metallurgy/Mineral Technology; Manufacturing Processes, and Materials Science and Technology. Manuscripts should present scientific solutions for technological problems. The three focus areas have a vertically lined multidisciplinarity, starting from mineral assets, their extraction and processing, their transformation into materials useful for the society, and their interaction with the environment.

**** Indexed by Scopus (2020) ****

More information about this series at http://www.springer.com/series/11054

Marcel Antonio Arcari Bassani ·
João Felipe Coimbra Leite Costa

Geostatistics with Data of Different Support Applied to Mining Engineering

Springer

Marcel Antonio Arcari Bassani
Mineral Exploration and Mining Planning Laboratory—LPM
Department of Mining Engineering
Federal University of Rio Grande do Sul
Porto Alegre, Rio Grande do Sul, Brazil

João Felipe Coimbra Leite Costa
Mineral Exploration and Mining Planning Laboratory—LPM
Department of Mining Engineering
Federal University of Rio Grande do Sul
Porto Alegre, Rio Grande do Sul, Brazil

ISSN 2364-3293 ISSN 2364-3307 (electronic)
Topics in Mining, Metallurgy and Materials Engineering
ISBN 978-3-030-80192-2 ISBN 978-3-030-80193-9 (eBook)
https://doi.org/10.1007/978-3-030-80193-9

This Springer imprint is published by the registered company Springer Nature Switzerland AG
The registered company address is: Gewerbestrasse 11, 6330 Cham, Switzerland

Marcel Antonio Arcari Bassani dedicates the book to his beloved wife Waleska.

João Felipe Coimbra Leite Costa dedicates to Luciana and Juan, his precious jewels.

Preface

This book aims to bring some new insights in using data of different support for mining applications. The integration of data of different support in Geostatistics is not a novelty in the geostatistical literature. However, most textbooks require a solid mathematical background, which is sometimes a barrier for many readers. In addition, many case studies about this topic shown in journal papers and conference proceedings are related to petroleum and remote sensing areas. In this book, we tried to keep the theory minimum and focus on case studies relevant to Mining Engineering.

The book is appropriate for professional engineers and geologists who work in the area of mineral resource estimation. Moreover, we believe most graduate and undergraduate students who have recently started studying Geostatistics will significantly benefit from the content presented.

The book is organized in eight chapters as follows:

- Chapter 1 provides an introduction to the topic of data of different support and reviews previous works;
- Chapter 2 discusses solutions to integrate data of different support into Geostatistics using kriging;
- Chapter 3 details the methods to perform geostatistical simulations with data of different support;
- Chapter 4 presents the change of support using the variogram for variogram regularization and deconvolution;
- Chapter 5 illustrates the solution of integrating data of different support in kriging in a case study;
- Chapter 6 presents a case study applied to continuous variables using Direct Sequential Simulation with data of different support;
- Chapter 7 is a case study of sequential indicator simulation with categorical data of different support;
- Chapter 8 closes with some conclusions.

The research presented in this book was developed in the Mineral Exploration and Mining Planning Laboratory (LPM) of the Federal University of Rio Grande do Sul (UFRGS). In this regard, the authors are grateful to all the LPM students and industry affiliates. Moreover, the LPM has benefited from extensive interaction with other geostatistical centers, such as those from the Stanford University, University of Alberta (UofA), and Instituto Superior Técnico (IST). Remarkably, the authors acknowledge the positive influence that Professors Clayton Deutsch (UofA), Jef Caers (Stanford), Alexandre Boucher (AR2Tech), Amílcar Soares (IST), and Leonardo Azevedo (IST) had on the LPM research group. In summary, all the people who have frequented the LPM made the laboratory a thriving research environment.

Porto Alegre, Brazil

Marcel Antonio Arcari Bassani
João Felipe Coimbra Leite Costa

Contents

Abbreviations

BSSIM	Block Sequential Simulation
ccdf	Conditional cumulative distribution function
cdf	Cumulative distribution function
DDH	Diamond drill holes
DSS	Direct Sequential Simulation
LHS	Left-hand-side
LMC	Linear Model of Coregionalization
REC14	Proportion of mass retained at 14 # sieve aperture
RHS	Right-hand-side
SGS	Sequential Gaussian Simulation
SIS	Sequential Indicator Simulation
SMU	Selective mining unity
TBS	Turning Bands Simulation

Chapter 1
Introduction

1.1 Data of Different Support

Geostatistical techniques are extensively used for mineral resources assessment, including grade estimation and uncertainty analysis. Support is the term used in Geostatistics to describe the size or volume of a sample. For example, suppose a grade sample obtained from diamond drill holes (DDH). In this case, the support is a function of the core's radius and length.

In Mining Engineering applications, data that have different support are often found. For instance, many mining operations contain datasets collected in different years, with different sampling protocols and core lengths. Another example is production data, including averaged grades over a mined out area. When the data have different support, this difference should be considered to build grade models.

The use of data of different support is well established in the petroleum industry. The petroleum industry routinely combines information from wells and seismic surveys. The wells have a vertical resolution of decimeters, whereas the seismic data have a vertical resolution of ten meters or more (Pyrcz and Deutsch 2014). The seismic data usually consist of a coarse grid where all the grid cells have the same size. Moreover, the seismic data often cover the entire area of interest.

This scenario rarely occurs in Mining Engineering applications. The data of different support are usually scattered in the area of interest. Moreover, the samples usually have different sizes. For instance, consider a dataset with data obtained from ten different sampling campaigns that used distinct sampling protocols. The samples are scattered over the area of interest and probably have different lengths.

One alternative to deal with samples of different lengths is to transform the original 3D dataset into a 2D dataset and work with the variable accumulation (Journel and Huijbregts 1978; Krige 1981; Bertoli et al. 2003; Marques et al. 2014). Accumulation is the product of grade and thickness. The problem is that this approach disregards the vertical mining selectivity.

Another alternative consists of compositing the samples and estimate using the composites (Rossi and Deutsch 2014). Compositing involves combining adjacent

M. A. Arcari Bassani and J. F. Coimbra Leite Costa, *Geostatistics with Data of Different Support Applied to Mining Engineering*, Topics in Mining, Metallurgy and Materials Engineering, https://doi.org/10.1007/978-3-030-80193-9_1

samples along one drill hole and calculating their average according to a specific length. However, compositing is a challenge when the length of the samples is highly variable. Compositing to a long length smooths the original grades, which may mask important short-scale geological structures. In contrast, compositing to a short length usually breaks the initial sample into several pieces with the same grade. The latter leads to an incorrect result and artificially diminishes the short-scale variability.

1.2 Previous Works

Many geostatistical books (Journel and Huijbregts 1978; Isaaks and Srivastava 1989; Goovaerts 1997) show how the kriging system may accommodate data of different support. In simple terms, the relationship between the data in kriging is obtained by covariances. The data representing a volume in the space are defined as "block data" (even though the block may have any shape), while the data representing a point in the space are called point data. The relationships between these two types of data are defined in the kriging system by point-to-block and block-to-block covariances.

The point-to-block covariances are primarily used for the block kriging algorithm (Journel and Huijbregts 1978; Isaaks and Srivastava 1989; Goovaerts 1997). Block kriging consists of estimating the average value of a volume using point samples. The kriging system is modified, so that the right-hand-side matrix uses point-to-block covariances. The result is that the average grade over a volume is estimated directly with one system of equations. Journel and Huijbregts (1978) proved that the block kriging estimate equals the linear average of point kriging estimates over the block volume. In summary, the point-to-block and block-to-block covariances preserve the linear relationship between the data of different support. Point-to-block and block-to-block covariances are linear average covariances calculated over points that discretize the blocks or volumes of interest.

When the data have different support and precision, Deutsch et al. (1996) affirm that this difference must be considered. In this context, the block cokriging approach is the most rigorous method to integrate these two sources of information. Similar to block kriging, block cokriging also uses average covariances to consider the difference in support between the data.

The main difference between block kriging and cokriging is that block kriging calculates all the covariances using the one variogram model, which refers to the variable of interest. This approach is the proper method when the data differ in support but not in precision. In contrast, the block cokriging approach uses the Linear Model of Coregionalization (MLC—Goovaerts 1997) to calculate the covariances. As a result, the average covariances would be calculated through the cross-variogram model. This book was focused on the difference in support between the samples, so block kriging was mainly used for the case studies.

Behrens et al. (1998) proposed the Sequential Gaussian Simulation (SGS—Isaaks 1990) with block kriging to incorporate the seismic data. The drawback is that the Sequential Gaussian Simulation uses a nonlinear transformation. The result is that the linear relationship between the data of different support is lost, consequently the block kriging approach with average covariances is inadequate.

Deutsch (2002) recommends avoiding nonlinear transformations when dealing with data of different support. The use of nonlinear transformations is a limitation for the Gaussian simulation algorithms such as Sequential Gaussian Simulation (SGS) and Turning Bands Simulation (TBS—Journel and Huijbregts 1978). These methods transform the original data into Gaussian values. In this context, the Direct Sequential Simulation (DSS—Journel 1994; Soares 2001) stands out as it uses the data directly.

Tran et al. (2001) used DSS with average covariances to integrate data of different support. As DSS does not use nonlinear transformations, the linear relationship between the data of different support is kept. The example shown by the authors combined data from wells, seismic surveys, and production tests. Yao and Journel (2000) also used DSS to build porosity models in an oil reservoir. The seismic information was incorporated by block cokriging.

The DSS that considers data of different support was implemented by Hansen and Mosegaard (2008) in the VISIM software. The software allowed use of volumetric data of irregular shapes for geostatistical simulations. A complete software package to integrate data of different support, called Bgeostats, was published by Liu and Journel (2009). The Bgeostats consisted of a series of plug-ins for SGeMS (Remy et al. 2008) and included algorithms for estimation and simulation.

The use of average covariances to integrate data of different support may also be used for categorical variables. In this context, Yao and Chopra (2000) utilized Sequential Indicator Simulation (SIS) with block cokriging to incorporate the lithology proportion derived from seismic to build simulated lithology models in a reservoir. The use of average covariances for simulation of categorical variables was implemented in the BlockSIS software (Deutsch 2006).

References

Behrens RA, MacLeod MK, Tran TT, Alimi AC (1998) Incorporating seismic attribute maps in 3D reservoir models. SPE Reservoir Eval Eng 1(02):122–126

Bertoli O, Vann J, Dunham S (2003) Two-dimensional geostatistical methods—theory, practice and a case study from the 1A shoot nickel deposit, Leinster, Western Australia. In: 5th International mining geology conference

Deutsch CV (2002) Geostatistical reservoir modeling. Oxford University Press, New York

Deutsch CV (2006) A sequential indicator simulation program for categorical variables with point and block data: BlockSIS. Comput Geosci 32(10):1669–1681

Deutsch CV, Srinivasan S, Mo Y (1996) Geostatistical reservoir modeling accounting for precision and scale of seismic data. In: SPE annual technical conference and exhibition. Society of Petroleum Engineers. https://doi.org/10.2118/36497-MS

Goovaerts P (1997) Geostatistics for natural resource evaluation. Oxford University Press, Oxford

Hansen TM, Mosegaard K (2008) VISIM: sequential simulation for linear inverse problems. Comput Geosci 34(1):53–76

Isaaks EH (1990) The application of Monte Carlo methods to the analysis of spatially correlated data. PhD Thesis, Stanford University

Isaaks HE, Srivastava MR (1989) An introduction to applied geostatistics. Oxford University Press, Oxford

Journel AG (1994) Modeling uncertainty: some conceptual thoughts. Geostatistics for the next century. Springer, Dordrecht, pp 30–43

Journel AG, Huijbregts CJ (1978) Mining geostatistics. Academic Press, New York

Krige DG, Krige DG (1981) Lognormal-de Wijsian geostatistics for ore evaluation. South African Institute of mining and metallurgy, Johannesburg

Liu Y, Journel AG (2009) A package for geostatistical integration of coarse and fine scale data. Comput Geosci 35(3):527–547

Marques DM, Rubio RH, Costa JFCL, Silva EMAD (2014) The effect of accumulation in 2D estimates in phosphatic ore. REM: Revista Escola de Minas 67(4):431–437

Pyrcz MJ, Deutsch CV (2014) Geostatistical reservoir modeling. Oxford University Press, New York

Remy N, Boucher A, Wu J (2008) Applied geostatistics with SGeMS: a user's guide. Cambridge University Press, New York

Rossi ME, Deutsch CV (2014) Mineral resource estimation. Springer, Berlin

Soares A (2001) Direct sequential simulation and cosimulation. Math Geosci 33(8):911–926

Tran TT, Deutsch CV, Xie Y (2001) Direct geostatistical simulation with multiscale well, seismic, and production data. In: SPE annual technical conference and exhibition. Society of Petroleum Engineers. https://doi.org/10.2118/71323-MS

Yao T, Chopra A (2000) Integration of seismic attribute map into 3D facies modeling. J Petrol Sci Eng 27(1–2):69–84

Yao T, Journel AG (2000) Integrating seismic attribute maps and well logs for porosity modeling in a west Texas carbonate reservoir: addressing the scale and precision problem. J Petrol Sci Eng 28(1–2):65–79

Chapter 2
Kriging with Data of Different Support

Support is the term used in the geostatistical literature to describe the volume or size of the data. For instance, the average grade of a mined pushback refers to a volume/support much larger than a sample that represents the average grade along one meter of a diamond drill hole. This difference in support must be considered if these two sources of information are used to build grade models.

This chapter reviews some fundamentals of kriging and details the use of data of different support for geostatistical applications. The use of average covariances to integrate data of different support is highlighted. An illustrative example is provided, and the effects of the data support in the estimates are discussed.

2.1 Overview of Kriging

Kriging was established for the problem of estimating a continuous attribute z at location $\mathbf{u}$ using a linear combination of the n surrounding samples $\{z(\mathbf{u}_i), i = 1, \ldots, n\}$. All types of kriging are variants of the general kriging estimator (Eq. 2.1):

$$z^*(\mathbf{u}) - m(\mathbf{u}) = \sum_{i=1}^{n} \lambda_i \cdot [z(\mathbf{u}_i - m(\mathbf{u}_i)], \tag{2.1}$$

where λ_i is the weight of the datum $z(\mathbf{u}_i)$. The values $m(\mathbf{u})$ and $m(\mathbf{u}_i)$ are the means of the attribute at locations $\mathbf{u}$ and $\mathbf{u}_i$, respectively. These means are called as trend components. Trends that are known should be treated deterministically and, thus, the weights are applied to the residuals, which are the difference between the attribute considered and its mean. What differs the variants of kriging is the way these trend components are considered.

The mathematical formulation of the kriging equations is well explained in classic geostatistical books (Journel and Huijbregts 1978; Isaaks and Srivastava 1989; Deutsch and Journel 1998; Goovaerts 1997). In this book, we quickly review the

M. A. Arcari Bassani and J. F. Coimbra Leite Costa, *Geostatistics with Data of Different Support Applied to Mining Engineering*, Topics in Mining, Metallurgy and Materials Engineering, https://doi.org/10.1007/978-3-030-80193-9_2

simple and ordinary kriging techniques and emphasize the use of data of different support in Geostatistics. Simple and ordinary kriging are the most used kriging algorithms in the mining industry. Simple kriging is widely used for simulation studies, while ordinary kriging is the main estimation technique for mineral resource estimation.

2.2 Simple Kriging

Simple kriging assumes that the mean m is known and constant over the entire area of interest. The simple kriging estimate is a linear combination of the samples and the mean m (Eq. 2.2):

$$z^*(\mathbf{u}) = \sum_{i=1}^{n} \lambda_i \cdot [z(\mathbf{u}_i) - m] + m$$
$$z^*(\mathbf{u}) = \sum_{i=1}^{n} \lambda_i \cdot z(\mathbf{u}_i) + \left[1 - \sum_{i=1}^{n} \lambda_i\right] \cdot m, \tag{2.2}$$

where $z^*(\mathbf{u})$ is the estimate, λ_i are the simple kriging weights, and $z(\mathbf{u}_i)$ are the data values.

The simple kriging weights are chosen, so that the error variance is minimized. The error variance is expressed using the random variables $Z^*(\mathbf{u})$ and $Z(\mathbf{u})$, which represent the estimated and true values at location $\mathbf{u}$. A detailed explanation of random variables is provided by Isaaks and Srivastava (1989). Equation 2.3 defines the error variance:

$$\sigma_E^2(\mathbf{u}) = \mathrm{Var}\{\ Z^*(\mathbf{u}) - Z(\mathbf{u})\}$$
$$\sigma_E^2(\mathbf{u}) = \mathrm{Var}\{\ Z^*(\mathbf{u})\} - 2\mathrm{Cov}\{Z^*(\mathbf{u}), Z(\mathbf{u})\} + \mathrm{Var}\{Z(\mathbf{u})\}, \tag{2.3}$$

Considering that the estimate $Z^*(\mathbf{u})$ is a linear combination of the data values (Eq. 2.2), the error variance may be expressed as a function of the estimation weights and covariances (Eq. 2.4):

$$\sigma_E^2(\mathbf{u}) = \sum_{i=1}^{n} \sum_{j=1}^{n} \lambda_i \cdot \lambda_j \cdot C(\mathbf{u}_i, \mathbf{u}_j) - 2 \cdot \sum_{i=1}^{n} \lambda_i \cdot C(\mathbf{u}_i, \mathbf{u}) + C(0), \tag{2.4}$$

where λ_i are the estimation weights, $C(\mathbf{u}_i, \mathbf{u}_j)$ is the covariance between the data located at $\mathbf{u}_i$ and $\mathbf{u}_j$, $C(\mathbf{u}_i, \mathbf{u})$ is the covariance between the datum located at $\mathbf{u}_i$ and the point $\mathbf{u}$ to be estimated, and $C(0)$ is the variance of the data. The terms $C(0)$, $C(\mathbf{u}_i, \mathbf{u})$, and $C(\mathbf{u}_i, \mathbf{u}_j)$ are calculated from a covariance model, which is usually built using the Linear Model of Regionalization (Journel and Huijbrets 1978; Goovaerts

1997). The weights that minimize the error variance are obtained by taking the partial first derivatives of Eq. 2.4 in relation to the weights and setting them to zero (Eq. 2.5):

$$\frac{\partial \sigma_E^2(\mathbf{u})}{\partial \lambda_i} = \sum_{i=1}^{n} \lambda_i \cdot C(\mathbf{u}_i, \mathbf{u}_j) - C(\mathbf{u}_i, \mathbf{u}) = 0$$
$$j = 1, \ldots, n, \tag{2.5}$$

which results in the simple kriging system (Eq. 2.6):

$$\sum_{i=1}^{n} \lambda_i \cdot C(\mathbf{u}_i, \mathbf{u}_j) = C(\mathbf{u}_i, \mathbf{u}) \quad j = 1, \ldots, n, \tag{2.6}$$

The substitution of Eq. 2.6 into the definition of the error variance (Eq. 2.4) results in the simple kriging variance (Eq. 2.7):

$$\sigma_E^2(\mathbf{u}) = C(0) - \sum_{i=1}^{n} \lambda_i \cdot C(\mathbf{u}_i, \mathbf{u}), \tag{2.7}$$

The simple kriging variance depends on the covariance model and the distance between the samples and the point $\mathbf{u}$ to be estimated. In general, as the samples get closer to the point to be estimated, their covariances with the point to be estimate $\mathbf{u}$ ($C(\mathbf{u}_i, \mathbf{u})$) increase and their weights increase. The result is that the simple kriging variance decreases.

2.3 Ordinary Kriging

The simple kriging estimate relies on a strong stationarity assumption, as the mean is considered known and constant over the entire area of study. In practice, datasets in the Earth Sciences do not have a constant mean. A location map of the samples often reveals areas of higher and lower values.

Ordinary kriging considers local variations of the mean, as it assumes that the mean is constant only inside the local neighborhood centered at the location $\mathbf{u}$ to be estimated. The local mean is considered unknown. Ordinary kriging filters the influence of the mean by forcing the kriging weights assigned to the surrounding data to sum one. In this context, the ordinary kriging estimator is written as a linear combination of the samples (Eq. 2.8):

$$z^*(\mathbf{u}) = \sum_{i=1}^{n} \lambda_i \cdot z(\mathbf{u}_i)$$

$$\sum_{i=1}^{n} \lambda_i = 1, \tag{2.8}$$

The ordinary kriging weights, similar to the simple kriging weights, also minimize the variance of the estimation error. The main difference is that the Lagrange formulism is used to add the constraint that the sum of the weights must be one. The ordinary kriging system is defined as follows (Eq. 2.9):

$$\begin{aligned} &\sum_{i=1}^{n} \lambda_i \cdot C(\mathbf{u}_i, \mathbf{u}_j) + \mu = C(\mathbf{u}_i, \mathbf{u}) \quad j = 1, \ldots, n \\ &\sum_{i=1}^{n} \lambda_i = 1, \end{aligned} \tag{2.9}$$

where μ is the Lagrange parameter. The substitution of Eq. 2.9 in the definition of the error variance (Eq. 2.4) results in the ordinary kriging variance (Eq. 2.10):

$$\sigma_E^2(\mathbf{u}) = C(0) - \sum_{i=1}^{n} \lambda_i \cdot C(\mathbf{u}_i, \mathbf{u}) - \mu, \tag{2.10}$$

2.4 Matrix Notation of Kriging

The matrix notation of the kriging system helps to understand the influence of data of different support in kriging and is reviewed. Using matrix notation, both the simple and ordinary kriging systems may be described using three matrixes:

- The left-hand-side (LHS) matrix **C**, which contains the covariances between the data;
- The weights vector **w**, which contains the kriging weights;
- The right-hand-side (RHS) matrix, **D**, which contains the covariances between the data and the point to be estimated.

Equation 2.11 describes the kriging system using matrix notation:

$$\mathbf{C} \cdot \mathbf{w} = \mathbf{D}, \tag{2.11}$$

The kriging weights are obtained by the multiplication of the inverse of the LHS matrix **C** by the RHS matrix **D**.

2.5 Block Kriging

In many applications, it is interesting to estimate the average value of some attribute Z over a volume. For instance, in the mining industry, the decision whether the material is ore or waste is based on the average value of a block (also called *selective mining unity*—SMU) that has a volume in the space. The block size is related to the selectivity of the mining equipment and data spacing.

The term block kriging refers to the estimation of average values over a line, surface, or volume (Journel and Huijbregts 1978; Isaaks and Srivastava 1989; Goovaerts 1997). As long as the average is linear, the estimation of the average value over a block (or volume) may obtained as the average of N point estimates. These point estimates occur at the N points that discretize the block V (Fig. 2.1). Traditionally, block kriging considers that the samples are at a quasi-point support, since the support of data is far smaller than the support of the block to be estimated.

Equation 2.12 defines the block estimate $z_V^*(\mathbf{u})$:

$$z_V^*(\mathbf{u}) = \frac{1}{N}\sum_{i=1}^{N} z^*(\mathbf{u}_i^{'}), \tag{2.12}$$

where $z_V^*(\mathbf{u})$ is the estimate of attribute z over the block V centered at location $\mathbf{u}$, $z^*(\mathbf{u}_i^{'})$ is the kriging estimate of attribute z at the discretizing point $\mathbf{u}_i^{'}$, and N is number of points used to discretize the block V.

This approach is computationally expensive since it demands the resolution of N systems of linear equations to obtain a single block estimate. A more efficient approach is provided by the block kriging system, which obtains directly the block estimate by solving just one system of linear equations. Equation 2.13 defines the

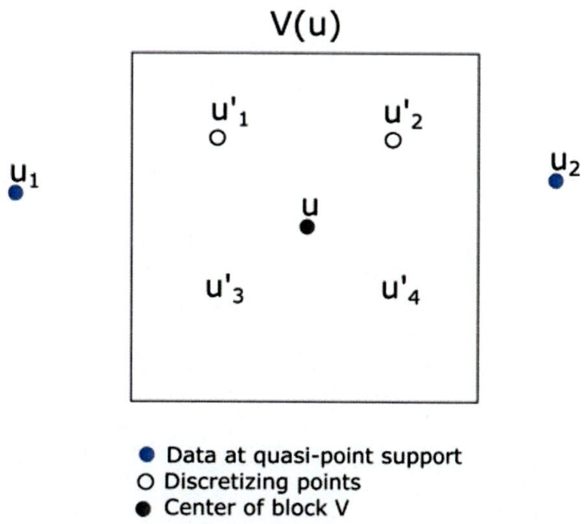

Fig. 2.1 Scheme of block V to be estimated with four discretizing points and two samples at quasi-point support

block kriging estimator:

$$z_V^*(\mathbf{u}) = \sum_{i=1}^{n} \lambda_i \cdot z(\mathbf{u}_i), \tag{2.13}$$

where λ_i refers to the weight related to the datum at location $\mathbf{u}_i$ and $z(\mathbf{u}_i)$ refers to the datum at location $\mathbf{u}_i$, and n is the number of data used for the estimation. The weights are obtained by the solving the block kriging system.

Similar to the traditional kriging system, the block kriging system contains three matrices: (i) the left-hand-side (LHS) matrix, which contains the covariances between the samples; (ii) the weight vector; and (iii) the right-hand side matrix, which contains the covariance between the samples and the block to be estimated. If the idea is to estimate a block (average value over a volume), the RHS matrix is modified to account for the volume to be estimated. In this case, the RHS matrix contains covariances point-to-block. Equation 2.14 defines the ordinary block kriging system:

$$\begin{aligned} &\sum_{i=1}^{n} \lambda_i \cdot C(\mathbf{u}_i, \mathbf{u}_j) + \mu = \overline{C}(\mathbf{u}_i, V) \quad j = 1, \ldots, n \\ &\sum_{i=1}^{n} \lambda_i = 1 \end{aligned} \tag{2.14}$$

The difference between the point and block kriging system is the construction of the RHS matrix. The RHS matrix of the "point" kriging system contains covariances point-to-point ($C(\mathbf{u}_i, \mathbf{u})$), while the RHS matrix of the block kriging system contains covariances point-to-block $\overline{C}(\mathbf{u}_i, V)$. The covariances point-to-block are obtained as a linear average of point-to-point covariances (Eq. 2.15):

$$\overline{C}(\mathbf{u}_i, V) = \frac{1}{N} \sum_{j=1}^{N} C(\mathbf{u}_i, \mathbf{u}_j'), \tag{2.15}$$

where N is the number of points that discretize the block V, $C(\mathbf{u}_i, \mathbf{u}_j')$ is the covariance point-to-point between the sample located at $\mathbf{u}_i$ and the discretizing point $\mathbf{u}_j'$. Figure 2.2 illustrates the calculation of the point-to-block covariance.

If the same samples are used to estimate all the N discretizing points, the estimate obtained by block kriging is the same as the average of point kriging estimates of the N discretizing points (Journel and Huijbregts 1978). The major benefit is that the block estimate is obtained directly by solving only one linear system of equations.

Equation 2.16 describes the ordinary block kriging variance:

$$\sigma_V^2(\mathbf{u}) = \overline{C}(V, V) - \sum_{i=1}^{n} \lambda_i \cdot \overline{C}(\mathbf{u}_i, V) - \mu, \tag{2.16}$$

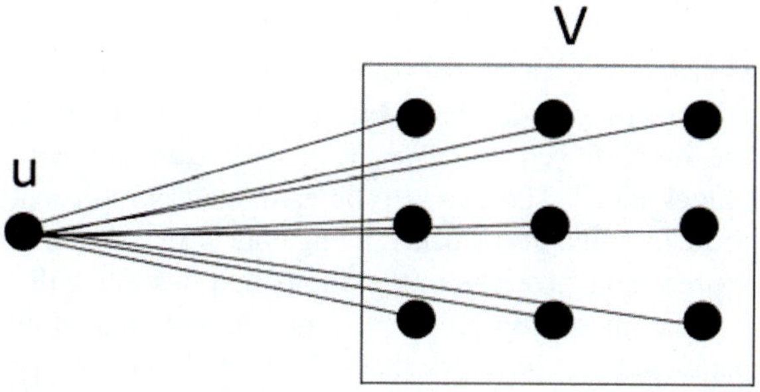

Fig. 2.2 Scheme for calculating point-to-block covariance between point *u* and block *V*. Each line represents a covariance point-to-point

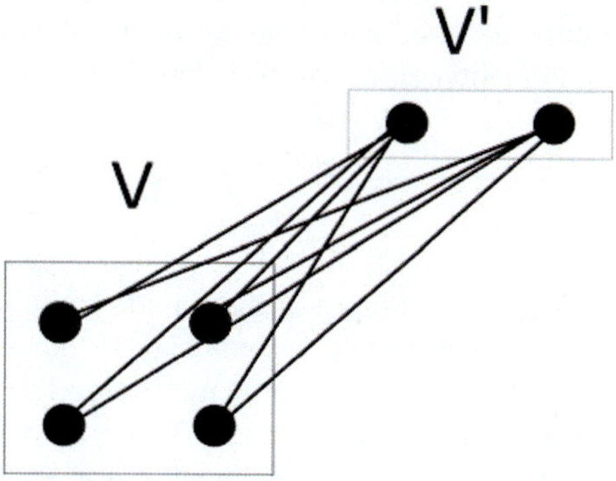

Fig. 2.3 Scheme of calculation of block-to-block covariance between block *V* and block *V*′. Each line represents a point-to-point covariance

where the block-to-block covariance[1] $\overline{C}(V, V)$, also called block variance, is approximated by the arithmetic average of point-to-point covariances defined between any point that discretize the block *V* (Eq. 2.17):

$$\overline{C}(V, V) = \frac{1}{N^2} \sum_{i=1}^{N} \sum_{j=1}^{N} C(\mathbf{u}_i^{'}, \mathbf{u}_j^{'}), \tag{2.17}$$

Figure 2.3 shows a scheme of the calculation of the block-to-block covariances.

[1] One complicating factor when calculating the block variance is the zero effect (see Journel and Huijbregts 1978, p. 96). By definition, the value of the covariance for the distance of zero equals the variogram sill. However, this procedure tends to overestimate the block variance. The solution for this consists of slightly shifting one discretizing point when calculating the covariance between two discretizing points that are exactly at the same location.

2.6 Kriging with Data of Different Support

Considering that the data are defined at different support, the kriging system may be adapted to consider these differences. The data consist of a set of values defined over the supports v_i $\{z(v_i); i = 1, \ldots, n\}$. Data defined over a volume are often called "block data." The block to be estimated is represented by V. Figure 2.4 illustrates the estimation of a block V using data with different volumes. The grades of the data represent the average of the points that discretize the volumes.

One interesting property is that the volumes of either the data or the block to be estimated may have any shape. These volumes are represented by sets of discretizing points. For example, in Fig. 2.4, the support v_1 is a line that represents the average grade along a line (the line mimics samples from diamond drill holes), while v_2 represents the average grade of a mined area. The estimate of $z^*(V)$ is a linear combination of the samples defined over the volumes v (Eq. 2.18):

$$z_V^*(\mathbf{u}) = \sum_{i=1}^{n} \lambda_i \cdot z(v_i), \tag{2.18}$$

When the support of the samples may not be considered as point or quasi-point support, the kriging system may be adapted to consider the support of the data. The adaptation consists of modifying the left-hand-side LHS, which contains the covariances between the data. When the data have different support, the LHS matrix is written using block-to-block covariances. The traditional block kriging system uses only point-to-point covariances in the LHS matrix (see Sect. 2.5). This property

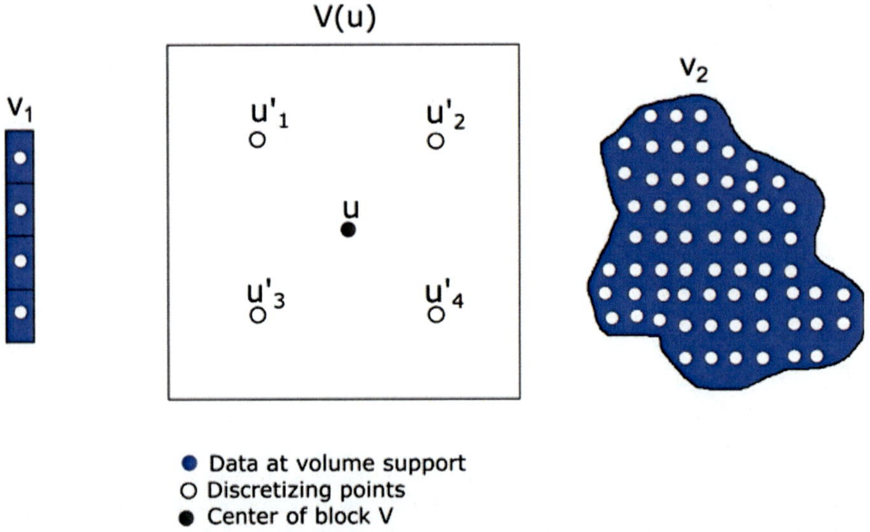

Fig. 2.4 Scheme of block to be estimated V using two samples of different support

is well described in the geostatistical literature (Journel and Huijbregts 1978; Isaaks and Srivastava 1989; Goovaerts 1997), but as far as the authors know has not been applied in the mineral industry. Equation 2.19 describes the ordinary kriging system with data of different support:

$$\sum_{i=1}^{n} \lambda_i \cdot \overline{C}(v_i, v_j) + \mu = \overline{C}(v_i, V) \quad j = 1, \ldots, n$$
$$\sum_{i=1}^{n} \lambda_i = 1, \tag{2.19}$$

where $\overline{C}(v_i, v_j)$ is the block-to-block covariance between the block datum v_i and the block datum v_j, and $\overline{C}(v_i, V)$ is the block-to-block covariance between the block datum v_i and the block to be estimated V. The block-to-block covariance $\overline{C}(v_i, v_j)$ is the average of point-to-point covariances between all the points that discretize the block v_i with all the points that discretize the block v_j (Eq. 2.20):

$$\overline{C}(v_i, v_j) = \frac{1}{N_i} \cdot \frac{1}{N_j} \sum_{\alpha=1}^{N_i} \sum_{\beta=1}^{N_j} C(\mathbf{u}_\alpha^{'}, \mathbf{u}_\beta^{'}), \tag{2.20}$$

where N_i is the number of points $\mathbf{u}_\alpha^{'}$ that discretize the block v_i and N_j is the number of point $\mathbf{u}_\beta^{'}$ that discretize the block v_j. $C(\mathbf{u}_\alpha^{'}, \mathbf{u}_\beta^{'})$ is the point-to-point covariance between the discretizing points $\mathbf{u}_\alpha^{'}$ and $\mathbf{u}_\beta^{'}$. Equation 2.21 describes the ordinary kriging variance using samples of different support:

$$\sigma_V^2(\mathbf{u}) = \overline{C}(V, V) - \sum_{i=1}^{n} \lambda_i \cdot \overline{C}(v_i, V) - \mu, \tag{2.21}$$

2.7 Illustrative Example

This example shows the effect of considering the support of the data in kriging estimates (Bassani et al. 2018). The goal is to estimate the attribute at point $\mathbf{u}$ using two samples. One sample is considered to have a quasi-point support, and the other is defined over a volume v in the space (Fig. 2.5). The volume of the block datum is centered at location $\mathbf{u}_1$, and the quasi-point datum is located at $\mathbf{u}_2$. The volume has equal size along the X and Z directions, and 16 discretizing points were used to calculate the average covariances. We compare kriging using only point-to-point covariances, which ignores the different support of the data, against kriging using average covariances. The estimations were performed using ordinary kriging.

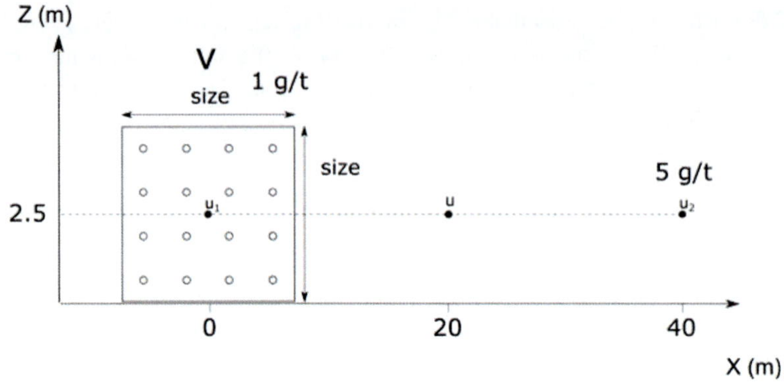

Fig. 2.5 Estimation at point **u** using a datum at block-support centered at location $\mathbf{u}_1$ and a datum at a quasi-point support at $\mathbf{u}_2$

2.7.1 Sensitivity Analysis

The estimate of the grade centered at **u** was carried out with different variogram models. The variogram model is defined at quasi-support. The effects on the weight of the block sample, on the estimate, and on the kriging variance, were measured. Since ordinary kriging was used, the weight of the quasi-point support sample is one minus the weight of the block sample.

Table 2.1 summarizes the sensitivity analysis. When one parameter was tested, the others remained unchanged for each comparative round. Four comparative rounds were performed to test four parameters (nugget effect, variogram range, variogram type, and block size). For example, the first comparative round tests the nugget effect (C_0). In this round, the variogram model consisted of the nugget effect tested plus a spherical structure with a fixed range of 60 m and contribution (C_1) of one minus the nugget effect tested.

Table 2.1 Summary of parameters tested in the sensitivity analysis

Comparative round	Parameter tested (variable)	Variogram model
1st	Nugget effect (C_0)	$\gamma(h) = C_0 + C_1 \cdot \text{Sph}\left(\frac{h}{60\,\text{m}}\right)$ $C_1 = 1 - C_0$
2nd	Range (a)	$\gamma(h) = 0.1 + 0.9 \cdot \text{Sph}\left(\frac{h}{a}\right)$
3rd	Variogram type (vtype): Gaussian, spherical, exponential	$\gamma(h) = 0.1 + 0.9 \cdot \text{vtype}\left(\frac{h}{60\,\text{m}}\right)$
4th	Block size (m)	$\gamma(h) = 0.1 + 0.9 \cdot \text{Sph}\left(\frac{h}{60\,\text{m}}\right)$

2.7.2 *Effect of Nugget Effect*

Figures 2.6a–c show the influence of the nugget effect on the weight of the block sample, the estimate, and the kriging variance, respectively. As the nugget effect decreases, the weight of the block sample decreases for kriging with samples of different support (Fig. 2.6a). A low nugget effect means that the attribute is spatially continuous, and the volume represented by the block sample is redundant in terms of information. The result is that the kriging system considers this redundancy and decreases the weight of the block sample (this effect is known as the declustering property of kriging). In the case of low nugget effect, kriging ignoring the difference in supports, i.e., considering all data as quasi-point samples and kriging with samples of different support tend to generate similar estimates (Fig. 2.6b).

As the nugget effect increases, kriging loses its declustering property (Journel and Huijbregts 1978). The block sample becomes less redundant, and its weight increases (Fig. 2.6a). Moreover, the difference between the estimates (performed by kriging considering all samples as quasi-point and kriging with samples of different support) increases as the nugget effect increases (Fig. 2.6b). Mathematically, this difference occurs because the nugget effect highly influences the covariances block-to-block

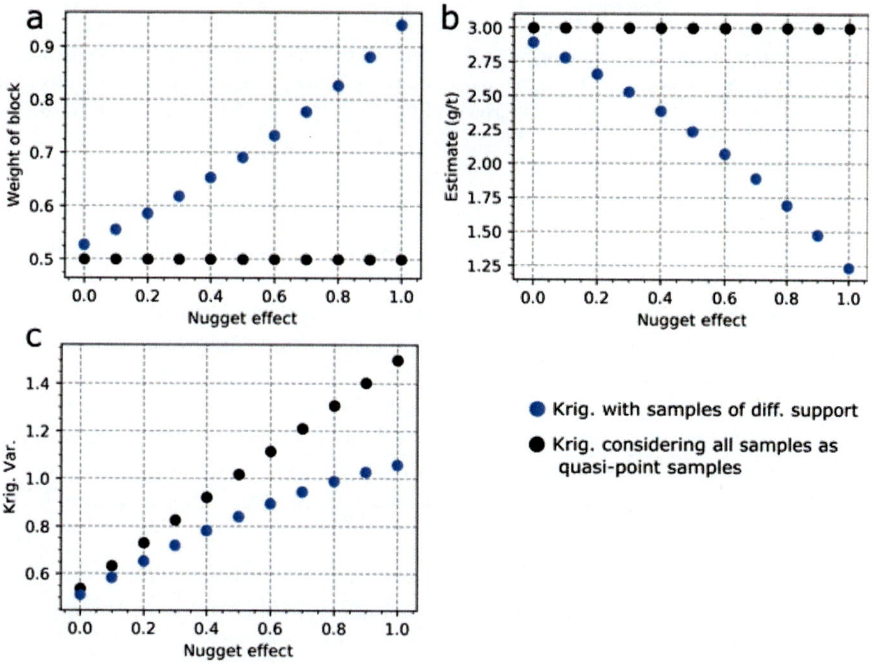

Fig. 2.6 Effect of the nugget effect on **a** the weight of the block sample, **b** on the estimate, and **c** on the kriging variance. The variogram model used has two structures: a nugget effect and a spherical model with range equal to 60 m

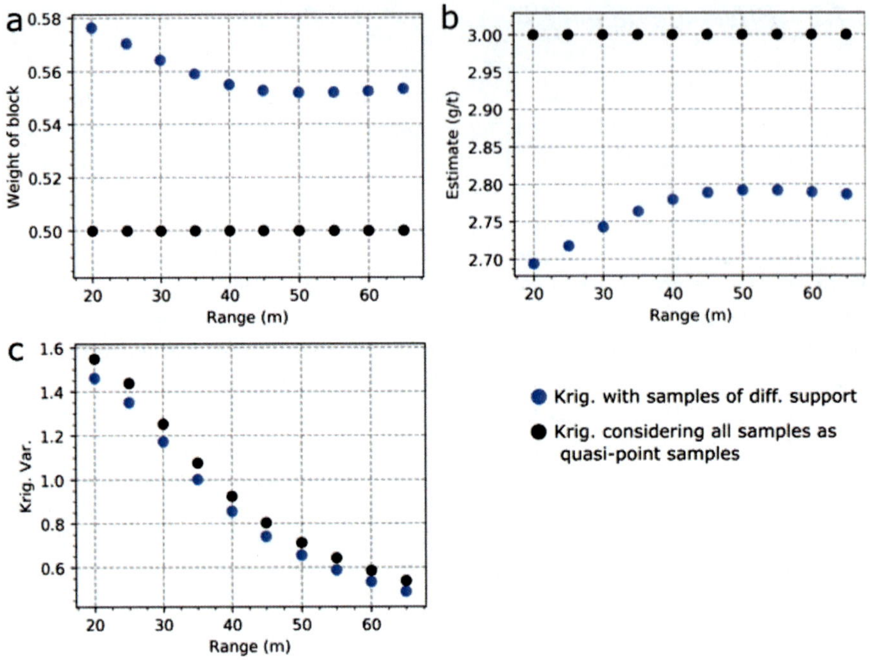

Fig. 2.7 Effect of the variogram range **a** on the weight of the block sample, **b** on the estimate, and **c** on the kriging variance. The variogram model used has a nugget effect of 0.10 and a spherical model with a contribution of 0.90. The distance from the samples to the point to be estimated is 20 m

and point-block. When the nugget effect increases, the covariances block-to-block and point-to-block used in the LHS matrix decreases. The result is that the inverse of the LHS matrix produces higher weights for the block sample.

Overall, the kriging variances are lower for kriging with samples of different support (Fig. 2.6c). The average volume over the block sample represents more information than the grade at the center of the block, which is the case of kriging with quasi-point samples. In this context, the kriging variance is lower when the volume of the block is considered in the kriging system. As the nugget effect decreases, the block sample is redundant, and the kriging variance is similar for both estimation methods (kriging considering all samples as quasi-point and kriging with samples of different support).

2.7.3 *Effect of Variogram Range*

Overall, the weight of the block sample does not change much as the variogram range increases (Fig. 2.7a). The variogram range did not affect much the covariances point-to-block and block-to-block used in the kriging with samples of different support. The covariances block-to-block and point-to-block are mainly affected when the range is small compared to the block size. In this case, the ranges are longer than the block size (the block size is 5 m while the shortest range considered is 20 m).

As the variogram range increases, the weight of the block sample decreases (Fig. 2.7a). A long range means that the attribute is more spatially continuous and, thus, the block sample becomes redundant, and its weight decreases. As a result, the estimates of the two methods (kriging with quasi-point samples and kriging with samples of different support) become more similar (Fig. 2.7b).

The kriging variance decreased as the variogram range increased (Fig. 2.7c). As the variogram range increases, the phenomenon is more spatially continuous, and the kriging variance is expected to be lower.

2.7.4 *Effect of Variogram Shape*

Figure 2.8 shows the effect of the variogram type on the weight of the block sample, on the estimate, and on the error variance. Overall, the variogram type did not influence much the weight of the block sample, on the estimate, and on the kriging variance. The Gaussian variogram is more spatially continuous than the spherical variogram, which is more spatially continuous than the exponential variogram. As the phenomenon becomes more spatially continuous, the weight of the block sample decreases (Fig. 2.8a), and the kriging variance decreases. In this context, the Gaussian variogram model resulted in the lowest weight for the block sample (Fig. 2.8a) and kriging variance (Fig. 2.8c).

2.7.5 *Effect of Support of the Samples*

Figure 2.9 shows the effect of the block size on the weight of the block sample, on the estimate, and on the kriging variance. In general, the weight of the block sample increases as its size increases (Fig. 2.9a). This result occurs because the block size directly affects the covariances block-to-block used by kriging with samples of different support. As the block size increases, the covariances block-to-block in the LHS matrix tend to decrease. As a result, the inverse of the LHS matrix generates higher weights for the block sample.

As the block size diminishes, the block datum acts similarly to a quasi-point support sample. As a result, its weight is similar to the weight obtained by kriging

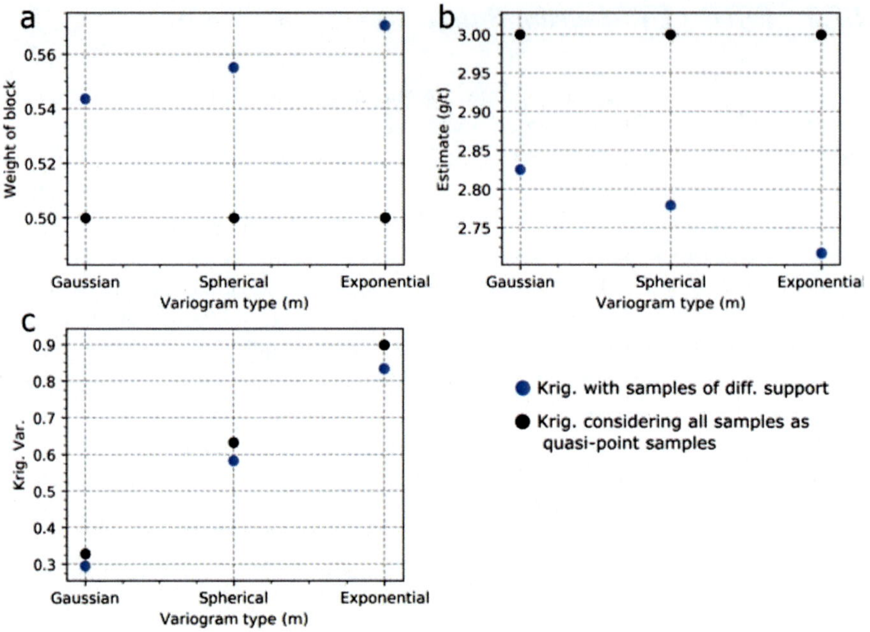

Fig. 2.8 Effect of the variogram type **a** on the weight of the block sample, **b** on the estimate, and **c** on the kriging variance. The variogram model used has a nugget effect of 0.10 and a contribution of 0.90 in the second structure with range equal to 60 m

considering all samples as quasi-points. The result is that, with small block sizes, the estimates obtained by the two methods are similar (Fig. 2.9b).

Overall, the kriging variance decreases as the block size increases. As the block size increases, part of the block becomes closer to the point to be estimated. The result is that the kriging variance decreases (Fig. 2.9c).

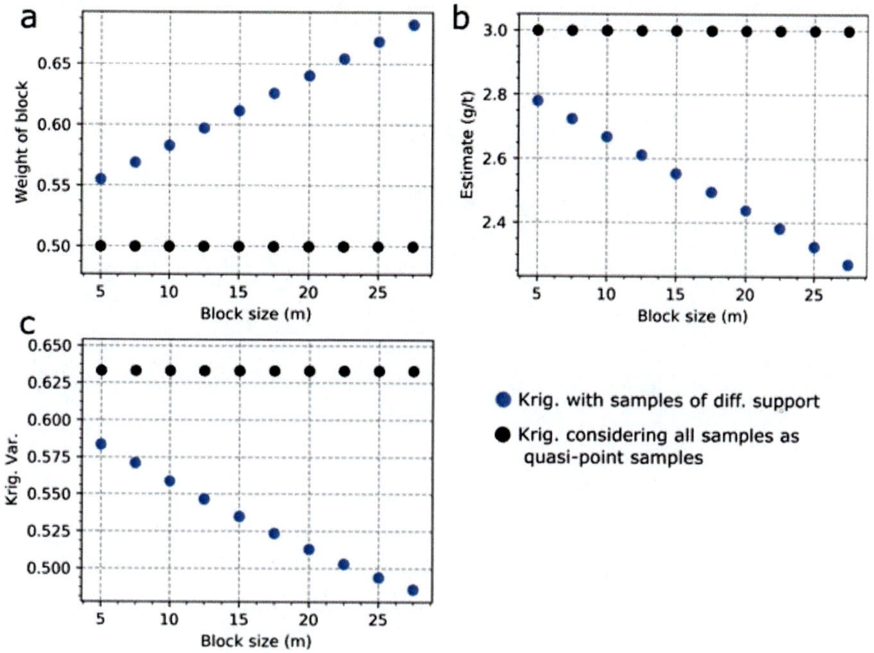

Fig. 2.9 Effect of the block size **a** on the weight of the block sample, **b** on the estimate, and **c** on the kriging variance. The variogram model used has a nugget effect of 0.10 and a contribution of 0.90 in the second structure with range equal to 60 m

References

Bassani MAA, Costa JFCL, Guaglianoni WC, Rubio RH (2018) Comparison between the indirect approach and kriging with samples of different support for estimation using samples of different length. Stoch Env Res Risk Assess 32(3):785–797

Deutsch CV, Journel AG (1998) GSLIB: geostatistical software library and user's guide. Oxford University Press, New York

Goovaerts P (1997) Geostatistics for natural resource evaluation. Oxford University Press, Oxford

Isaaks HE, Srivastava MR (1989) An introduction to applied geostatistics. Oxford University Press, Oxford

Journel AG, Huijbregts CJ (1978) Mining geostatistics. Academic Press, New York

Chapter 3
Simulation with Data of Different Support

3.1 Overview of Geostatistical Simulations

The estimates performed by kriging are built, so that each estimate considered individually (without considering the neighboring estimates) is "best." Best because the estimate minimizes the variance of the estimation error. However, when several estimated locations are considered together, such as an estimated grade model, the estimates do not represent the true spatial variability of the modeled variable. The map of the estimates has less variability than the modeled variable. The variance of the estimates is much lower than the variance of the data. This effect is called smoothing.

Conversely, geostatistical simulations generate a set of models with the same statistical and spatial characteristics of the data. The simulated models are known as realizations. The realizations, obtained by most algorithms, reproduce both the histogram and variogram of the data. The set of realizations also provides a measure of spatial uncertainty. For instance, areas where the simulated values differ a lot from one realization to another characterize areas of high uncertainty. The opposite is also true, i.e., low uncertainty areas are the areas where the simulated values are nearly the same at all realizations.

Estimated models should be used carefully when the variability of the data impacts the results where the models are applied. For instance, the ore tonnage of a deposit is often calculated based on the proportion of the blocks from the grade model above a given cut-off. The variance of the block model profoundly influences this proportion. A smooth grade model obtained by kriging leads to a biased tonnage estimation for ore. As a result, the ore tonnage derived from an estimated model differs from the actual ore tonnage. Geostatistical simulation overcomes this limitation of kriging. As the simulated models reproduce the variability of the phenomenon, the ore tonnage predicted by simulated models tends to be closer to the true tonnage (which is only known after complete extraction of the deposit) than the tonnage obtained by a kriged model.

M. A. Arcari Bassani and J. F. Coimbra Leite Costa, *Geostatistics with Data of Different Support Applied to Mining Engineering*, Topics in Mining, Metallurgy and Materials Engineering, https://doi.org/10.1007/978-3-030-80193-9_3

3.2 Data of Different Support in Geostatistical Simulation

3.2.1 Continuous Variables

Many geostatistical simulation methods are found in the literature (Goovaerts 1997; Deutsch and Journel 1998; Chilès and Delfiner 1999). For continuous variables, the most common methods are the Sequential Gaussian Simulation (SGS), Sequential Indicator Simulation (SIS), and Turning Bands Simulation (TBS). The problem is that these simulation algorithms use a nonlinear transformation of the data, which is not adequate to integrate data of different support. A nonlinear transformation distorts the linear relationship between data of different support. In this context, the Direct Sequential Simulation (DSS) (Journel 1994; Oz et al. 2003; Soares 2001) is the appropriate method to integrate data of different support. DSS uses the data in their original units to condition the simulations and, thus, keeps the linear relationship between the data of different support.

3.2.2 Categorical Variables

In the case of the simulation of categorical variables such as rock type, the suitable method to integrate data of different support is the Sequential Indicator Simulation (SIS) (Deutsch 2006) with the use of average covariances in the kriging system. Although SIS is not convenient for data of different support in case of continuous variables, the method works well for categorical ones. This suitability occurs because the average of indicators of categorical variables represents the mean probability of a given rock type over a volume in space. Many Gaussian methods are available for categorical variables, such as the Truncated Gaussian and Pluri-Gaussian Simulation (Xu and Journel 1993; Emery 2007; Armstrong et al. 2011). However, the Gaussian transformation used by these methods does not keep the linear relationship between the data of different support.

3.3 Direct Sequential Simulation with Data of Different Support

The Direct Sequential Simulation (DSS) algorithm is based on the theorem shown by Journel (1994). Journel (1994) demonstrated that, if the local cumulative distribution function (cdf) used at the sequential simulation has a mean and variance that equal the simple kriging estimate and variance, then the variogram is reproduced. The reproduction of the variogram does not depend on the shape of the local cdf.

The Direct Sequential Simulation algorithm consists of the following steps:

(i) Choose randomly a location **u** in the simulation grid and search for the nearby data and previously simulated nodes;
(ii) Calculate the mean and variance of the local conditional cumulative distribution function (ccdf) by simple kriging using the data and previously simulated nodes. The mean corresponds to the simple kriging estimate while the variance corresponds to the simple kriging variance;
(iii) Define a local ccdf whose mean and variance are equal to the simple kriging estimate and variance, respectively;
(iv) Draw a value from the local ccdf and add it to the dataset;
(v) Repeat steps i–iv until all the grid nodes are simulated.

The integration of data of different support in DSS involves kriging with samples of different support (using average covariances, as explained in Chap. 2) to obtain the mean and variance of the local ccdfs. There are many methods to sample from the local ccdf (Nowak and Srivastava 1997; Soares 2001; Oz et al. 2003). Among these methods, the Soares approach (Soares 2001) and the ccdf lookup table (Oz et al. 2003) tend to generate better results in terms of histogram reproduction. These two methods use the global reference histogram to sample the local ccdfs.

3.3.1 Sampling from the Local CCDF

The main methods to sample from the local ccdf are the (i) Soares approach (Soares 2001) and (ii) ccdf lookup table (Oz et al. 2003). These two methods use the normal score transform to sample from the global histogram. In this context, the normal score transform is reviewed before the sampling methods are explained.

Normal score transformation

The normal score transformation φ transforms the original data z into Gaussian values y. The Gaussian values y have a standard Gaussian distribution and are called normal scores. The normal score transformation (Deutsch and Journel 1998) is obtained by matching the quantiles of the cumulative distribution function (cdf) of the original data $F(z)$ with the quantiles of the standard Gaussian cdf $G(y)$.

Graphically, $F(z)$ may be interpreted as a function whose input is a value of the original data z and the result is the corresponding cdf, which is between 0 and 1 (Fig. 3.1). Similarly, the function $G(y)$ calculates the cdf associated to a normal score y.

Equation 3.1 defines the normal score transformation φ:

$$y = \varphi(z) = G^{-1}[F(z)], \tag{3.1}$$

where G^{-1} is the quantile function of the standard Gaussian distribution. The function G^{-1} has as input a cdf value between 0 and 1 and returns the corresponding normal score value (Fig. 3.2):

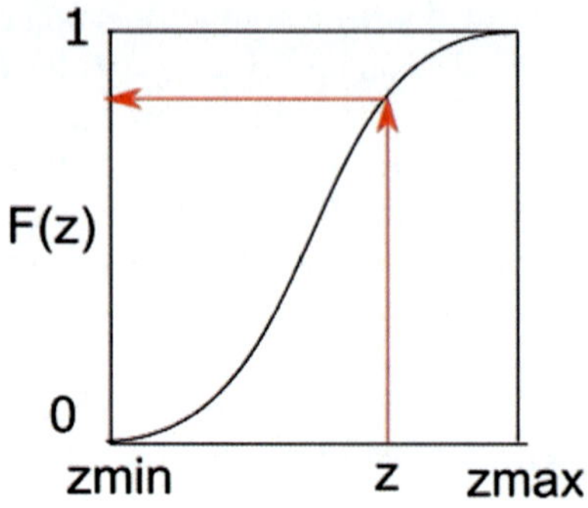

Fig. 3.1 Scheme of the cumulative distribution function of the original data

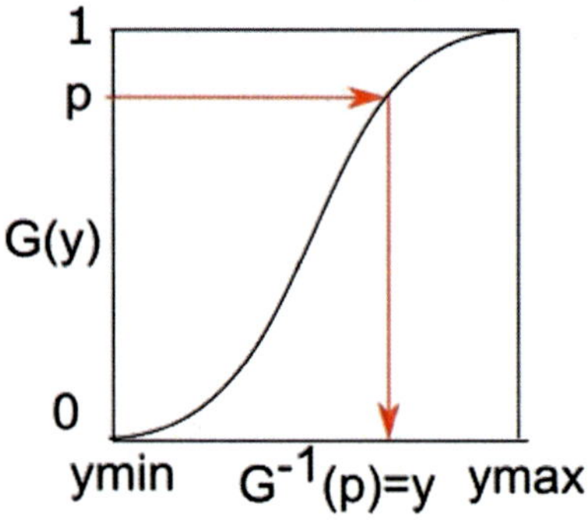

Fig. 3.2 Scheme of the quantile function of the standard Gaussian cdf

The normal score transformation may be seen as a quantile–quantile table. Each quantile in the space of the original data matches a quantile in the Gaussian space (Fig. 3.3).

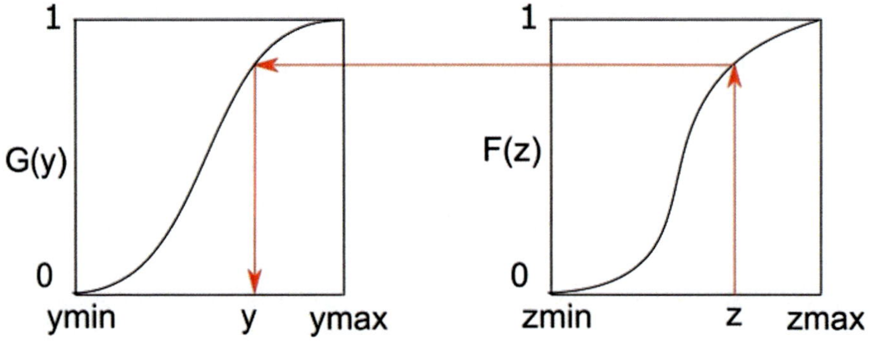

Fig. 3.3 Scheme of the normal score transformation

Soares approach

The Soares approach (Soares 2001) to sample from the local ccdf is shown in Fig. 3.4 and involves the following steps:

(i) Calculate at the location **u** the mean $z^*(\mathbf{u})$ and variance $\sigma^2(\mathbf{u})$ of the local ccdf by simple kriging using the original and previously simulated data in the

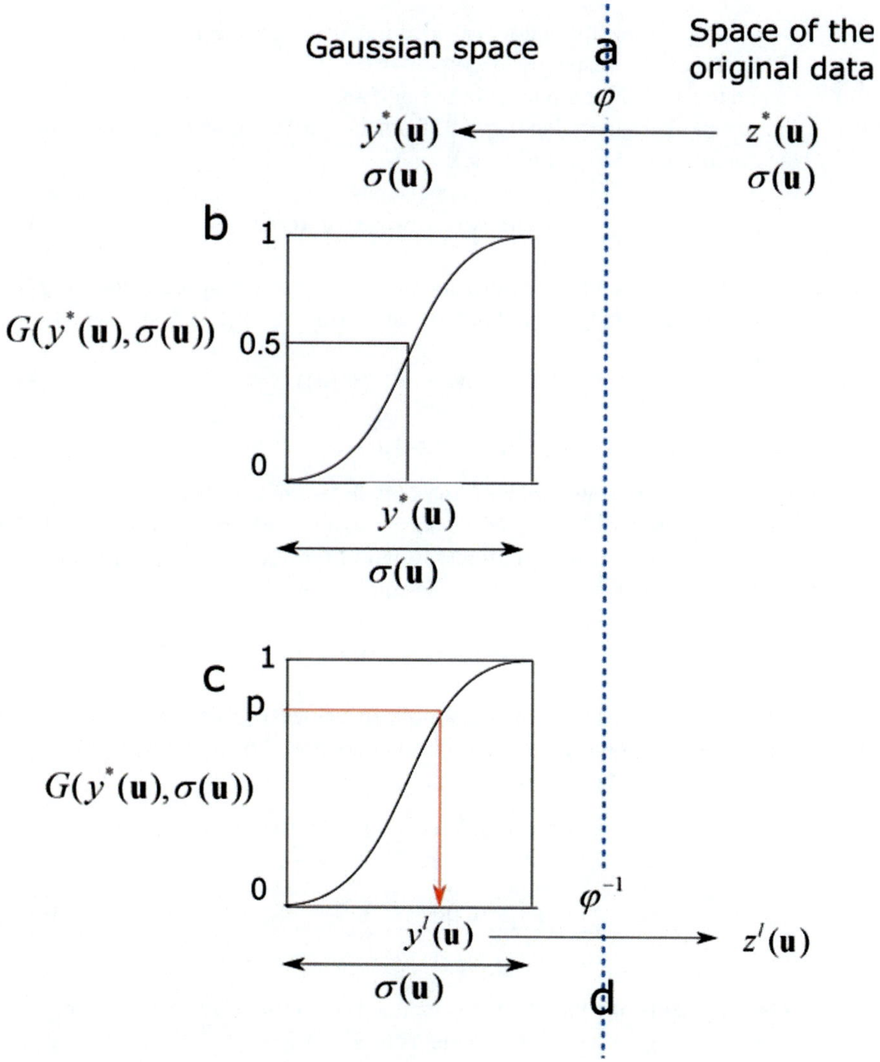

Fig. 3.4 Scheme of a simulated value z using the Soares approach: **a** normal score transform, **b** Gaussian cdf with mean and variance equals to $y^*(\mathbf{u})$ and $\sigma^2(\mathbf{u})$, **c** sampling from the local cdf, and **d** normal score transform back-transform

original units. $z^*(\mathbf{u})$ and $\sigma^2(\mathbf{u})$ are the simple kriging estimate and variance, respectively. $\sigma^2(\mathbf{u})$ is calculated using the standardized variogram of the original data, so that the sill is one;

(ii) Identify the Gaussian value $y^*(\mathbf{u})$ (Eq. 3.2) that matches the mean of the local ccdf $z^*(\mathbf{u})$ using the normal score transformation φ (Fig. 3.4a):

$$y^*(\mathbf{u}) = \varphi(z^*(\mathbf{u})) \tag{3.2}$$

(iii) Build a Gaussian distribution at location $\mathbf{u}$ whose mean and variance are $y^*(\mathbf{u})$ and $\sigma^2(\mathbf{u})$, respectively (Fig. 3.4b);

(iv) Generate a random number p between 0 and 1;

(v) Simulate a Gaussian value (Eq. 3.3) from the local Gaussian distribution using the random number p (Fig. 3.4c):

$$y^l(\mathbf{u}) = G^{-1}(y^*(\mathbf{u}), \sigma(\mathbf{u}), p) \tag{3.3}$$

(vi) Transform the simulated Gaussian value $y^l(\mathbf{u})$ to the space of the original data using the normal score back-transformation φ^{-1} (Eq. 3.4) (Fig. 3.4d):

$$z^l(\mathbf{u}) = \varphi^{-1}(y^l(\mathbf{u})) \tag{3.4}$$

(vii) Add the simulated value $z^l(\mathbf{u})$ to the dataset.

The steps i–vii are repeated until all the grid nodes are simulated.

The method proposed by Soares performs a nonlinear transformation of the estimate of the original data. This transformation may result in a degree of bias, which must be corrected. The following bias may occur (Eq. 3.5):

$$E[z^l(\mathbf{u})] \neq z^*(\mathbf{u}) \tag{3.5}$$

In other words, the average of the simulated values $E[z^l(\mathbf{u})]$ may differ from the estimate of the original data $z^*(\mathbf{u})$. In this context, Soares (2001) proposes a correction that involves the following steps:

(i) Estimate the average of the simulated value $E[z^l(\mathbf{u})]$ using Eq. 3.6:

$$E^*[z^l(\mathbf{u})] = \frac{1}{N}\sum_{i=1}^{N} z^l_{\mathrm{MC}}(\mathbf{u}) \tag{3.6}$$

where $z^l_{\mathrm{MC}}(\mathbf{u})$ is one the N values obtained by Monte Carlo Simulation.

(ii) Correct the original simulated value using Eq. 3.7:

$$z^{l'}(\mathbf{u}) = z^l(\mathbf{u}) + \left[z^*(\mathbf{u}) - E^*[z^l(\mathbf{u})]\right] \tag{3.7}$$

Soares (2001) states that the impact of this correction is higher when the global histogram (used for the normal score transformation) has classes with low frequency.

Ccdf lookup table

Oz et al. (2003) propose to use a table of local ccdfs, which is called ccdf lookup table, to sample the simulated value. The process to obtain the ccdf lookup table involves the following steps:

(i) Build a list of all possible local Gaussian distributions. These Gaussian distributions are obtained by the combination of all possible means and standard deviations. In general, the values for the means are evenly spaced values between minus 3.5 and 3.5. For the standard deviation, the values are evenly spaced between zero and one. The combination of a mean m with a standard deviation σ results in a Gaussian distribution $G(m, \sigma)$;

(ii) For each Gaussian distribution generated at step i, a series of quantiles p are used to simulate normal scores values y^l (Eq. 3.8). Then, each simulated normal score value y^l is transformed to the space of the original units z^l using the normal score back-transform (Eq. 3.9):

$$y^l(p) = G^{-1}(m, \sigma, p) \tag{3.8}$$

$$z^l(p) = \varphi^{-1}(y^l(p)) \tag{3.9}$$

The result is a set of simulated values in the space of the original units. This set comprises a ccdf in the space of the original units. For each Gaussian distribution, there is a corresponding ccdf in the space of the original units.

(iii) Calculate the mean and variance of the ccdf in the space of the original units and store into a table;

The result is a ccdf lookup table in the space of the original units, with its mean and variance. During DSS, the local ccdf used is that whose mean and variance equal the simple kriging estimate and variance, respectively.

Figure 3.5 shows how a local ccdf in the space of the original units is built starting from a local Gaussian distribution using only four quantiles. In practice, one hundred or more quantiles are used.

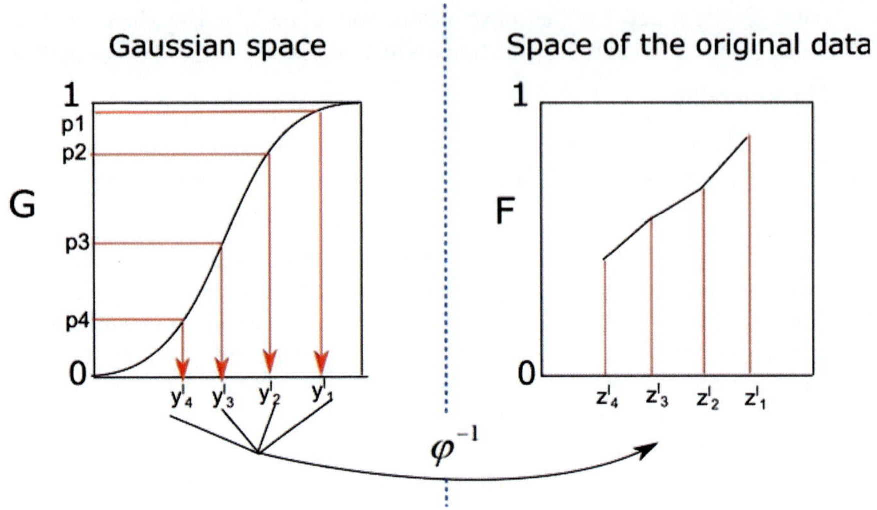

Fig. 3.5 Scheme of obtaining a cdf in the space of the original units by transforming a series of quantiles in the Gaussian space

3.4 Sequential Indicator Simulation with Data of Different Support

3.4.1 Indicators for Categorical Variables

Consider K categories that are mutually exclusive (only one category may prevail at a given location). The indicator approach requires that the dataset is transformed into a series of K indicators. Equation 3.10 defines how the indicators are obtained:

$$i(\mathbf{u}_i; k) = \begin{cases} 1 \text{ if category } k \text{ prevails at location } \mathbf{u}_i \\ 0 \text{ otherwise} \end{cases}, \tag{3.10}$$

where $i(\mathbf{u}_i; k)$ is the indicator for category k at location $\mathbf{u}_i$. The indicator represents the probability that a given category prevails at a given location. In the case of *hard data*,[1] the indicator is 1 if the category prevails or zero if does not prevail. For instance, consider that a rock type is classified as either ore or waste. A sample that is ore results in an indicator of 1 for the ore category. In other words, there is 100% probability that the sample is ore.

[1] The term hard data refers to data that do not have sampling errors in the geostatistical literature.

3.4.2 Indicator Kriging

Indicator kriging for categorical variables estimates the probability that category k prevails at location $\mathbf{u}$. Many kriging algorithms may be used (simple kriging, ordinary kriging, cokriging, simple kriging with local mean). These kriging algorithms are well detailed in the literature (Goovaerts 1997; Deutsch and Journel 1998; Chilés and Delfiner 1999). Here, we focus on the integration of data of different support for categorical variables.

The simple kriging estimator assumes that the mean probability p_k is known and constant over all the area of interest. Equation 3.11 defines the simple indicator kriging estimator:

$$i(\mathbf{u}; k)^* = p_k + \sum_{i=1}^{n} \lambda_i \cdot [i(\mathbf{u}_i; k) - p_k] \tag{3.11}$$

where $i(\mathbf{u}_\alpha; k)^*$ is the estimate (obtained by simple kriging) of the probability that category k prevails at location $\mathbf{u}$, λ_i is the simple kriging weight associated to the indicator $i(\mathbf{u}_i; k)$, and p_k is the mean probability for category k. The weights are obtained by the resolution of the simple kriging system.

Equation 3.12 defines the ordinary kriging estimator:

$$i(\mathbf{u}; k)^* = \sum_{i=1}^{n} \lambda_i \cdot i(\mathbf{u}_i; k) \tag{3.12}$$

where $i(\mathbf{u}_\alpha; k)^*$ is the estimate of the probability that category k prevails at location $\mathbf{u}$, λ_i is the ordinary kriging weight associated to the indicator. The weights are obtained by the resolution of the simple kriging system.

The integration of data of different support for indicator kriging involves using average covariances to build the indicator kriging system, as explained in Chap. 2. The LHS and RHS matrices are built using covariances point-to-block and block-to-block, so that the difference of support between the samples is considered.

3.4.3 Sequential Indicator Simulation with Data of Different Support

The sequential indicator with data of different support (Deutsch 2006) assumes that the data defined over a volume represents the average probability that a given category prevails over this volume. The data that represent average values over volumes or areas are called block data. For instance, consider that the proportion of ore over a volume of $5 \times 5 \times 2$ m^3 is 60%. In this case, the indicator data for this volume is 0.60.

The sequential indicator simulation with data of different support involves the following steps:

(i) Choose randomly a location **u** in the simulation grid and search for nearby data and previously simulated nodes;
(ii) Calculate the probability that each category k prevails at location **u** by indicator kriging $[i(\mathbf{u}; k)^*]$. Simple or ordinary kriging may be used. The indicator kriging system is built using average covariances so that the difference of support between the samples is considered;
(iii) Define any order of the K categories and build a cdf-type function by adding the probabilities estimated at step ii (Eq. 3.13):

$$F(\mathbf{u}; k')^* = \sum_{k=1}^{k'} i(\mathbf{u}; k)^* \tag{3.13}$$

(iv) Draw a value from the local ccdf and add it to the dataset.

The steps i–iv are repeated until all the grid nodes are simulated.

References

Armstrong M, Galli A, Beucher H, Loc'h G, Renard D, Doligez B, Eschard R, Geffroy F (2011) Plurigaussian simulations in geosciences. Springer Science & Business Media, Berlin
Chilés JP, Delfiner P (1999) Geostatistics: modeling spatial uncertainty. Wiley, New York
Deutsch CV (2006) A sequential indicator simulation program for categorical variables with point and block data: blockSIS. Comput Geosci 32(10):1669–1681
Deutsch CV, Journel AG (1998) GSLIB: geostatistical software library and user's guide. Oxford University Press, New York
Emery X (2007) Simulation of geological domains using the plurigaussian model: new developments and computer programs. Comput Geosci 33(9):1189–1201
Goovaerts P (1997) Geostatistics for natural resource evaluation. Oxford University Press, Oxford
Journel AG (1994) Modeling uncertainty: some conceptual thoughts. Geostatistics for the next century. Springer, Dordrecht, pp 30–43
Nowak MS, Srivastava RM (1997) A geostatistical conditional simulation algorithm that exactly honours a predefined grade-tonnage curve. In: Geostatistics Wollongong vol 96, pp 669–682
Oz B, Deutsch CV, Tran TT, Xie Y (2003) DSSIM-HR: A FORTRAN 90 program for direct sequential simulation with histogram reproduction. Comput Geosci 29(1):39–51
Soares A (2001) Direct sequential simulation and cosimulation. Math Geol 33(8):911–926
Xu W, Journel AG (1993) GTSIM: gaussian truncated simulations of reservoir units in a W Texas carbonate field. SPE 27412:3–6

Chapter 4
Change of Support in the Variogram

The use of average covariances for geostatistical simulation requires the variogram model at point support. However, obtaining the variogram at point support is a challenge when the samples have completely different supports. The goal of this chapter is to show how to change the support of the variogram. Changing the support of the variogram means that we can obtain a variogram at block support starting from a variogram at point support and vice-versa.

The technical terms for the change of support in the variogram are variogram regularization and deconvolution. Variogram regularization means obtaining a variogram at block support from a variogram at point support (this process is also called variogram upscaling). Variogram deconvolution is the opposite: obtaining a variogram at point support from a variogram at block support (this process is also called variogram downscaling). The methods for variogram regularization and deconvolution are shown in this chapter.

4.1 Experimental Variogram for Data of Different Support

Equation 4.1 defines the experimental variogram for point-support samples:

$$\hat{\gamma}(\mathbf{h}) = \frac{1}{2N(\mathbf{h})} \sum_{i=1}^{N(\mathbf{h})} [z(\mathbf{u}_i) - z(\mathbf{u}_i + \mathbf{h})]^2, \quad (4.1)$$

where $N(\mathbf{h})$ is the number of pairs for samples separated by the vector **h**, $z(\mathbf{u}_i)$ is the datum located at $\mathbf{u}_i$, and $z(\mathbf{u}_i + \mathbf{h})$ is the datum located at $\mathbf{u}_i + \mathbf{h}$. Similarly, Eq. 4.2 defines the experimental variogram for samples of different support (Goovaerts 2008):

M. A. Arcari Bassani and J. F. Coimbra Leite Costa, *Geostatistics with Data of Different Support Applied to Mining Engineering*, Topics in Mining, Metallurgy and Materials Engineering, https://doi.org/10.1007/978-3-030-80193-9_4

$$\hat{\gamma}(\mathbf{h}) = \frac{1}{2N(\mathbf{h})} \sum_{i=1}^{N(\mathbf{h})} \left[z_{v_i}(\mathbf{u}_i) - z_{v_{i+\mathbf{h}}}(\mathbf{u}_i + \mathbf{h}) \right]^2 \tag{4.2}$$

The use of samples of different support to calculate the experimental variogram may be questioned (Journel 1986). In theory, samples with different support have different variances. These different variances lead to different sills for the variogram model. In this context, the practitioner should select a subset of similar support samples to calculate the experimental variogram. The problem occurs when there are few data of similar support. In this case, the experimental variogram becomes noisy and does not show spatial continuity. From a practical view, the samples of different support aid to infer the spatial continuity model and may be used for the experimental variogram. The idea is to use all available information to infer the spatial continuity model as best as possible.

4.2 Variogram Regularization for Blocks of Regular Shape

Under the stationarity hypothesis, the point-support variogram and the variogram at block support are related by Eq. 4.3 (Journel and Huijbregts 1978, p. 77):

$$\gamma_v(\mathbf{h}) = \bar{\gamma}(v, v_{\mathbf{h}}) - \bar{\gamma}(v, v), \tag{4.3}$$

where $\gamma_v(\mathbf{h})$ is the variogram at block support at a distance $\mathbf{h}$, $\overline{\gamma}(v, v_{\mathbf{h}})$ is the average variogram between the block v and the block v separated by a distance $\mathbf{h}(v_{\mathbf{h}})$. The term $\overline{\gamma}(v, v)$ is the average variogram within the block v and is commonly called *gammabar*. Equation 4.4 shows the calculation of the *gammabar* $\overline{\gamma}(v, v)$:

$$\bar{\gamma}(v, v) = \frac{1}{N^2} \sum_{i=1}^{N} \sum_{j=1}^{N} \gamma(\mathbf{u}_i^{'}, \mathbf{u}_j^{'}), \tag{4.4}$$

where N is the number of discretizing points of the block v. Equation 4.5 shows the calculation of the variogram between blocks $\overline{\gamma}(v, v_{\mathbf{h}})$:

$$\bar{\gamma}(v, v_{\mathbf{h}}) = \frac{1}{N_i N_j} \sum_{i=1}^{N_i} \sum_{j=1}^{N_j} \gamma(\mathbf{u}_i^{'}, \mathbf{u}_j^{'}), \tag{4.5}$$

where N_i is the number of discretizing points of the block v and N_j is the number of discretizing points of the block $v_{\mathbf{h}}$. The presented relations assume that all the blocks have the same geometric shape. Goovaerts (2008) showed a general formula of variogram regularization, which is suitable for blocks of irregular shapes. This formula is shown in Sect. 4.3.

4.3 Variogram Regularization for Blocks of Irregular Shape

When all the blocks have the same shape, the term $\overline{\gamma}(v, v)$ is constant. However, when the blocks have different shapes, the term $\overline{\gamma}(v, v)$ depends on the distance **h** and on the shapes of the volumes considered. Equation 4.6 shows the variogram regularization for blocks with different shapes and sizes (Goovaerts 2008):

$$\gamma_v(\mathbf{h}) = \bar{\gamma}_{\mathbf{h}}(v, v_{\mathbf{h}}) - \bar{\gamma}_{\mathbf{h}}(v, v), \tag{4.6}$$

The term $\bar{\gamma}_{\mathbf{h}}(v, v)$ is the arithmetic average of *gammabars* for blocks separated by a distance **h** (Eq. 4.7):

$$\bar{\gamma}_{\mathbf{h}}(v, v) = \frac{1}{2N(\mathbf{h})} \sum_{i=1}^{N(\mathbf{h})} \left[\bar{\gamma}(v_i, v_i) + \bar{\gamma}(v_{i+\mathbf{h}}, v_{i+\mathbf{h}})\right], \tag{4.7}$$

where $N(\mathbf{h})$ is the number of pairs of blocks separated by a distance **h**. $\bar{\gamma}(v_i, v_i)$ is the gammabar, calculated by Eq. 4.4.

Equation 4.8 defines the calculation of $\bar{\gamma}_{\mathbf{h}}(v, v_{\mathbf{h}})$:

$$\bar{\gamma}_{\mathbf{h}}(v, v_{\mathbf{h}}) = \frac{1}{N(\mathbf{h})} \sum_{i=1}^{N(\mathbf{h})} \bar{\gamma}(v_i, v_{i+\mathbf{h}}), \tag{4.8}$$

where $\overline{\gamma}_{\mathbf{h}}(v, v_{\mathbf{h}})$ is the arithmetic average of variograms between blocks separated by the distance **h**. The term $\bar{\gamma}(v_i, v_{i+\mathbf{h}})$ is obtained by Eq. 4.5.

4.4 Variogram Deconvolution

Journel and Huijbregts (1978, p. 90) proposed a method to obtain the point-support variogram considering that the samples are measured at a volume v. This process is called variogram deconvolution. The method involves the following steps:

1. Define an initial point-support variogram model $\gamma(\mathbf{h})$ using the experimental variogram at block support $\hat{\gamma}_v(\mathbf{h})$;
2. Compute the theoretical variogram model at block support $\gamma_v(\mathbf{h})$ using Eq. 4.3 and compare it against the experimental points $\hat{\gamma}_v(\mathbf{h})$;
3. Adjust the point-support variogram model's parameters $\gamma(\mathbf{h})$ until the theoretical model $\gamma_v(\mathbf{h})$ fits the experimental points $\hat{\gamma}_v(\mathbf{h})$.

This method has been successfully applied by Goovaerts (2008) in a case study where the blocks had different sizes and shapes. Often, the central gap to integrate data of different support in kriging is inferring the point-support variogram model.

The method proposed by Journel and Huijbregts (1978) and reviewed here shows how to fill this gap.

References

Goovaerts P (2008) Kriging and semivariogram deconvolution in the presence of irregular geographical units. Math Geosci 40(1):101–128
Journel AG (1986) Geostatistics: models and tools for the earth sciences. Math Geol 18(1):119–140
Journel AG, Huijbregts CJ (1978) Mining geostatistics. Academic Press, New York

Chapter 5
Case Study of Kriging with Data of Different Support

5.1 Dataset Presentation

The dataset contains 686 drill holes located on a relatively regular grid of 200 m × 200 m spacing along the East (X) and North (Y) directions. This dataset was used by Bassani and Costa (2016). The original Z coordinates were converted into stratigraphic coordinates. The stratigraphic coordinate is the Z coordinate of the sample's centroid minus the Z coordinate of the top of the seam used as a reference. Several authors discuss the importance of stratigraphic coordinates for geostatistical modeling in tabular deposits (Deutsch 2002; Pyrcz and Deutsch 2014; Rubio et al. 2015).

The variable of interest is the fraction of the total sample mass retained at the 14 # sieve aperture (REC14). This variable represents the proportion of coarse fragments and is essential to predict the amount of concentrated ore after the washing process. REC14 is an additive variable, like grades, so the methods presented here are also adequate for grades.

REC14 was sampled roughly at the nominal length of 0.5 m. To demonstrate the technique, which regards samples of different lengths, half of the 686 drill holes were selected randomly. In these selected drill holes, REC14 was composited by the total intercepted seam. The composite length corresponds to the ore thickness, and the composite grade represents the average REC14 over this thickness. The final dataset contains 343 drill holes with one sample intercepting the total ore thickness (blue points in Fig. 5.1) plus 343 drill holes with samples whose length is nearly 0.5 m (red points in Fig. 5.1). The dataset imitates a dataset whose samples came from different campaigns with different sampling lengths.

Figure 5.2a shows the length-weighted histogram of REC14. The distribution is roughly symmetric around the mean, and the coefficient of variation is low (0.19). The coefficient of variation is much lower than one, indicating that the distribution does not contain extreme values that predominantly affect the mean. The drill holes are regularly spaced, so these statistics are representative of the study area. Figure 5.2b shows the QQ-plot between the original and declustered data. The data

M. A. Arcari Bassani and J. F. Coimbra Leite Costa, *Geostatistics with Data of Different Support Applied to Mining Engineering*, Topics in Mining, Metallurgy and Materials Engineering, https://doi.org/10.1007/978-3-030-80193-9_5

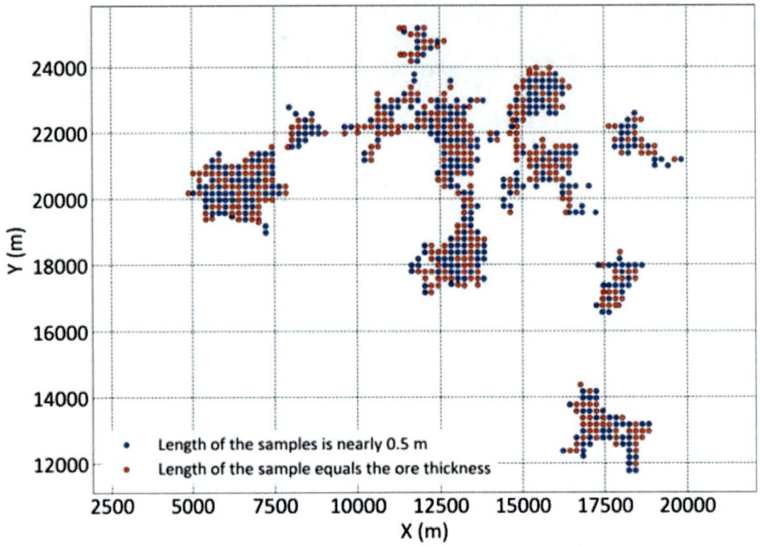

Fig. 5.1 Location map of the samples

were declustered using the nearest neighbor estimate, which approximates the polygonal declustering method (Isaaks and Srivastava 1989). The points in Fig. 5.2b are close to the line $y = x$ and show that the two distributions are similar, as expected.

Figure 5.2c shows the histogram of the length of the samples. The length of the samples ranges between 0.25 and 7.88 m. Since about 80% of the samples are short, whose lengths are between 0.25 and 0.75 m, the geomodeler may feel motivated to retain only these short samples (all of them at equivalent support) for estimation. However, keeping only the short samples results in discarding 50% of the drill holes. This loss of information is likely to worsen the estimates.

5.2 Variogram Analysis and Modeling

The point-support variogram model was obtained by the variogram deconvolution method described in Chap. 4. Figure 5.3a, b shows the experimental variogram along with the horizontal and vertical directions. Both the quasi-point (samples with a length of 0.5 m) and block samples (samples longer than 0.5 m) were used for the experimental variogram, which is called block experimental variogram (blue dots in Fig. 5.3). The block variogram (dashed green line in Fig. 5.3) was obtained by the regularization of the point-support variogram model (black line in Fig. 5.3). The goal is to minimize the difference between the block experimental variogram and the block variogram.

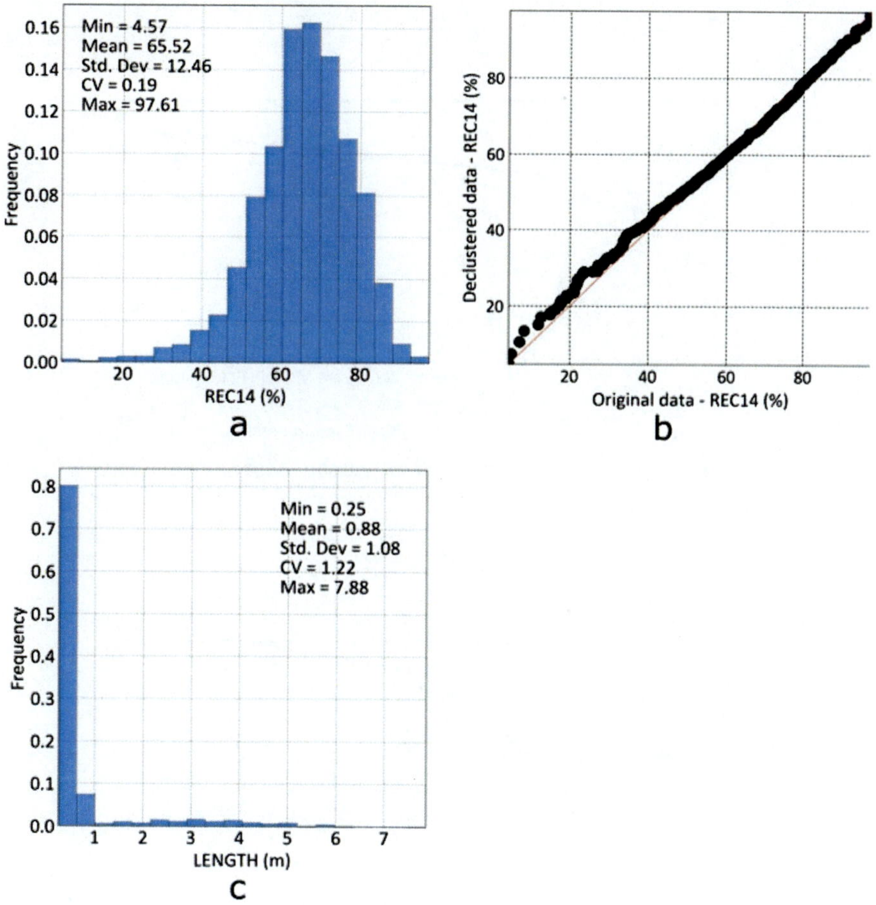

Fig. 5.2 **a** Histogram of REC14, **b** QQ-plot between the original and declustered data, and **c** histogram of the length of the samples

Equation 5.1 describes the point-support variogram model of REC14. The variogram is standardized so that the sill equals one.

$$\gamma(\mathbf{h}) = 0.11 + 0.59 \cdot \mathrm{Sph}\left(\frac{\mathrm{NS}}{300\,\mathrm{m}}, \frac{\mathrm{EW}}{300\,\mathrm{m}}, \frac{\mathrm{vert.}}{3\,\mathrm{m}}\right) + 0.30 \cdot \mathrm{Sph}\left(\frac{\mathrm{NS}}{5000\,\mathrm{m}}, \frac{\mathrm{EW}}{5000\,\mathrm{m}}, \frac{\mathrm{vert.}}{7\,\mathrm{m}}\right) \tag{5.1}$$

The variable of interest is isotropic on the horizontal plane, and the direction of minor spatial continuity is vertical. This variogram model is consistent with the geology of the deposit. Figure 5.4 shows a location map of the samples colored

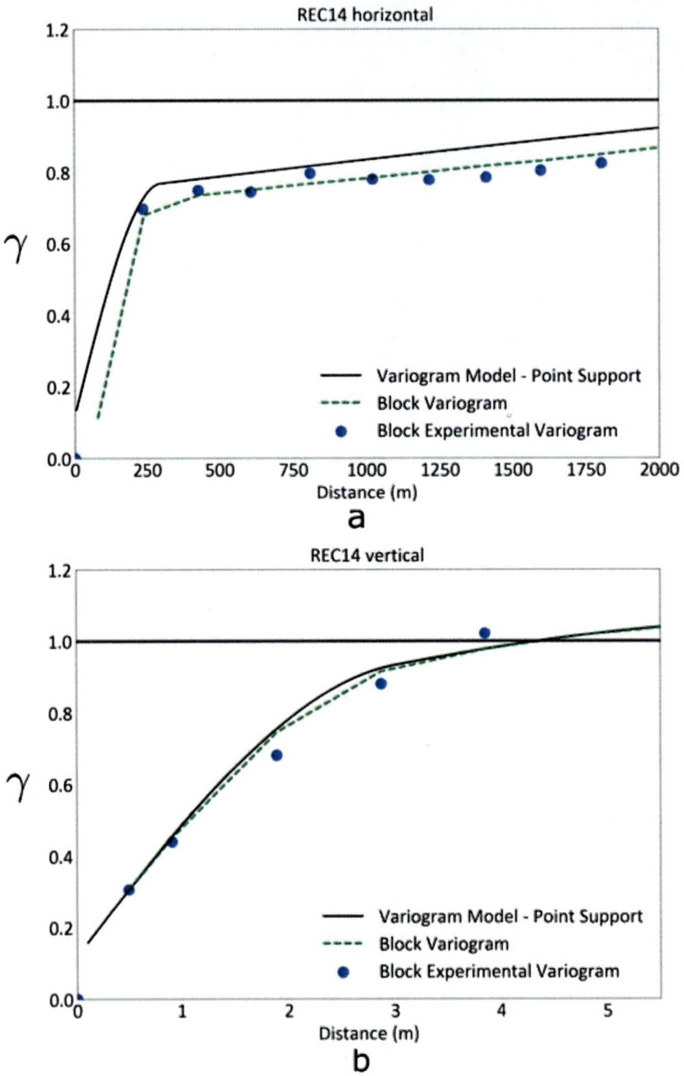

Fig. 5.3 Experimental variogram, variogram model at point support, and block variogram along with the **a** horizontal and **b** vertical directions

by the variable of interest. The samples do not show an evident anisotropy and are consistent with the variogram model.

The variable of interest has a short range along the vertical direction. If the attribute is spatially discontinuous (i.e., presents high nugget effect and short range), kriging with samples of different support results in higher weights to the long samples compared to kriging using point covariances. This effect is observed in the simple example shown in Chap. 2. As the ratio length of the sample/range of the variogram

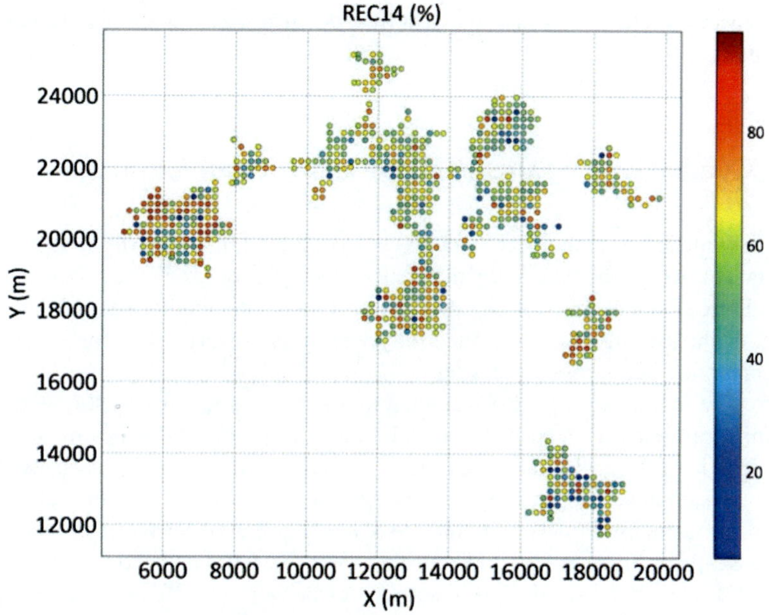

Fig. 5.4 Location map of the samples colored by REC14

increases, the covariance block-to-block tends to decrease, and the kriging weights increase.

5.3 Estimation and Validation

Ordinary kriging with samples of different support was used to estimate REC14. The drill hole samples were discretized along the direction of the drill hole (vertical direction). The discretization spacing corresponds to 0.5 m, which is the length of the shortest samples. This length represents quasi-point support. These discretization points were used to calculate the covariances point-to-block and block-to-block between the samples.

The block model consists of blocks whose dimensions are 50 m × 50 m × 0.5 m along the *X*, *Y*, and *Z* directions. The block discretization was set to 5 × 5 × 1. The estimates were limited for the blocks inside the geological model.

The model was checked by visual inspection and swath plots. The visual inspection consists of plotting and analyzing the grade model and the data using the same color scale. The data should justify the grade model.

The swath plot starts by dividing the area of interest into a series of swaths. Within each swath, the average of the estimates is compared against the average of the declustered data. In this case, the data were declustered using the nearest neighbor estimate.

5.4 Comparison Against Kriging Using Point Covariances

Kriging with different support samples was compared against kriging using point covariances, which disregards the difference of support between the samples. In this case, all the samples are considered to have quasi-point support. The left-hand-side matrix of the kriging system is filled only with point covariances. The comparison was performed with cross-validation.

Cross-validation allows the practitioner to compare the actual and estimated values to minimize their differences. Cross-validation starts by removing one sample at a particular location. Then, the estimate is performed at that location using the remaining samples. The procedure is repeated for all the locations of the dataset. The same parameters used to estimate the block model are used for the cross-validation. Last, the estimation error (the difference between the estimated and true values), the mean absolute error, and the mean squared error were calculated. The mean estimation error measures the accuracy of the estimates. The mean absolute error, the mean squared error, and the standard deviation of the error are used to measure the precision of the estimates.

5.5 Results

5.5.1 Estimates

Figure 5.5 shows a plan view of the estimates. The West part of the block model shows high grades, matching the high-grade samples in this region (see Fig. 5.4).

Figure 5.6 shows a plan view of the block model and the samples with the same color scale. The high-grade blocks are close to the high-grade samples, as expected.

The swath plots show that the block model reproduced the trend of the data along the *X*, *Y*, and *Z* directions (see Figs. 5.7a–c). Along with the three directions, the local mean of the estimates is similar to the local mean of the declustered data. The swath plot also shows that the estimates are not always above or always below the declustered data.

The swath plot along the *Z* direction (Fig. 5.7c) showed discrepancies between the declustered mean and the mean of the estimates for the lowest elevations. This discrepancy occurred because the stratigraphic coordinates are based on the *Z* coordinate of the seam's top. Figure 5.8 shows a cross section of the drill holes with the stratigraphic coordinates. The areas of lower elevations have fewer samples than the

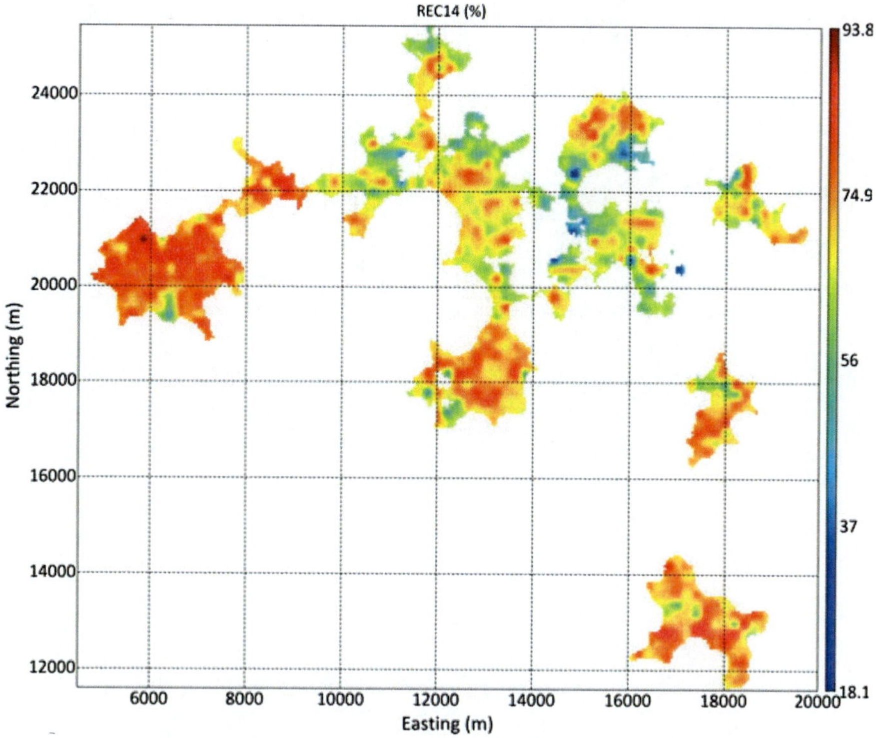

Fig. 5.5 Block model estimates

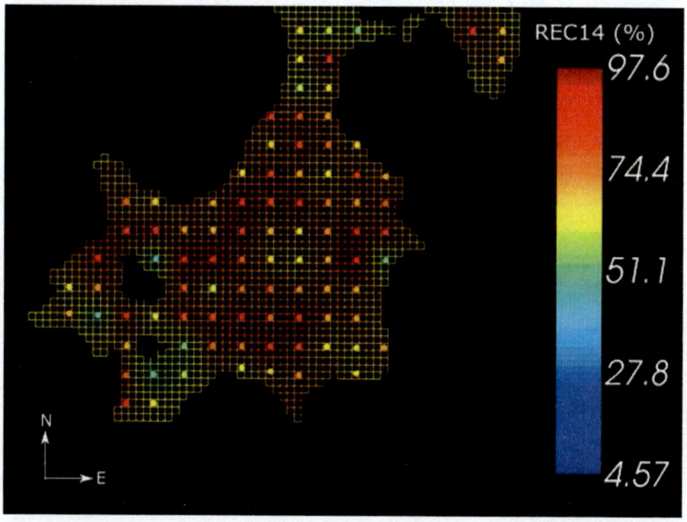

Fig. 5.6 Block model with the samples

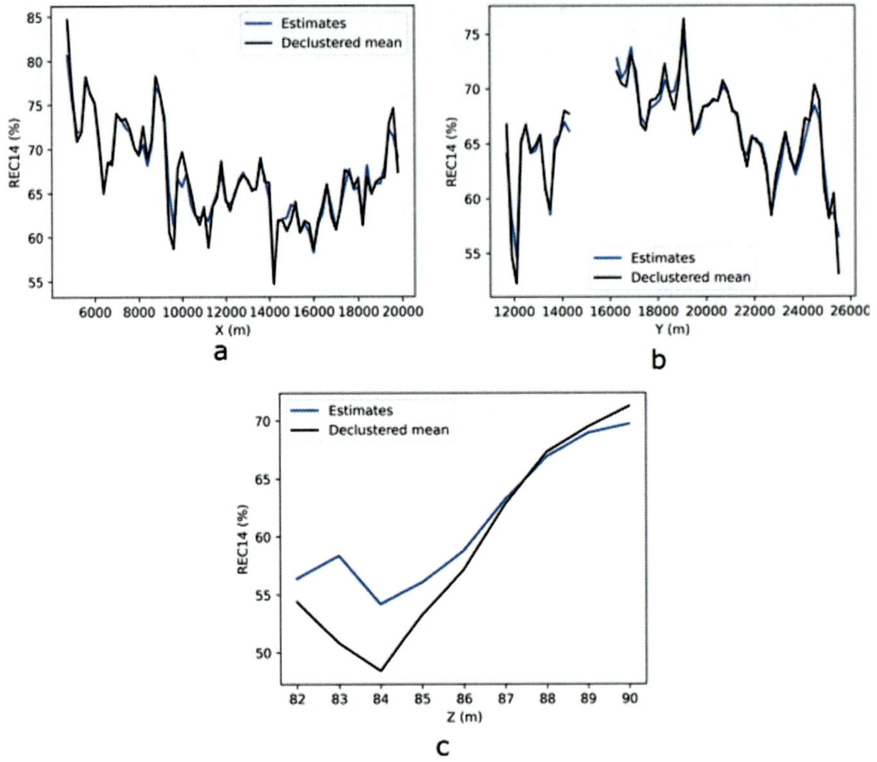

Fig. 5.7 Swath plot along with the **a** *X*, **b** *Y*, and **c** *Z* directions

areas of higher elevations. This lack of data in the areas of lower elevations resulted in discrepancies in the swath plot.

5.5.2 *Comparison with Kriging Using Point Covariances*

Table 5.1 shows the cross-validation results for kriging with samples of different support and kriging using point covariances. The estimates obtained by kriging with samples of different are more accurate, as the mean error is closer to zero. Moreover, kriging with samples of different support resulted in higher precision. For this method, the standard deviation of the error, the mean absolute error, and the mean squared error are lower. The mean squared error obtained by kriging with samples of different support is 4.5% lower than the mean squared error obtained by kriging using point covariances.

The results emphasize the capacity of kriging to handle samples of different support. Practitioners often overlook that the kriging system may handle data of

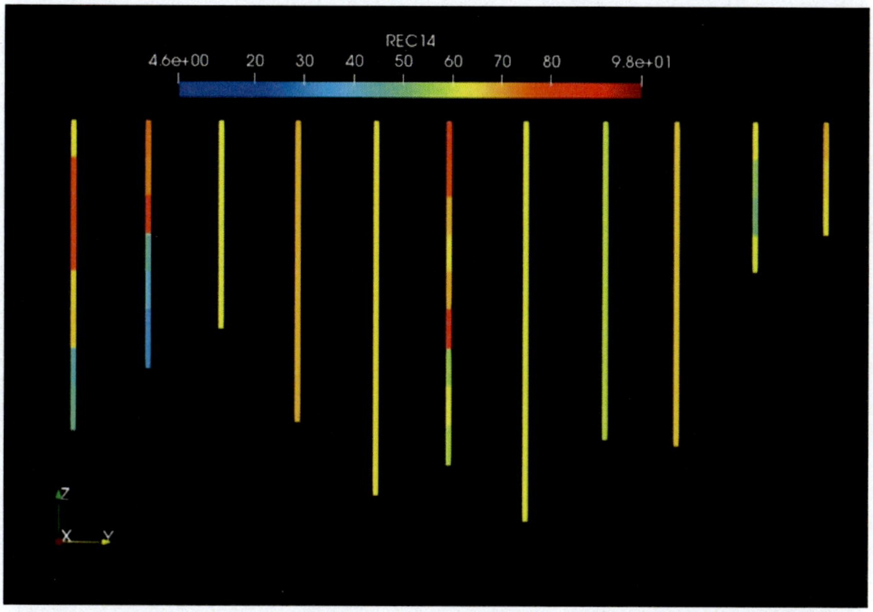

Fig. 5.8 Cross section of the drill holes with stratigraphic coordinates. The vertical exaggeration is 20:1

Table 5.1 Comparative table between kriging with samples of different support and kriging using point covariances

	Kriging with samples of different support (using average covariances)	Kriging using point covariances
Count	2342	2342
Mean error (%)	0.29	0.45
Std. Dev. of the error (%)	9.48	9.69
Mean absolute error (%)	7.06	7.32
Mean squared error ($\%^2$)	89.97	94.02

different support (Journel 1986). In the example, kriging was used to estimate grades in a mineral deposit using samples with different lengths.

In comparison against kriging using point covariances, kriging with samples of different support produced more accurate and precise estimates. When the samples have different supports, this difference should be considered in the kriging system through average covariances.

References

Bassani MAA, Coimbra Leite Costa JF (2016) Grade estimation with samples of different length. Appl Earth Sci 125(4):202–207

Deutsch CV (2002) Geostatistical reservoir modeling. Oxford University Press, New York

Isaaks HE, Srivastava MR (1989) An introduction to applied geostatistics. Oxford University Press, Oxford

Journel AG (1986) Geostatistics: models and tools for the earth sciences. Math Geol 18(1):119–140

Pyrcz MJ, Deutsch CV (2014) Geoestatistical reservoir modeling. Oxford University Press, New York

Rubio RH, Koppe VC, Costa JFCL, Cherchenevski PK (2015) How the use of stratigraphic coordinates improves grade estimation. Rem: Revista Escola de Minas 68(4):471–477

Chapter 6
Case Study of Direct Sequential Simulation with Data of Different Support

The majority of the theory about using data of different support has already been covered in the previous chapters. This chapter shows a case study of Direct Sequential Simulation (DSS) to integrate data of different support in a mining context. The result is a series of equally probable grade models that may be directly applied for stochastic mine planning. Also, these grade models provide a measure of uncertainty. The chapter highlights some practical aspects of DSS with data of different support. Moreover, the impact of the data support on uncertainty is discussed.

6.1 Dataset Presentation

The dataset used is the same as in Chap. 5, which shows a thorough exploratory data analysis. In this chapter, we emphasize the dataset's main characteristics used for Direct Sequential Simulation. The dataset is obtained from a bauxite deposit, and the variable of interest is the proportion of coarse fragments, called REC14.

The dataset contains 686 drill holes. At 343 drill holes, the sample length is 0.5 m along the vertical direction (blue points in Fig. 6.1). The other 343 drill holes have one sample that spans the entire ore thickness (red points in Fig. 6.1). In this case, the sample grade is the average grade over the ore thickness. This situation mimics a mine operation whose sampling protocols changed along its lifetime.

Direct Sequential Simulation (DSS) requires a reference distribution at the support to be simulated (Soares 2001; Oz et al. 2003). The simulated values are drawn from the region of the reference distribution whose mean and variance match the simple kriging estimate and variance. Early attempts of DSS did not use the reference distribution (Caers 2000). However, the methods that sample from the reference distribution have shown better histogram reproduction (Soares 2001).

In this case, we are interested in creating high-resolution models, so the simulations are performed at point support. Although Sequential Simulation algorithms may be adapted to simulate grades directly at block support (Godoy 2003; Boucher

M. A. Arcari Bassani and J. F. Coimbra Leite Costa, *Geostatistics with Data of Different Support Applied to Mining Engineering*, Topics in Mining, Metallurgy and Materials Engineering, https://doi.org/10.1007/978-3-030-80193-9_6

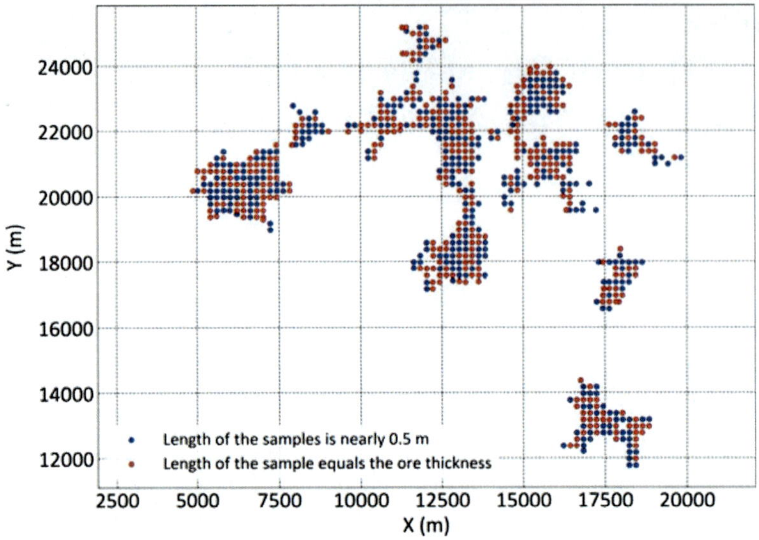

Fig. 6.1 Location map of the samples

and Dimitrakopoulos 2009; Emery and Ortiz 2011), we preferred to use point-support simulation to add more flexibility. The point-support simulations may be block averaged according to any volume. This property is attractive when the mine planner wants to measure the grade uncertainty over an irregular volume. The simulated volume grade is the average of simulated-point values inside this volume (Journel and Huijbregts 1978). Another benefit of point simulation is that the grade models may be directly compared against the data. Block simulation usually requires changing the support of the histogram and variogram. This change of support diminishes the variance and smooths the extreme values, which may mask some valuable information.

Figure 6.2a shows the length-weighted histogram of the variable of interest considering all the samples. In contrast, Fig. 6.2b shows the histogram considering only the samples with nearly 0.5 m of length, called point-support histogram (the 0.5 m samples represent a "point" in the space as they are the shortest sampling interval). The standard deviation for the histogram considering only the point samples is higher, as expected. Also, the mean of the point-support histogram (Fig. 6.2b) is very similar to the mean of the mix-support histogram (Fig. 6.2a). This similarity indicates that the point-support samples are not preferentially sampled in either a high-grade or low-grade zone.

Cell declustering was performed to verify whether some clusters could affect the statistics of the point-support samples. In this context, Fig. 6.3 shows the QQ-plot between the equal-weighted and declustered histogram considering only the point samples. The QQ-plot shows that the two distributions are very similar, as the points follow the line y = x. This result shows that the equal-weighted histogram of REC14

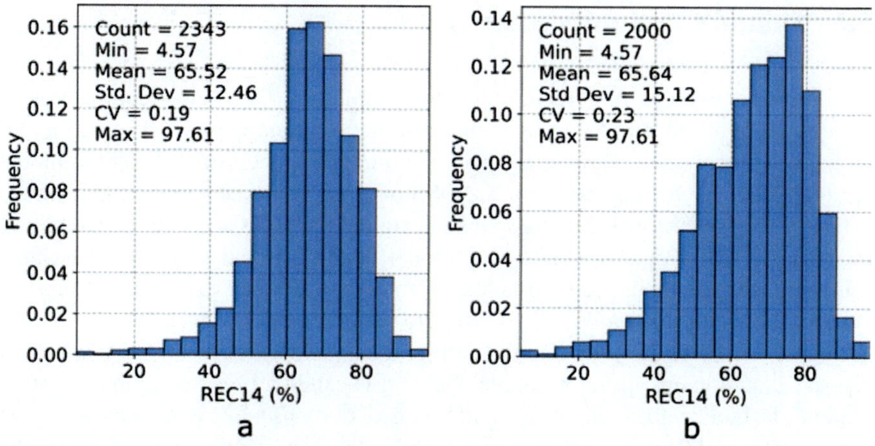

Fig. 6.2 **a** Length-weighted histogram of REC14 considering all the samples and **b** equal-weighted histogram of REC14 considering only the point samples

Fig. 6.3 QQ-plot between the equal-weighted and declustered histogram of REC14 considering only the point samples

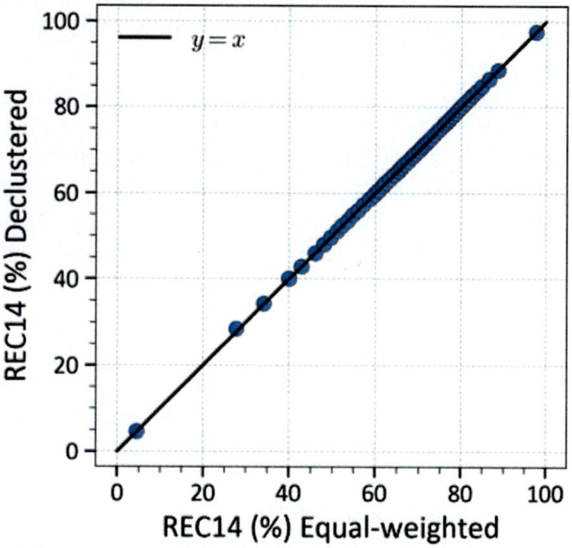

is representative of the area of interest. Thus, this histogram was used as a reference distribution for DSS.

This particular dataset has many point-support samples to obtain a point-support reference distribution (Fig. 6.2b). When the dataset exhibits a small number of point samples, the user may apply change-of-support techniques to obtain a point-support histogram. Several change-of-support methods are presented in the literature, such as the discrete Gaussian model (Journel and Huijbregts 1978), Affine (Isaaks and Srivastava 1989), and indirect lognormal (Isaaks and Srivastava 1989).

6.2 Spatial Continuity Analysis

Similar to the histogram, Direct Sequential Simulation (DSS) also requires a variogram model at the support to be simulated, that is, point support. The variogram model was obtained by variogram deconvolution and is the same variogram model used in Chap. 5. One advantage of the DSS over the Gaussian simulation methods is that kriging is performed in the space of original units. As a result, the variogram model used for DSS is the variogram model of the original variable. The Gaussian simulation methods require that the original data are transformed into Gaussian values. The variogram model is then fitted to these Gaussian values.

Figure 6.4 shows the point-support variogram model. The experimental variogram was calculated using all the available data and is called a block experimental variogram. The block variogram is obtained by variogram upscaling of the point-support variogram model (Eq. 4.6 in Chap. 4). Fitting the block variogram to the experimental variogram is not straightforward and required some trial and error. Increasing the nugget effect of the point-support variogram elevates the point-support variogram line. However, a higher nugget effect increases the variogram within blocks, which decreases the block variogram. This result occurs because the block variogram is the variogram between blocks minus the variogram within blocks (see Chap. 4, Eqs. 4.6–4.8).

The direction of major spatial continuity is horizontal and agrees with the sedimentary process that formed the deposit. The direction of minor spatial continuity is vertical. This spatial continuity pattern is widely observed at sedimentary formations such as coal deposits and oil reservoirs. Equation 6.1 shows the variogram model. The variogram is standardized so that the sill equals one. The standardized variogram model is used during DSS to define the sample the global histogram using the Soares approach (see Chap. 3).

$$\gamma(\mathbf{h}) = 0.11 + 0.59 \cdot \mathrm{Sph}\left(\frac{\mathrm{NS}}{300\,\mathrm{m}}, \frac{\mathrm{EW}}{300\,\mathrm{m}}, \frac{\mathrm{vert.}}{3\,\mathrm{m}}\right) + 0.30 \cdot \mathrm{Sph}\left(\frac{\mathrm{NS}}{5000\,\mathrm{m}}, \frac{\mathrm{EW}}{5000\,\mathrm{m}}, \frac{\mathrm{vert.}}{7\,\mathrm{m}}\right) \tag{6.1}$$

The geostatistical simulations are more sensitive to the variogram model than kriging estimates. In this context, we presented a more detailed variogram discussion here than in Chap. 5, which covers kriging estimates.

6.3 Geostatistical Simulations

The simulations were performed using Direct Sequential Simulation (DSS) with data of different support, which is known as Block Sequential Simulation—BSSIM (Liu and Journel 2009). The variable of interest is simulated on a high-resolution grid.

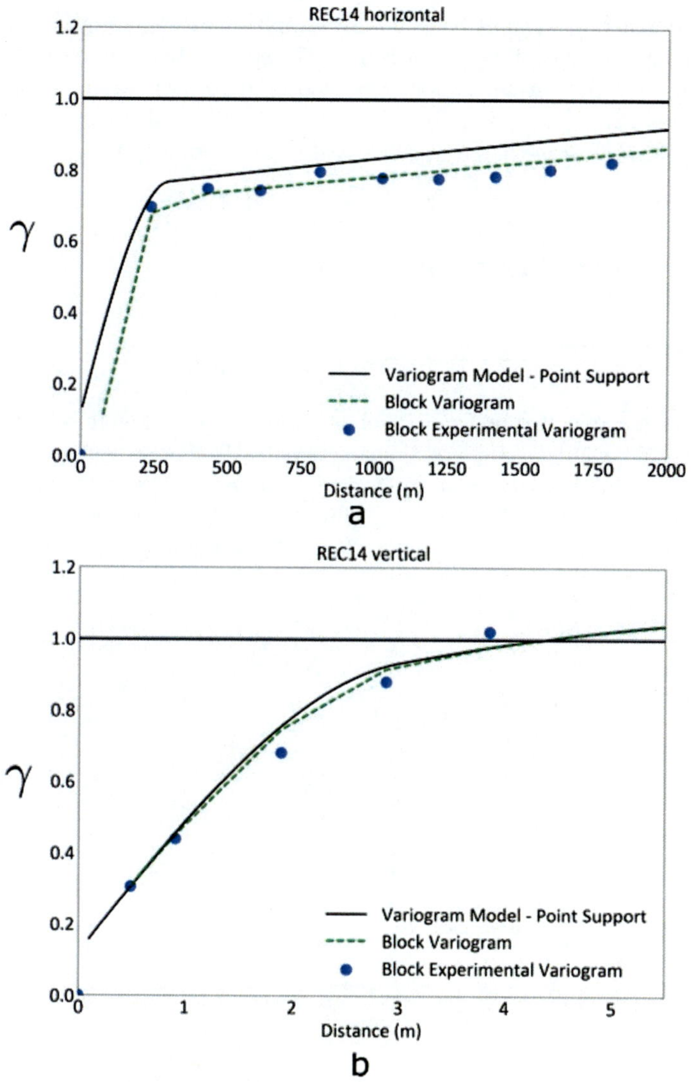

Fig. 6.4 Experimental variogram, variogram model at point support, and block variogram along with the **a** horizontal and **b** vertical directions

The grid spacing is 12.5 m × 12.5 m × 0.5 m along the X, Y, and Z directions. The number of nodes along the X, Y, and Z directions is 1240, 1108, and 17, respectively. The entire grid was simulated, and then the simulated values outside the geological domain were clipped. In this case study, twenty-five realizations were built. The software for DSS consists of an executable that was published by Azevedo and Soares (2017). The sampling method for the local cumulative distributions is the Soares approach (Soares 2001).

The difference of support between the samples was considered in the kriging system through block-to-block covariances. The spacing between the discretization points is 0.50 m. For instance, a sample whose length is 1.5 m is represented by three discretization points. The discretization points are used to calculate the block-to-block and point-to-block covariances used in the kriging system.

6.4 Results and Discussion

6.4.1 Realizations

Figure 6.5a, b shows two realizations obtained by DSS. Although both models are different, they have the same spatial characteristics. The high- and low-grade areas are in the same regions for both models. For comparison purposes, an estimated model was built with ordinary kriging (Fig. 6.5c). The estimated model shows a smoother transition between the grades due to the smoothing effect of kriging. On

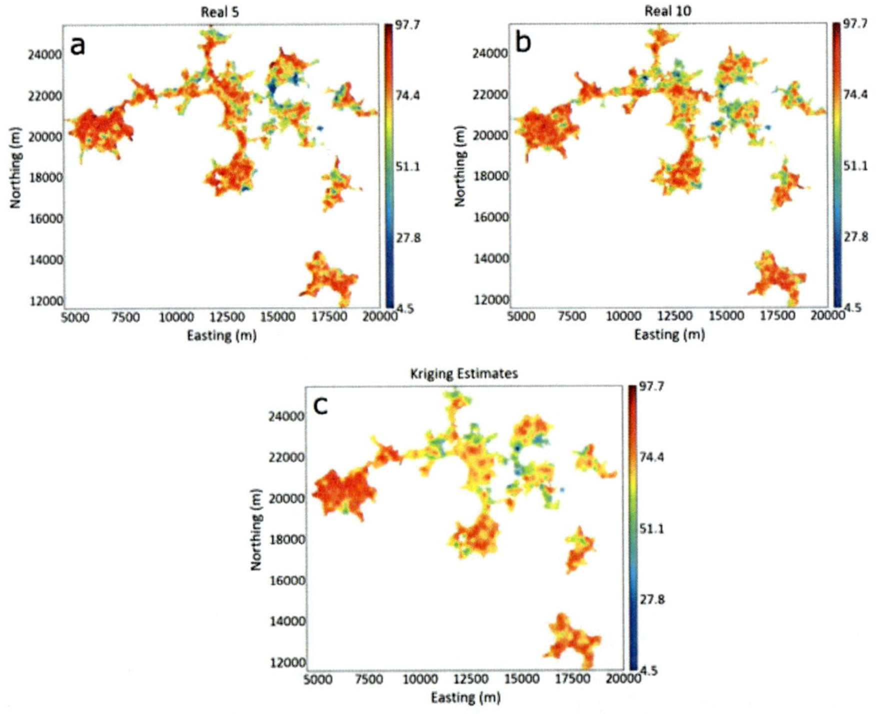

Fig. 6.5 Maps of the **a** fifth and **b** fifteenth realizations, and **c** map of kriging estimates

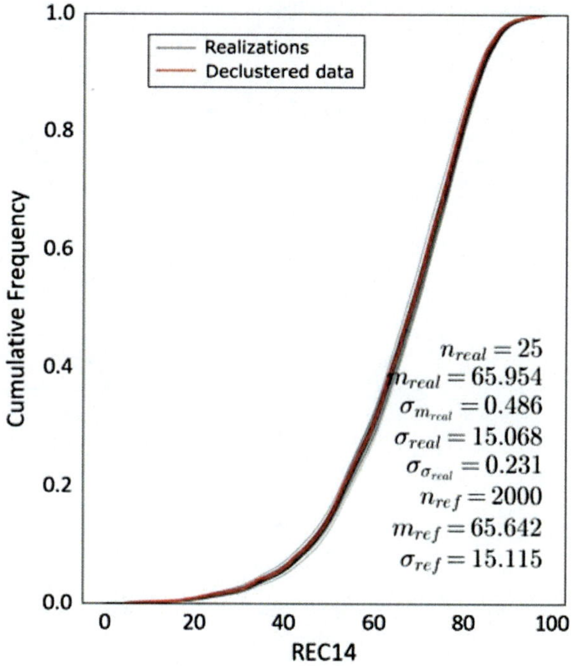

Fig. 6.6 Histogram reproduction of the realizations

the other hand, the realizations show higher variability and characterize the deposit's true spatial variability.

6.4.2 *Histogram Reproduction*

Figure 6.6 shows the histogram reproduction of the realizations. The histograms of the realizations are close to the histogram of the declustered data (see Fig. 6.6). This result shows that the realizations reproduced the distribution of the declustered data. Also, the histograms of the realizations are close to each other. This proximity indicates low global uncertainty. This result is expected as the area of interest is densely sampled.

6.4.3 *Variogram Reproduction*

Figure 6.7 compares the experimental variogram of the realizations against the variogram model used to conditioning the realizations. The experimental variograms of the realizations were standardized so that the sill is one. The results show that the realizations reproduced the variogram model, as expected.

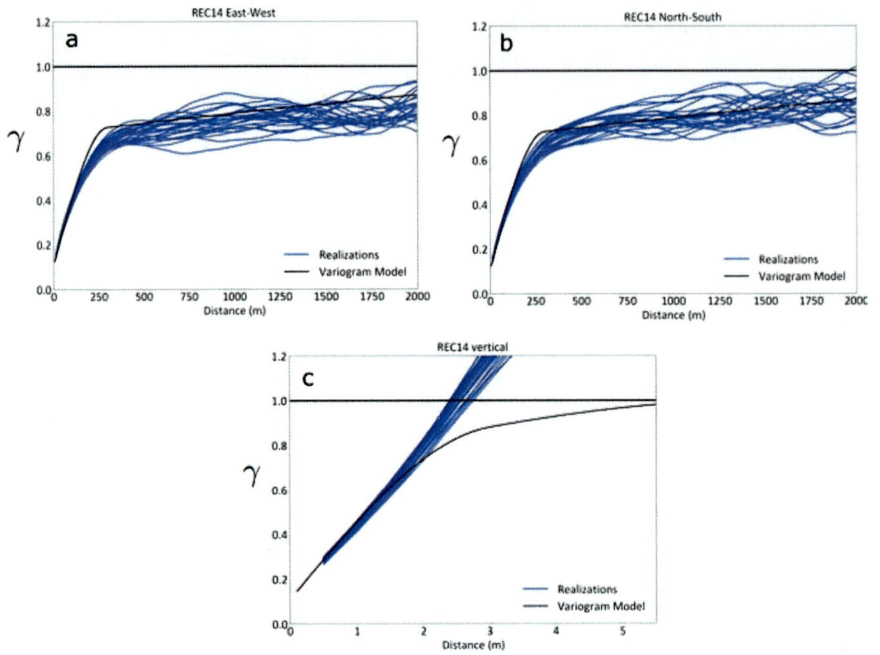

Fig. 6.7 Variogram reproduction along the **a** East–West, **b** North–South, and **c** vertical directions

6.4.4 Post-processing

Figure 6.8 shows the E-type of the realizations. The E-type consists of the average over all the realizations. As expected, the E-type shows a smooth map that resembles the kriging estimates (Fig. 6.5c).

Figure 6.9a shows the map of the standard deviation, which represents a measure of uncertainty. The standard deviation shows, for each grid node, the standard deviation over all the realizations. The regions where the standard deviation is low are the regions where all the realizations have similar simulated values. In other words, all the possible values are similar, and the practitioner knows reasonably well whether this region has either high or low grade (low uncertainty). In contrast, the regions with a high standard deviation mean the opposite: high uncertainty. As expected, the regions closer to the data points have lower uncertainty.

Figure 6.9b amplifies a region to demonstrate the effect of data support on uncertainty. Overall, the regions closer to the samples whose length is nearly 0.5 m have a lower standard deviation than the regions closer to the samples whose length corresponds to the ore thickness. This result agrees with the results shown by Horta et al. (2014) and is intuitive. For instance, a sample representing the average grade over 2 m contains less information than four aligned samples of 0.5 m. The four samples inform all the grades of the four 0.5 m intervals. In contrast, the 2 m sample does not inform if the grade at the first 0.5 m differs from that of the last 0.5 m. In this context,

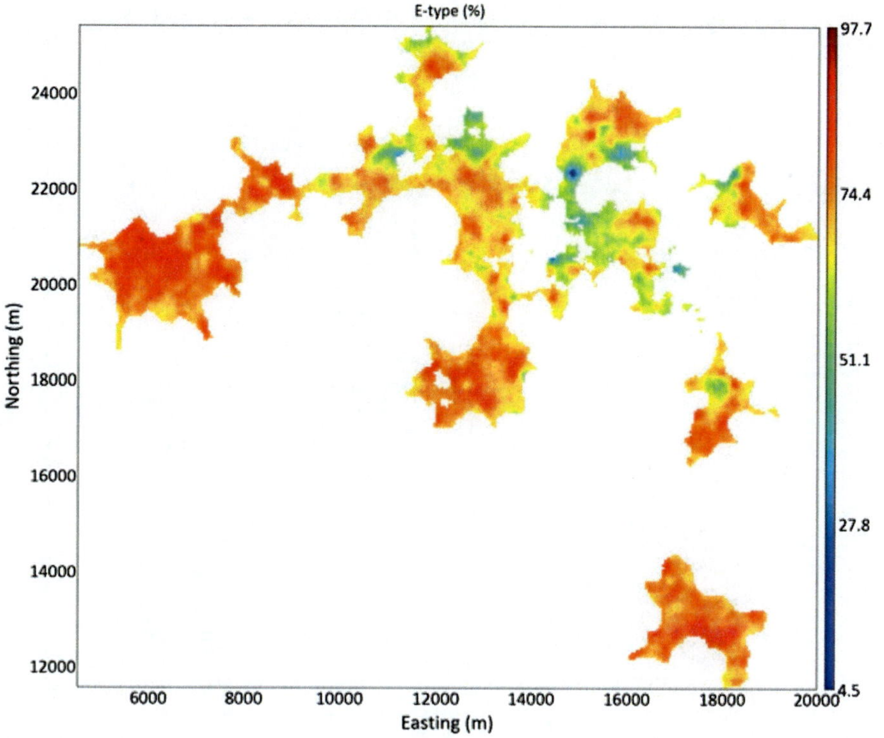

Fig. 6.8 Map of the average over all the realizations (E-type)

Direct Sequential Simulation with samples of different support detected this loss of information, represented by the areas of higher uncertainty.

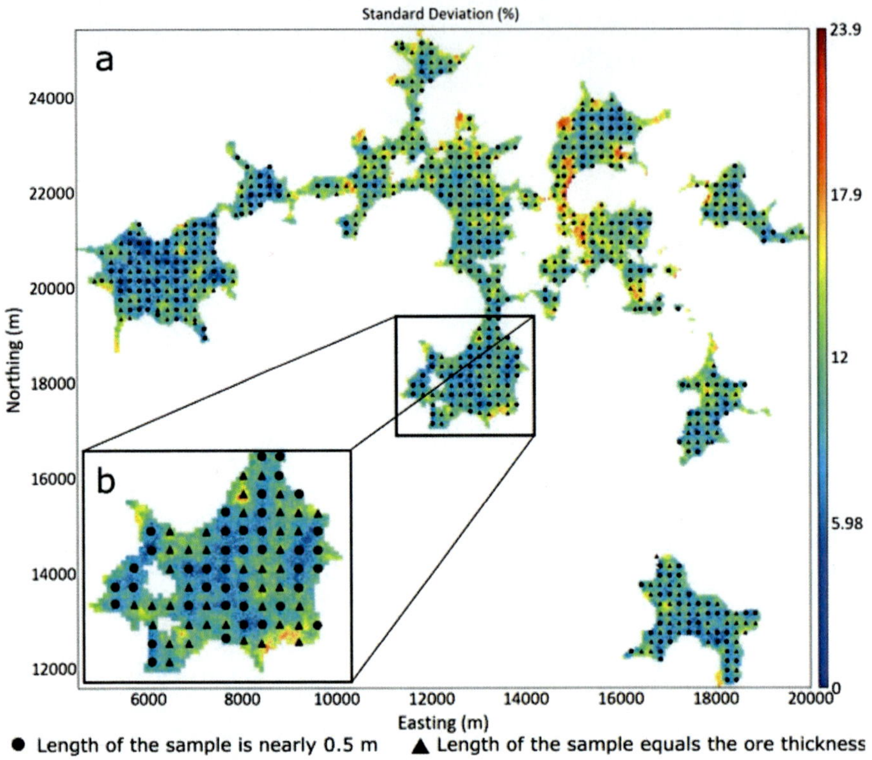

Fig. 6.9 **a** Map of the standard deviation over all the realizations and **b** detailed view of a region of the deposit

References

Azevedo L, Soares A (2017) Geostatistical methods for reservoir geophysics. Springer, Switzerland

Boucher A, Dimitrakopoulos R (2009) Block simulation of multiple correlated variables. Math Geosci 41(2):215–237

Caers J (2000) Adding local accuracy to direct sequential simulation. Math Geol 32(7):815–850

Emery X, Ortiz JM (2011) Two approaches to direct block-support conditional co-simulation. Comput Geosci 37(8):1015–1025

Godoy M (2003) The effective management of geological risk in long-term scheduling of open pit mines. PhD, University of Queensland

Horta A, Pereira MJ, Gonçalves M, Ramos T, Soares A (2014) Spatial modelling of soil hydraulic properties integrating different supports. J Hydrol 511:1–9

Isaaks E, Srivastava M (1989) An introduction to applied geostatistics. Oxford University Press, New York

Journel AG, Huijbregts CJ (1978) Mining geostatistics. Academic Press, New York

Liu Y, Journel AG (2009) A package for geostatistical integration of coarse and fine scale data. Comput Geosci 35(3):527–547

Oz B, Deutsch CV, Tran TT, Xie Y (2003) DSSIM-HR: a FORTRAN 90 program for direct sequential simulation with histogram reproduction. Comput Geosci 29(1):39–51

Soares A (2001) Direct sequential simulation and cosimulation. Math Geol 33(8):911–926

Chapter 7
Case Study of Sequential Indicator Simulation with Data of Different Support

This chapter illustrates the application of Sequential Indicator Simulation with data of different support to measure the geological complexity of a mineral deposit. We define the geological complexity based on the number of rock types and how connected they are in a mining operation. For example, an ore formation that contains a single rock type has minimum geological complexity. In contrast, an ore formation that contains five different rock types intermingled has high complexity.

Managing geological complexity is a crucial factor for the successful operation of mining activities. Consider, for instance, two rock types with two different processing routes. Mixing these rock types is likely to produce suboptimal results in terms of metallurgical recovery. The areas where these two rocks are in contact have high geological complexity. These areas may be prioritized for infill drilling so that their contacts become more accurate than they are currently.

This chapter shows how Sequential Indicator Simulation with data of different support may be applied to measure the geological complexity in an iron ore formation. The case study uses data derived from an iron ore deposit located in the Quadrilaterro Ferrifero region in Brazil.

7.1 Dataset Presentation

The dataset consists of diamond drill holes (DDH) with rock type information and the interpreted geological model. There are four rock types: (i) waste, (ii) hard itabirite, (iii) friable itabirite, and (iv) hematitite. These four rock types play different roles in the mining operation. Waste does not have economic value and is sent directly to the waste dump. The itabirites represent low-grade ore whose Fe contents range from 30 to 60% roughly. The hematitite has the highest Fe contents, exceeding 60%.

Figure 7.1 shows the drill hole samples and the interpreted geological model. The block size of the interpreted block model is 10 m × 10 m × 10 m along the *X*, *Y*, and *Z* directions, respectively. The length of the drill hole samples is 10 m. The hematitite

M. A. Arcari Bassani and J. F. Coimbra Leite Costa, *Geostatistics with Data of Different Support Applied to Mining Engineering*, Topics in Mining, Metallurgy and Materials Engineering, https://doi.org/10.1007/978-3-030-80193-9_7

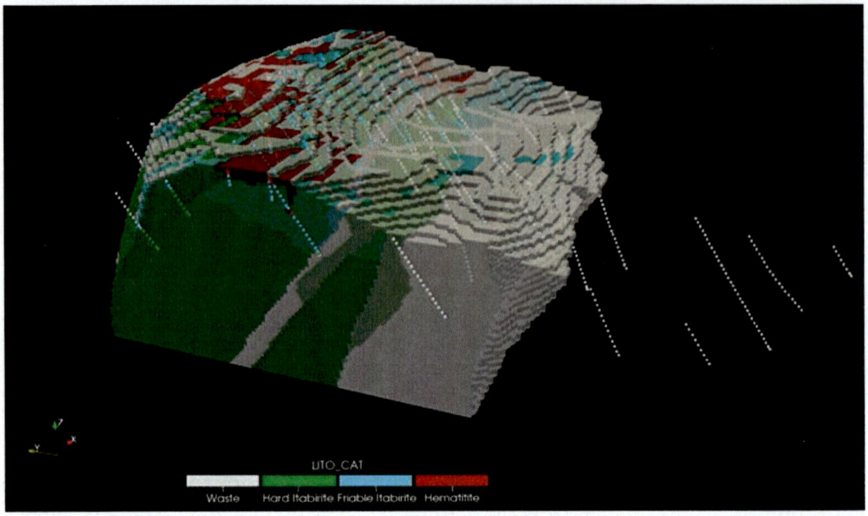

Fig. 7.1 Drill hole samples and interpreted geological model. Note that the opacity of the interpreted geological model was reduced so that the drill holes are visible

and friable itabirite occur in the upper parts of the deposit in the form of lenses. In contrast, the hard itabirite and waste are more continuous and cover the majority of the deposit. Another interesting aspect is a waste dyke that crosses in the middle of the ore formation.

The interpreted geological model was built using interpreted cross sections (this method is also referred to as an explicit model). In this method, the geologist manually defines the contours of the mineralization in a series of cross sections or slices (Abzalov 2016; Sinclair and Blackwell 2002). Then the cross sections are linked together to create a wireframe or geological volume. Lastly, the block model is flagged according to the wireframes. Figure 7.2 shows a cross section of the interpreted block model together with the drill holes. Folding events caused the curvature of the deposit.

We selected four benches of the upper part of the deposit to measure its geological complexity. The upper part of the iron ore formation has four rock types intermingled, while the lower part consists of two very continuous rock types (hard itabirite and waste). Moreover, the upper part of the deposit contains the rock types with the highest Fe contents. Figure 7.3a shows the interpreted block model with low opacity with the selected benches highlighted. Figure 7.3b shows the selected benches of the interpreted block model and the drill hole samples inside the benches' volume.

Figure 7.4 shows plan views of the four benches. The four benches cover an area of 460×530 m^2 and have a total height of 40 m (10 m per bench). The areas where the rock type changes as the bench elevation decreases represent a challenge. Iron ore mining often uses large excavation equipment to enhance productivity. The drawback is that the drilling and blasting operations move large ore volumes and mix ore from two consecutive benches. If the rock types of the consecutive benches are

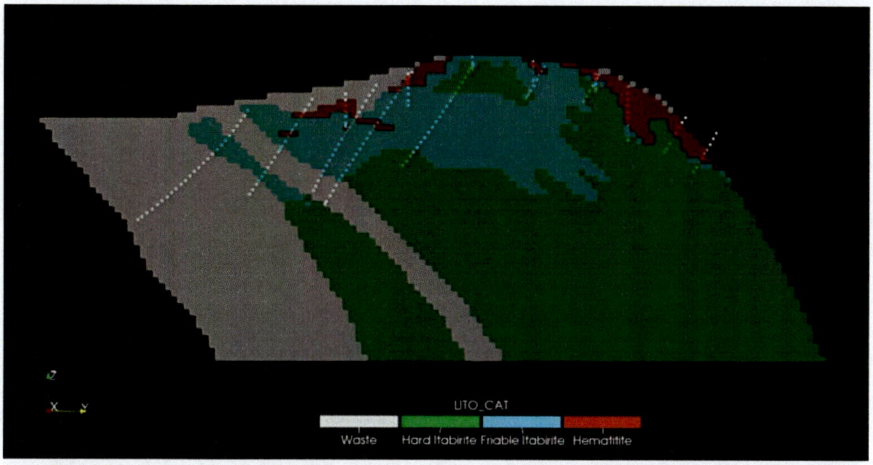

Fig. 7.2 Vertical section of interpreted block model and drill hole samples. The opacity of the block model was reduced to improve the visibility of the drill hole samples

different, the blasted ore will consist of a mixture of these rock types. This mixture leads to suboptimal use of the processing plant. For instance, the Fe contents of the hematitite ore may be lowered due to the itabirites and waste mixture.

Figure 7.5 shows the global proportions of the rock types for the drill hole samples and the interpreted block model in the study area. The proportions of the interpreted block model are the reference proportions to be reproduced by the simulations. The highest proportion occurs for the itabirites, while hematitite has the lowest proportion.

7.2 Methodology

The proposed methodology combines the information from the interpreted block model and the drill hole samples. The primary assumption is that the rock type proportions of the interpreted block model at larger scales are well known, but the exact limits of the rock types at block scale are not. The drill hole samples represent information without error so that the simulations perfectly reproduce these samples.

The algorithm uses the larger-scale proportion of the interpreted block model and the drill hole samples to simulate the rock types. The result is a set of equally probable simulated models that honor both the large-scale proportions and drill hole samples.

The methodology consists of the following steps:

1. Calculate the average proportion of each rock type over a large scale;
2. Codification of the rock type variable into indicators;
3. Spatial continuity analysis of the indicators;
4. Sequential Indicator Simulation with data of different support;
5. Post-processing.

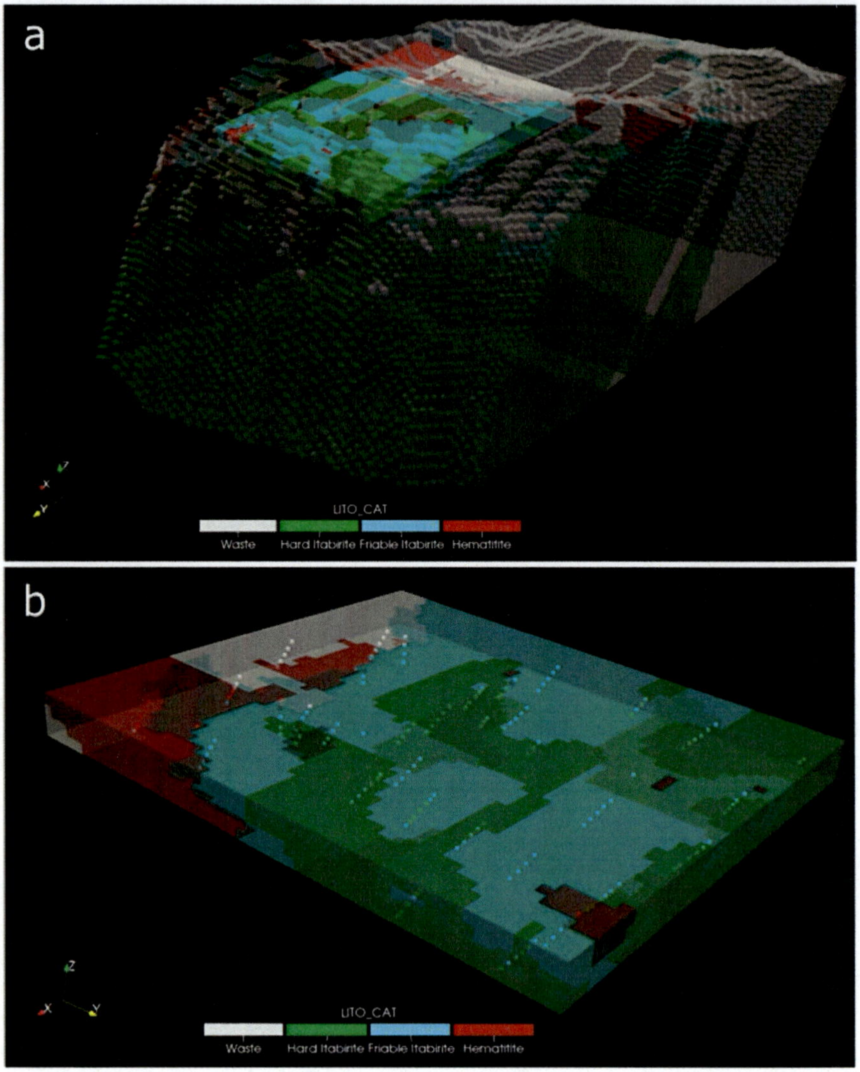

Fig. 7.3 **a** Interpreted block model with low opacity with the four benches selected as the study area and **b** four benches selected as the study area and drill hole samples

7.3 Average Proportions and Indicators

The average proportion was calculated as the vertical proportion of each rock type over the four benches of study. The vertical proportions were obtained from the interpreted block model. If an interpreted block model is not available, one could replace it by calculating the moving window averages of the drill hole samples.

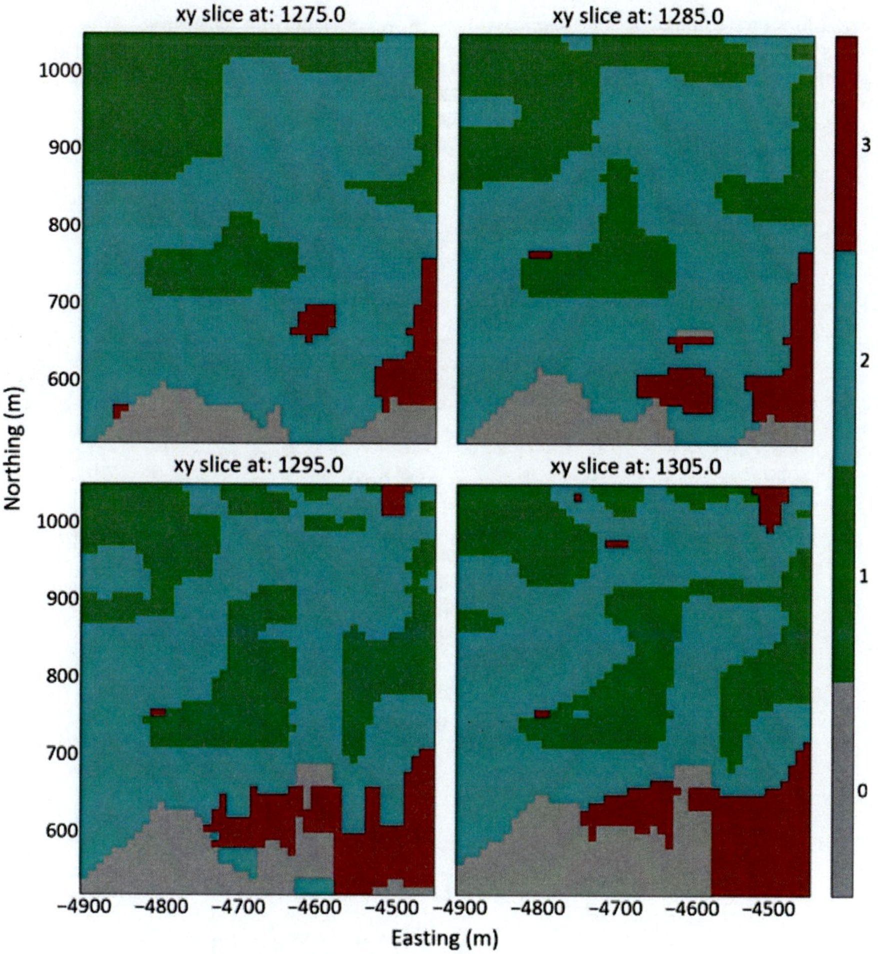

Fig. 7.4 Horizontal slices of the four benches selected as the study area. In gray the waste, in dark green the hard itabirite, in light green the friable itabirite, and in red the hematitite

Figure 7.6 shows the vertical proportions of the four rock types. The regions where one single rock type is present at the four benches have a vertical proportion of one (this is the case for the hard itabirite in the northwestern corner of the study area).

The original dataset with rock type information was coded as indicators. The indicators for the rock types were obtained by Eq. 7.1:

$$i(\mathbf{u}_i; k) = \begin{cases} 1 \text{ if category } k \text{ prevails at location } \mathbf{u}_i \\ 0 \qquad\qquad \text{otherwise} \end{cases} \tag{7.1}$$

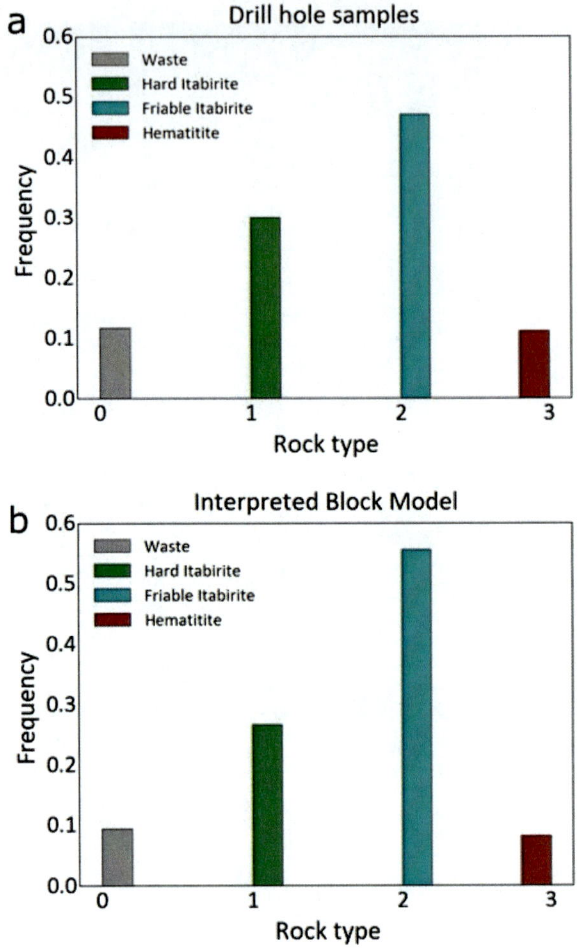

Fig. 7.5 **a** Proportion of rock types in the drill hole samples and **b** the interpreted block model inside the study area

where $i(\mathbf{u}_i; k)$ is the indicator for category k at the location $\mathbf{u}_i$. Figure 7.7 shows the original dataset with rock type information, while Fig. 7.8 the dataset coded as indicators for the four rock types.

7.4 Spatial Continuity Analysis

Figures 7.9, 7.10, 7.11 and 7.12 show the experimental indicator variograms and the variogram models for the main directions of anisotropy. The experimental variograms were calculated using all the drill hole samples. The experimental values were standardized so that the sill equals one.

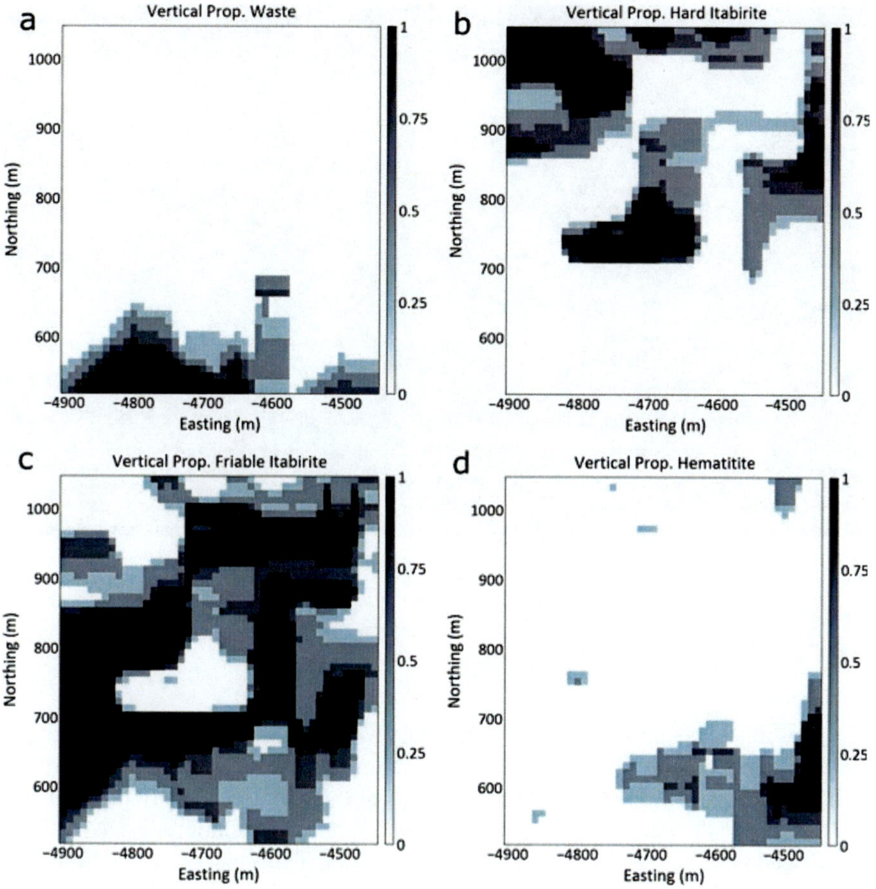

Fig. 7.6 Average vertical proportions of **a** waste, **b** hard itabirite, **c** friable itabirite, and **d** hematitite obtained from the interpreted block model

The main directions of anisotropy found were NS/Dip 30°, N90°E, and N180°E/Dip 60°, which agree with the geological structures. The direction NS/Dip 30° aligns with the North Flank of the fold, which has a shallow dip. The direction N90°E follows the strike of the ore formation, which is the EW direction. The direction N180°E/Dip 60° is perpendicular to the other two main directions and is roughly the down-the-hole direction.

Table 7.1 describes the indicator variogram models of the four rock types. The nugget effect was set to 0.05 for all the cases. Although some experimental variograms exhibited a large short-scale variability, the nugget effect was kept small as rock types are usually continuous for small distances.

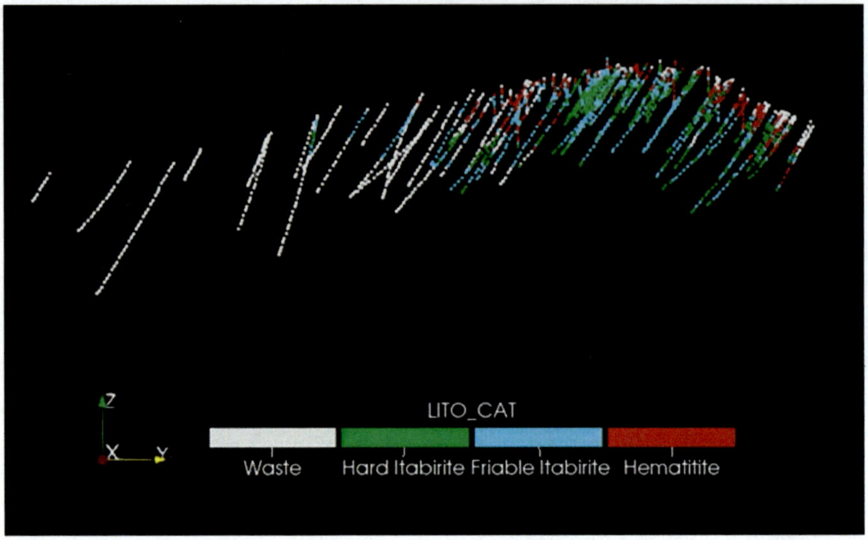

Fig. 7.7 Drill hole samples colored by rock type

7.5 Geostatistical Simulations

The geostatistical simulations were performed with the software BlockSIS (Deutsch 2006). Fifty realizations were run. The conditioning information consisted of the drill hole samples and the vertical proportions of the rock types. Only the drill hole samples inside the study area were used to condition the simulations. Each indicator was kriged with its corresponding variogram model. This characteristic strength of the Sequential Indicator Simulation allows the geomodeler to inform different spatial continuities for different rock types.

The simulation grid corresponds to the four benches of the study area to avoid extrapolation. The grid size has $46 \times 53 \times 4$ cells along the X, Y, and Z directions while the cell size is 10 m $\times$ 10 m $\times$ 10 m, which mimics the resolution of the interpreted block model. A higher-resolution grid was not used as the drill hole samples have a nominal length of 10 m. Keeping the same vertical size of the grid cells and the drill hole samples avoids loss of information in sequential simulation algorithms (Rossi and Deutsch 2014).

The difference in support between the vertical proportions and the drill hole samples was considered by using block-to-block and point-to-block covariances in the indicator kriging system. In this case study, we assumed that the vertical proportions do not have errors and are considered hard data in the kriging system. Thus, the block kriging option was chosen in BlockSIS (Deutsch 2006). If the interpreted block model is considered uncertain (soft data), the correct approach is to choose the block cokriging option in BlockSIS.

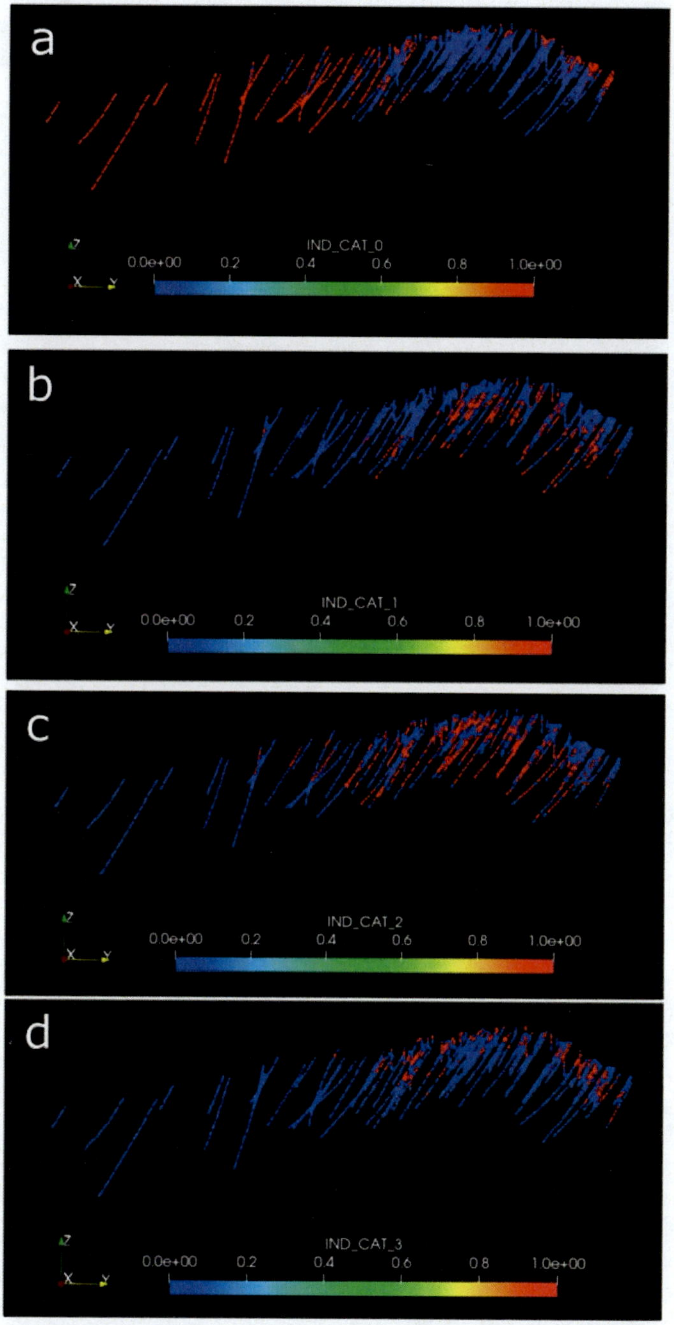

Fig. 7.8 Drill hole samples colored by indicators of **a** waste, **b** hard itabirite, **c** friable itabirite, and **d** hematitite

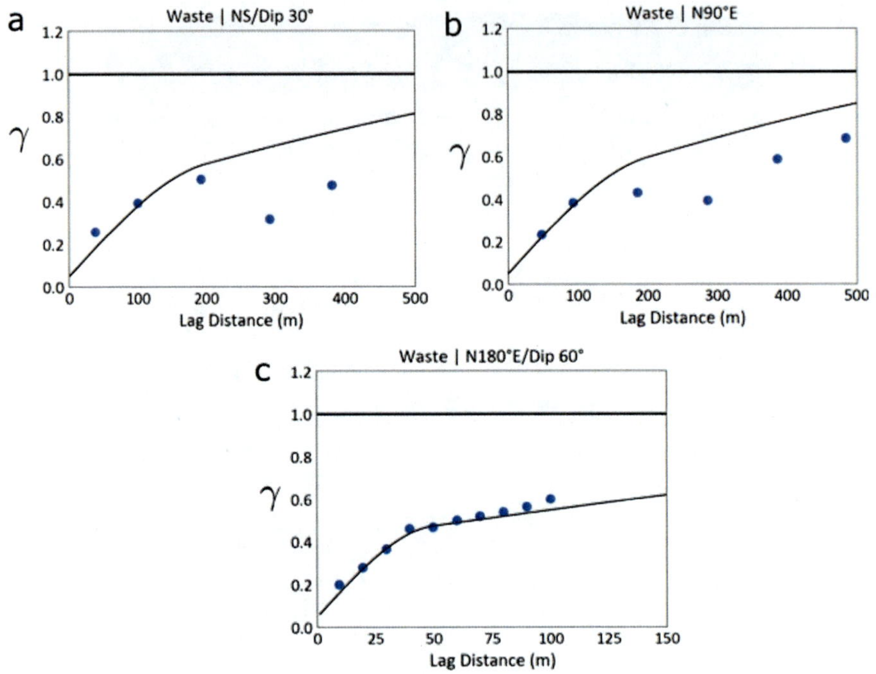

Fig. 7.9 Experimental indicator variogram and variogram model of waste along with the directions of **a** major, **b** medium, and **c** minor spatial continuity

The block kriging option considers only the difference in support between the drill hole samples and the vertical proportions. In contrast, the block cokriging option considers the difference in both support and precision between the two types of data. In comparison to block cokriging, block kriging increases the weights assigned to the interpreted block model. Thus, the realizations are highly constrained by the vertical proportions and resemble the interpreted block model. On the other hand, the block cokriging option diminishes the influence of the vertical proportions on the realizations. In general, the realizations built by block kriging show less variability and uncertainty than the block cokriging counterparts. The decision of which kriging option to choose depends on the reliability of the interpreted block model. Moreover, a sensitivity analysis may be performed to check which option results in the more realistic model. These vertical proportions allow the geomodeler to use the interpreted block model as conditioning information. For instance, consider that a vertical proportion of hard itabirite over the four benches is 0.50. In this case, out of the four simulated blocks that comprise the four benches, only two simulated blocks are flagged as hard itabirite. In summary, the use of vertical proportions aids the simulation algorithm to create models that agree with the geologists' expertise portraited in the interpreted block model.

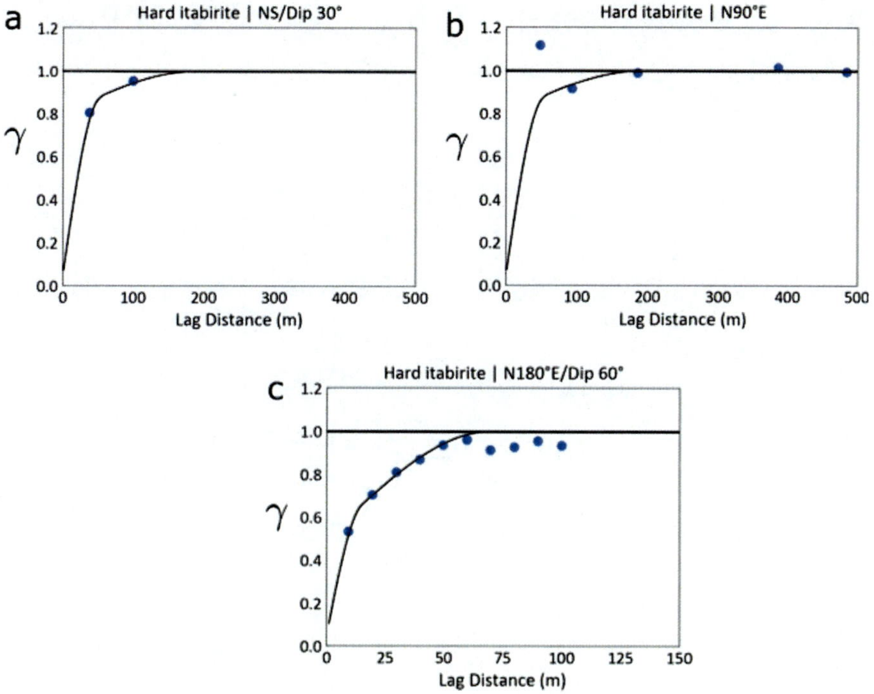

Fig. 7.10 Experimental indicator variogram and variogram model of hard itabirite along with the directions of **a** major, **b** medium, and **c** minor spatial continuity

7.6 Results and Discussion

7.6.1 Realizations

Figure 7.13 shows the interpreted block model and two realizations. The realizations have more variability than the interpreted block model, which is a smooth representation of reality. Moreover, the realizations are consistent with the interpreted block model. In other words, the regions where one particular rock type prevails in the interpreted block model are the same regions where this rock type prevails over all the realizations. For instance, the northwestern corner of the four benches has a high proportion of hard itabirite in the interpreted block model (see Fig. 7.13a). This high proportion of hard itabirite is present at the first and thirty-ninth realizations (Fig. 7.13b, c).

Figure 7.14 shows a scatter plot between the global rock type proportions of the interpreted block model against the global rock type proportions of the first realization. The results show that the rock type proportions were perfectly reproduced, as the points are close to the line $y = x$. The excellent reproduction of rock type proportions is an advantage of the Sequential Indicator Simulation with data

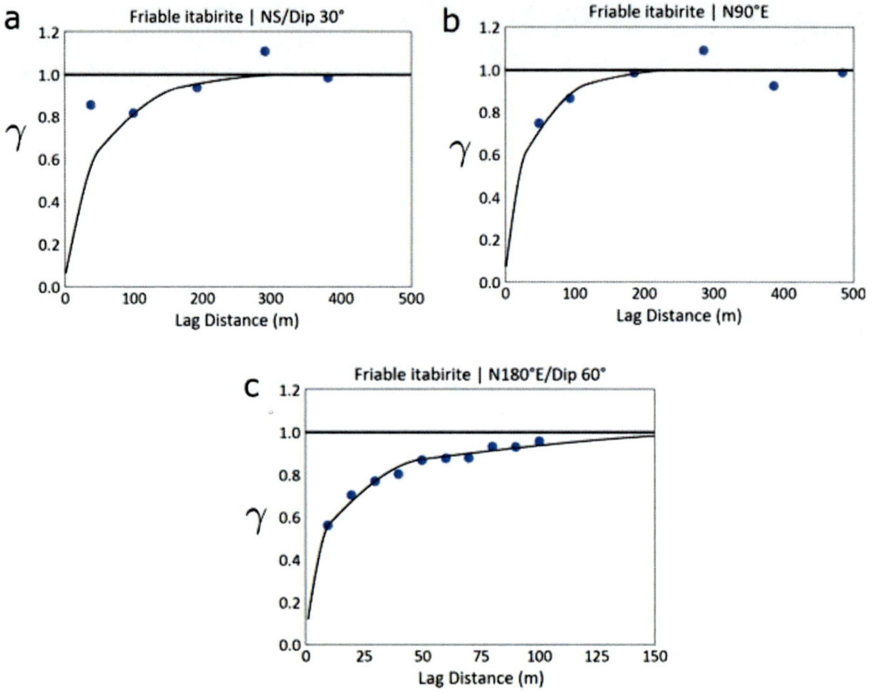

Fig. 7.11 Experimental indicator variogram and variogram model of friable itabirite along with the directions of **a** major, **b** medium, and **c** minor spatial continuity

of different support compared to the conventional approach of Sequential Indicator Simulation, which uses only the drill hole samples. The conventional approach of Sequential Indicator Simulation often fails to reproduce the proportions of the rock types that represent a small fraction of the study volume (Soares 1998).

Figure 7.15 shows the vertical proportions of each rock type of the first realization. The vertical proportions of the first realization resemble the vertical proportions of the interpreted block model (Fig. 7.6).

7.6.2 Post-processing

Figure 7.16 shows the probability of each category calculated over the fifty realizations, called E-type. For instance, consider a grid cell that is simulated as hematitite in twenty-five realizations out of the fifty. This particular grid cell has a 0.50 probability of being hematitite. In summary, the E-type model informs, for any grid cell, the probability that a specific rock type prevails in this cell. The E-type model agrees with the interpreted block model (Fig. 7.3b). For instance, the

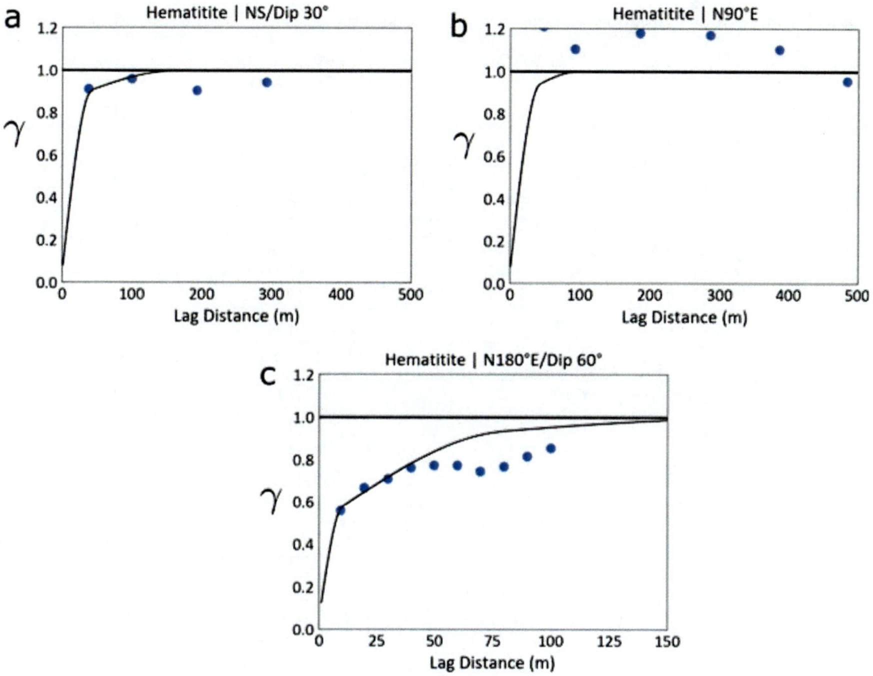

Fig. 7.12 Experimental indicator variogram and variogram model of hematitite along with the directions of **a** major, **b** medium, and **c** minor spatial continuity

highest probabilities of being waste occur in the same regions where the waste is the dominant rock type (considering its close vicinity) in the interpreted block model.

Figure 7.17 shows the uncertainty, calculated as one minus the highest probability (Goovaerts 1997). The highest probability is informed by the E-type model (Fig. 7.16). A region where a particular rock type is dominant has little uncertainty. For instance, the probability of being hard itabirite in the northwestern region of the model is roughly one (see Fig. 7.16b). In other words, there is no question that this region is hard itabirite, so its uncertainty is zero (Fig. 7.17). The highest uncertainty occurs when the four rock types have the same probability of 0.25, which results in an uncertainty of 0.75. The uncertainty model shows that the highest uncertainties occur around the contact between rock types, as expected. These results are consistent with those found by Boucher et al. (2014) and Carvalho et al. (2014).

From a practical perspective, the uncertainty model informs the geological complexity of the deposit. The geological complex areas are the areas near the contacts and present two challenges for the mining operation: (i) the contacts are uncertain, and (ii) the risk of dilution and ore loss is exacerbated. The correct destination of a block depends on its rock type. When the rock type of a particular block is uncertain, choosing the correct processing route for this block is difficult. Moreover, these geological complex areas are prone to dilution and ore loss. The

Table 7.1 Indicator variogram models for the four rock types

Rock type	Structure	Contribution	Range NS/Dip 30° (m)	Range N90°E (m)	Range N180°E/Dip 60° (m)
Waste	Nugget Effect	0.05	0	0	0
	Spherical	0.35	200	200	50
	Spherical	0.60	1000	900	600
Hard itabirite	Nugget Effect	0.05	0	0	0
	Spherical	0.45	50	50	15
	Spherical	0.30	60	60	70
	Spherical	0.20	190	190	70
Friable itabirite	Nugget Effect	0.05	0	0	0
	Spherical	0.40	50	30	10
	Spherical	0.35	170	120	50
	Spherical	0.20	330	250	200
Hematitite	Nugget Effect	0.05	0	0	0
	Spherical	0.45	40	40	10
	Spherical	0.35	45	45	80
	Spherical	0.15	170	100	200

blasting operations and equipment excavation are unable to extract the geological volumes near the contacts perfectly. The result is waste mixed with ore (dilution) and ore that is not extracted (ore loss).

The correct identification of the complex geological areas allows the mine planning team to take the required actions to mitigate these effects. For instance, these areas may be prioritized for infill drilling so that their uncertainty is reduced. Another action is to use high-selectivity equipment to mine these complex geological areas to avoid dilution and ore loss.

Further use of the geological uncertainty model involves mineral resources classification. The geological uncertainty may be considered when the mineral resources are classified as measured (low uncertainty), indicated (medium uncertainty), and inferred (high uncertainty). In this case, the regions with high geological uncertainty would be classified as inferred, with medium uncertainty as indicated, and with low uncertainty as measured.

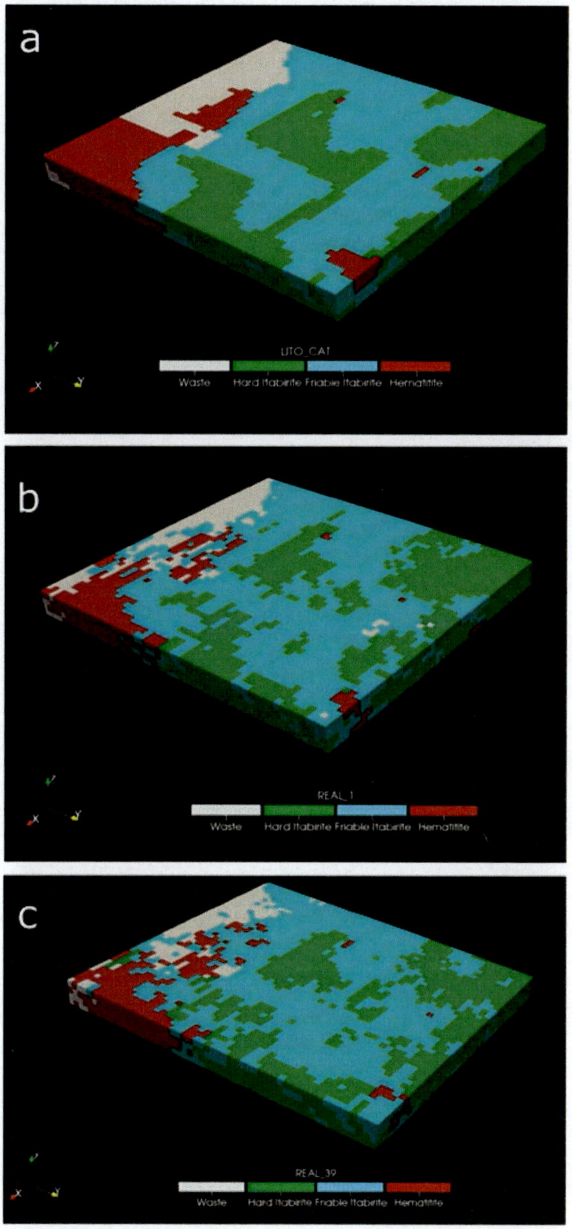

Fig. 7.13 **a** interpreted block model, **b** first realization, and **c** thirty-ninth realization

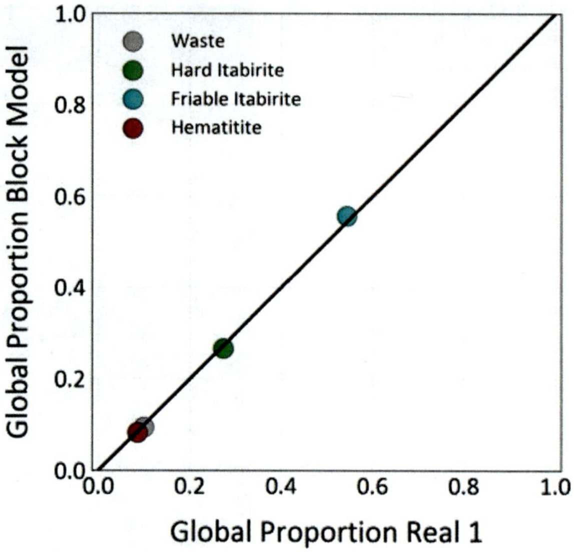

Fig. 7.14 Global proportion of rock types for the interpreted block model and first realization

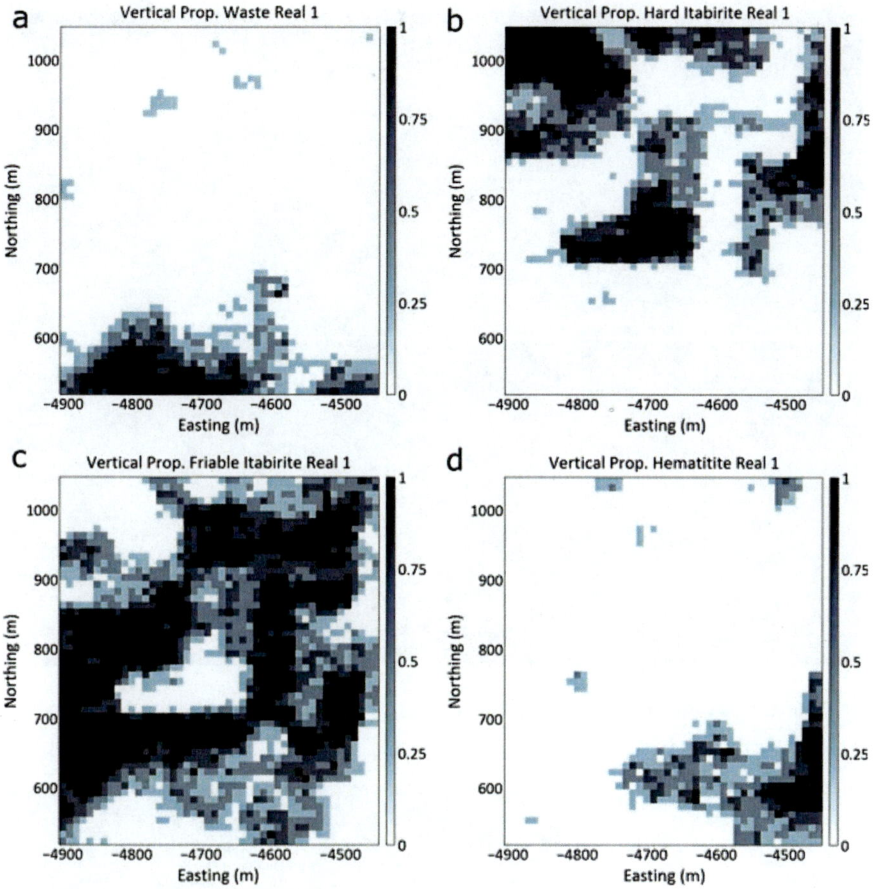

Fig. 7.15 Average vertical proportions of **a** waste, **b** hard itabirite, **c** friable itabirite, and **d** hematitite of the first realization

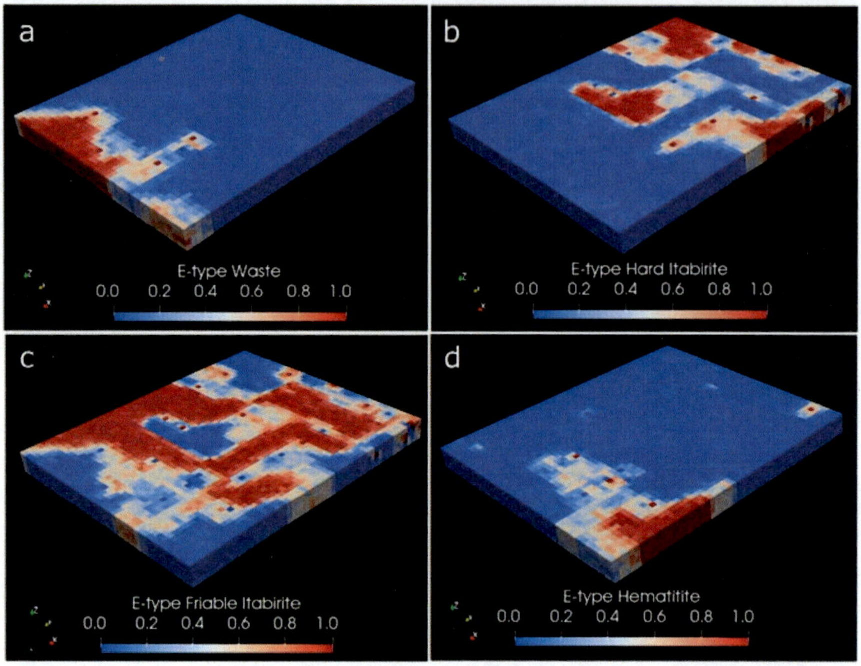

Fig. 7.16 E-type of **a** waste, **b** hard itabirite, **c** friable itabirite, and **d** hematitite

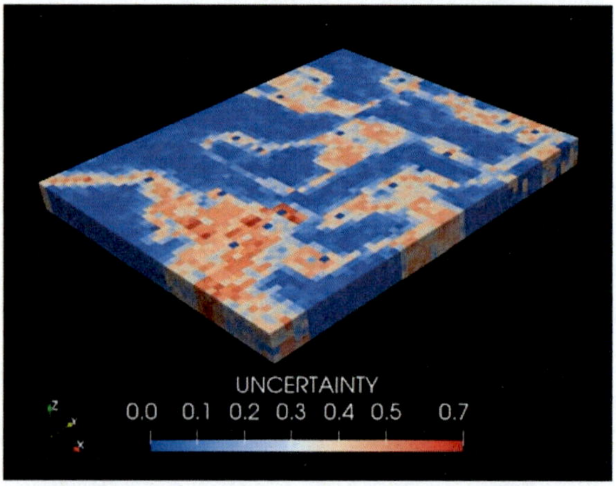

Fig. 7.17 Uncertainty model

References

Abzalov M (2016) Applied mining geology, vol 12. Springer International Publishing, Switzlerand

Boucher A, Costa JF, Rasera LG, Motta E (2014) Simulation of geological contacts from interpreted geological model using multiple-point statistics. Math Geosci 46(5):561–572

Carvalho J, Correia P, Menezes S, Peixoto C, Soares A (2014) Uncertainty assessment of the orebodies geometry by using block indicator simulation. Mathematics of planet earth. Springer, Berlin, pp 731–734

Goovaerts P (1997) Geostatistics for natural resource evaluation. Oxford University Press, Oxford

Deutsch CV (2006) A sequential indicator simulation program for categorical variables with point and block data: BlockSIS. Comput Geosci 32(10):1669–1681

Rossi ME, Deutsch CV (2014) Mineral resource estimation. Springer, Berlin

Sinclair AJ, Blackwell GH (2006) Applied mineral inventory estimation. Cambridge University Press, London

Soares A (1998) Sequential indicator simulation with correction for local probabilities. Math Geol 30(6):761–765

Chapter 8
Conclusions

Geostatistics provides an extensive toolbox to build grade models required for mining planning. This extensive toolbox often is not fully explored, and practitioners tend to use the more traditional algorithms such as ordinary kriging. The result is those geostatistical algorithms that allow the integration of data of different support are disregarded.

In this book, we emphasize the importance of using the data as they are. In other words, if two types of data have different support, this difference must be considered. Most importantly, Chapter 2 allows the practitioner to understand when the difference of support impacts more or less the estimates based on the variogram model.

We also tried to improve the clarity of this topic by focusing on practical applications. The use of data of different support in Geostatistics is usually taught in Geostatistics courses for graduate students. However, the classroom environment typically uses simplified examples. The primary goal is to facilitate the students' understanding of the fundamentals. In this book, we show case studies with data obtained from real mineral deposits.

The case studies covered the major applications of grade models used in Mining Engineering. Chapter 5 shows an application of grade estimation using ordinary kriging with data of different support. Ordinary kriging is the main interpolation engine for mineral resources estimation. Chapter 6 shows the application of Direct Sequential Simulation to measure grade uncertainty when the data have different support. Geostatistical Simulation builds equally probable models that are used for stochastic mine planning. In addition, the simulated models provide a measure of uncertainty. Chapter 7 applies the Sequential Indicator Simulation with data of different support to measure the geological uncertainty. Often, the geological uncertainty is the highest uncertainty present in a mineral resources model. The technique presented in Chapter 7 integrates the information provided by the interpreted block model and the drill hole samples for categorical variables. The interpreted block model represents the mine geologists' expertise.

M. A. Arcari Bassani and J. F. Coimbra Leite Costa, *Geostatistics with Data of Different Support Applied to Mining Engineering*, Topics in Mining, Metallurgy and Materials Engineering, https://doi.org/10.1007/978-3-030-80193-9_8

Springer Proceedings in Earth and Environmental Sciences

The series Springer Proceedings in Earth and Environmental Sciences publishes proceedings from scholarly meetings and workshops on all topics related to Environmental and Earth Sciences and related sciences. This series constitutes a comprehensive up-to-date source of reference on a field or subfield of relevance in Earth and Environmental Sciences. In addition to an overall evaluation of the interest, scientific quality, and timeliness of each proposal at the hands of the publisher, individual contributions are all refereed to the high quality standards of leading journals in the field. Thus, this series provides the research community with well-edited, authoritative reports on developments in the most exciting areas of environmental sciences, earth sciences and related fields.

Gisele Barbosa dos Santos ·
Miguel Fernandes Felippe · Roberto Marques Neto
Editors

Geomorphology of Brazil: Complexity, Interscale and Landscape

XIII SINAGEO (National Symposium of Geomorphology)

Springer

Editors
Gisele Barbosa dos Santos
Department of Geosciences
Universidade Federal de Juiz de Fora
Juiz de Fora, Brazil

Miguel Fernandes Felippe
Department of Geosciences
Universidade Federal de Juiz de Fora
Juiz de Fora, Brazil

Roberto Marques Neto
Department of Geosciences
Universidade Federal de Juiz de Fora
Juiz de Fora, Brazil

ISSN 2524-342X ISSN 2524-3438 (electronic)
Springer Proceedings in Earth and Environmental Sciences
ISBN 978-3-031-05177-7 ISBN 978-3-031-05178-4 (eBook)
https://doi.org/10.1007/978-3-031-05178-4

This Springer imprint is published by the registered company Springer Nature Switzerland AG
The registered company address is: Gewerbestrasse 11, 6330 Cham, Switzerland

Contents

Introduction: Geomorphology at the Start of the Twenty-First Century

Andrew S. Goudie

1 Introduction

Geomorphology is the study of the Earth's surface and the processes which shape it (Goudie and Viles 2010a, b). It is largely carried out by geologists and geographers. However, it is also an interdisciplinary discipline that has linkages to hydrology, archaeology, environmental history, engineering, ecology, and climatology. The discipline's recent history has been reviewed by Burt et al. (2008), Goudie (2016a), Gardner (2020), and Burt et al. (2022), while the role of various national schools has been recounted by Walker and Grabau (1993). Geomorphology has also become increasingly international in scope, as evidenced by the establishment of the International Association of Geomorphologists in 1989, and by the participation of geomorphologists in the meetings of the EGU.

The purpose of this chapter is to highlight some of the major features of Geomorphology at the start of the twenty-first century.

2 Development of Techniques

In recent decades there has been an explosion of techniques that have become available to geomorphologists (Goudie 1990). These have (i) allowed improved field measurements (e.g., through the use of GPS and data loggers), (ii) improved surveying of landform distribution and morphometry (through remote sensing, LIDAR, GIS, unmanned aerial vehicles, etc.) (Eckardt 2022), (iii) geophysical

A. S. Goudie (✉)
School of Geography, University of Oxford, South Parks Road, Oxford OX1 3QY, UK
e-mail: andrew.goudie@stx.ox.ac.uk

G. Barbosa dos Santos et al. (eds.), *Geomorphology of Brazil: Complexity, Interscale and Landscape*, Springer Proceedings in Earth and Environmental Sciences,
https://doi.org/10.1007/978-3-031-05178-4_1

techniques to permit two- or three-dimensional views of the materials and structures which make up the landscape (e.g., Ground Penetrating Radar and resistivity surveys), (iv) superior analyses of geochemical properties of materials, by, for example, the use of XRD and XRF, (v) assessment of the hardness of rocks (Viles et al. 2011), (vi) absolute dating of landforms and deposits by means of isotopes, optically stimulated luminescence, Caesium-137, cosmogenic nuclides, and thermochronology, including fission track dating (Anderson 2022), (vii) experimentation in the laboratory and under real field conditions, using programmable environmental cabinets, wind tunnels, rainfall simulators, dust and sand traps, electronic sensors, etc. (Church 2022), (viii) detailed environmental reconstruction (especially by miscellaneous types of core analysis), (ix) statistical analyses of large sets of data by means of computers, (x) and computer-based modelling (Church 2010; Martin 2022). Without all these technical developments, geomorphology would be a very different discipline from the one it has become.

3 Development of Landscapes Over Time

A major concern of geomorphologists for much of the past two centuries was the study of the long-term development of landscapes in response to climate changes and tectonic history. This involved the study of cycles of erosion, the establishment of denudation chronologies, and the analysis of landscape development in response to changes in climate (e.g., Büdel 1982) and base levels. Although in the second half of the twentieth century this historical/evolutionary approach became less dominant in the discipline, it has recently been re-energized because of the availability of a suite of new dating techniques (e.g., optical dating, cosmogenic radionuclides) and techniques for environmental reconstruction (e.g., by coring) (Anderson et al. 2013; Anderson 2022). There has been a renewed burst of interest in the role of Late Cenozoic environmental changes at a wide range of temporal scales. Quaternary geomorphology is a vibrant field, not least in lower latitudes, where the impact of pluvial and arid phases has been fundamental for understanding landscapes, including those of Brazil (de Paula Barros and Junior 2020; Mescolotti et al. 2021).

Longer-term studies of landform evolution have also blossomed because of an interest in plate tectonics, continental drift, sea-floor spreading, epeirogeny, and orogeny (Summerfield 2000). This enables one to explain such phenomena as drainage-basin evolution at a continental scale (e.g. Goudie 2005), the evolution of great escarpments on passive margins, and the distribution of volcanoes around the world.

Geomorphologists have also been much concerned with shorter-term environmental history and it is here that their work overlaps with that of environmental historians (see Hudson et al. 2008) and geoarchaeologists. Geoarchaeology is a fertile field of research with its own journals. Working with prehistorians and archaeologists, geomorphologists have investigated the effects of climatic, tectonic, and sea-level changes on human societies (e.g., Flemming 1999) and have assessed the relationship of archaeological sites to geomorphological settings, including dunes (Allchin et al. 1978), arroyos, colluvium, calcareous tufas, caves, lakes and lunette dunes, coastal erosion and construction, deltas, old river systems, badlands and alluvial deposits (Vita-Finzi 1967).

4 Rocks and Relief

Understanding the impact of rock types of landscapes is a fundamental component of Geomorphology. Thus the study of the links between rocks and relief has a long history during which studies have been performed on the links between particular rock types and landscape patterns. These have included studies of landforms on limestones and dolomites (Ford and Williams 2007), sandstones and conglomerates (Young et al. 2009), and granites (Migoń 2007). However, notwithstanding Yatsu's (1966) exhortation, quantification of the links between rock properties and landforms remains an under-researched part of Geomorphology. There was indeed a great school of rock control work that arose in Japan (Ouchi 1996), notably by Suzuki and colleagues (see Suzuki et al. 2000 for a history of this work). As Goudie (2016b) argued, rock properties occur at a range variety of scales, from large lineaments and fractures (which are measured in the field) through to individual rock micro-pores (which are calculated in the laboratory). At the mega-scale, there are discontinuities—faults, joints, and bedding planes. At the mesoscale, rock strength can be determined both in the laboratory and in the field by measuring such properties as abrasiveness and abradability, compressive, shear and tensile strengths, penetrometer resistance, surface hardness, and Young's Modulus of Elasticity. At a smaller scale, rocks can be tested in the laboratory to establish their resistance to weathering (particularly frost and salt action) and to assess the role of such factors as their porosities and water absorption capacities A new technique for the small-scale analysis of materials that was first developed in the 1930s but evolved from the 1960s onwards, was Scanning Electron Microscopy (SEM) (Whalley 1978).

Techniques such as the assessment of rock mass strength (RMS) (Selby 1980) and the study of the relationships between rock pore characteristics and resistance to weathering (e.g., Yu and Oguchi 2009) are indicative of the progress that is now being achieved. Rock hardness determination has developed as a research field (Viles et al. 2011), involving the use of the Schmidt Hammer (SH) (Goudie 2006), the Equotip, the Grindosonic, and dilatometric and sonometric techniques. Efforts have been made to relate this to such diverse phenomena as slope forms and instability, the morphology of shore platforms, glacial trough geometry, river channel dimensions, valley forms, the formation of cavernous weathering features, and inselberg development (Duszyński et al. 2022).

5 Processes

There are two main types of Earth surface processes: exogenic and endogenic (Goudie 2016a). The former refers to those processes (weathering, erosion, sediment transport, etc.) that are ultimately fuelled by the Sun's energy and which operate via the climate system. Particularly since the 1960s geomorphology has concerned itself with these exogenic processes, often at a reductionist level and using quantitative techniques, such as computational fluid dynamics (Lane et al. 1999). As a result of the quantitative revolution, great efforts have been made to measure such processes as grain entrainment and solute movements in small catchments. Classic and influential examples of this genre are those by Leopold et al. (1964) on rivers, by Carson and Kirkby (1972) on slopes, by Drewry (1986) on glaciers, by Washburn (1979) on the cryosphere, by Yatsu (1988) on weathering, by Masselink and Hughes (2003) on coasts, and by Gillette (1977) on aeolian processes. Discussions on recent developments in three of the biggest components of geomorphology—rivers, coasts, and slopes—are provided by Ferguson et al. (2022), Spencer and French (2022), and Kirkby (2022), respectively.

On the other hand, the latter refers to volcanic and tectonic processes powered by energy derived from the inside of the Earth. These exogenic processes operate over long time scales and over great regional extents. Since the 1960s, they have received increased attention as a consequence of the emergence of the plate tectonics paradigm (Burbank and Anderson 2011), and have contributed to a greater understanding of the global pattern of phenomena such as volcanoes, rift valleys, mountain ranges, and guyots (Summerfield 1991, 2022). They have also helped us to understand rates of denudation and fluvial incision in areas of active orogeny (Whipple and Meade 2006; Whittaker et al. 2007). Moreover, current work has shown that climate, as well as the erosional development of the landscape, feeds back into the ongoing tectonic processes (Whipple 2009). As Dadson (2010, p. 390) remarked 'the results from coupled geomorphic and geodynamic models suggest that climate-driven erosion is of first-order significance in the evolution of mountain belts across a range of time scales'. The study of rates of chemical weathering and physical denudation under different climatic and tectonic conditions has been boosted by a concern with how these processes relate to global carbon cycle (Goudie and Viles 2012). Exogenic process geomorphology and the new models of long-term landscape evolution, associated with new ideas on plate tectonics and novel geochronometric techniques, need to be combined more effectively than they have been in the past (Summerfield 2005; Bishop 2007).

More traditional evaluations of the links between climate conditions and the nature and rate of geomorphological processes persist as an active area of research, with syntheses of geomorphological phenomena in different climatic zones being produced, including those of humid tropical environments (Thomas 1994), glaciated areas (Benn and Evans 2010), deserts (Goudie 2013), and periglacial and permafrost regions (French 2017).

6 Living Landscapes

Organic agencies are crucial for understanding landform development. Thus since the 1980s, there has been a burgeoning concern with establishing links between plants, animals, and geomorphology, and the terms biogeomorphology (Viles 1988), zoogeomorphology (Butler 1995), ecogeomorphology, and geoecology have been used. Coombes (2016b) and Viles (2020) undertook citation analyses, which showed that publications in biogeomorphology increased exponentially during the 1990s.

Biogeomorphology is 'an approach to geomorphology which explicitly considers the role of organisms' (Viles 1988, p. 1), or as Coombes (2016a) defined it, 'Biogeomorphology is the scientific study of interactions and feedbacks between living and non-living parts of the landscape'. Viles recognized that there are two linked foci in biogeomorphology: 'The influence of landforms/geomorphology on the distributions and development of plants, animals, and microorganisms', and 'The influence of plants, animals and microorganisms on earth surface processes and the development of landforms.' The whole spectrum of biological life-forms is involved in biogeomorphological interactions, from bacteria and fungi affecting weathering and mineral precipitation to elephants excavating wallows, to cows causing ground compaction, to the effects of a large forest on the behavior of river catchments (Viles 2004).

Undoubtedly during the evolution of life, the impact of organisms on geomorphological processes has also evolved, and, for example, the Palaeozoic development of plant life about 440 million years ago would have dramatically changed the channel activity of rivers (Ielpi and Lapôtre 2020). Likewise, Algeo and Scheckler (1998) argued that the evolution of trees and seed plants and the appearance of multi-storied forests in the Devonian led to an intensification of soil formation and increased fluvial solute fluxes.

Recent work has tried to provide quantitative measures of relief complexity and to link this to biodiversity. Landforms have been seen as important components of habitat, particularly in river floodplains (e.g., Graf 2001; Bennett and Simon 2004). Moreover, geomorphological processes enhance an area's biodiversity by introducing dynamism and creating new habitats (Viles et al. 2008). Plants and animals are not merely passive occupiers of the Earth's surface. They play an active and key role in many geomorphological processes and can create unique landforms (beaver dams, coral and serpulid reefs, termitaria, phytogenic dunes, animal dens, ant mounds, etc.). Above all, biological influences can either accelerate or retard the rate of operation of exogenic processes. Organisms such as ants, notwithstanding their small size, achieve a remarkable amount of geomorphological work (Viles et al. 2021). Processes, including tree fall and root penetration, have major effects on slope forms, shallow landslides, and creep, while vegetation cover influences rainfall interception, infiltration rates, runoff, and sub-surface flow, temperature characteristics, and wind action. Riparian vegetation impacts upon river channel forms, flood plains, and bank erosion. Vegetation cover is also a crucial factor in controlling wind velocities and turbulence at the ground surface and in reducing wind erosion, dust storm generation, and sand dune movements. The combined effects of erosion reduction and accretion enhancement can be termed 'bioprotection' (Carter and Viles 2005), but conversely, organisms can accelerate erosion, a process which is called 'bioerosion'.

7 Submarine Geomorphology

Using an array of new techniques, geomorphologists have started to discover a great deal about the ocean floors (Micallef et al. 2022) and extra-terrestrial landscapes (Conway 2022).

For a long time, the former remained largely unexplored directly by humans, apart from some submarine-based expeditions, but their major features have now been mapped through ship and satellite-based remote sensing. Together, these have been used to create global topographic maps or digital elevation models (DEMs) of the ocean floor. Sidescan sonar and 3D seismic survey are among the techniques that allow the creation of 'images' of surface and sub-surface materials. Many large-scale features have been found which reflect the impact of glacial action (Ottesen and Dowdeswell 2009), tectonics, mass movements, and other processes. One particularly productive area of recent research has been the identification and interpretation of subsea mass movements, for landslides, creep phenomena, flows, slumps, slides, and falls are all common on the seafloor (Micallef et al 2007; 2009; 2018).

These mass movements can be hazardous to humans (Innocenti et al. 2021) and so this is a major research frontier for applied geomorphologists (Moore et al. 2018). Landslides in fjords, in the Gulf of Mexico (Fan et al. 2020), and on the flanks of oceanic islands, such as the Canaries, can generate tsunamis (Coppo et al. 2009), though this is not always the case (Løvholt et al. 2017). In addition, turbidity currents can pose challenges for engineering structures such as oil platforms (Clare et al. 2020). Submarine geomorphology also has implications for finding and developing hydrocarbons in places like the Congo and Angola Fans in the Atlantic off western Africa (Anka et al. 2010), the delta of the Nile (Li et al 2021), and the South China Sea (Wang et al. 2021a, b).

8 Extra-Terrestrial Geomorphology

Today, planetary geomorphology, thanks to the pioneering work of people like Greely and colleagues (Greeley and Iversen 1985), is a flourishing area of study (Baker 2008; Diniega et al. 2021; Conway 2022). Much work has been undertaken on Mars. Among the many Martian phenomena for which analogs have been sought on Earth, are wind scouring, yardangs and ventifacts, mass movements, flood deposits and alluvial fans, dunes, sand ripples, saltation phenomena, wind streaks, haloclasty and split rocks, chemical coatings on rocks groundwater-sapping features, relief inversion, coastal sabkhas, and dust events (Bhardwaj et al. 2021). This has stimulated research on a number of landforms and processes and has also led to research in a number of

Earth's drylands, including the Namib (Bourke and Goudie 2009), the Western Desert of Egypt (El-Baz and Maxwell 1982), Australia (Mann et al. 2004), the sandstone terrains of Utah (Chan et al. 2011) and the Qaidam Basin of the Tibetan Plateau (Xiao et al. 2017).

Titan is the largest of Saturn's moons and following the Cassini mission, which was launched in 1997 and remained active until 2017, we now know much more about its characteristics. The images sent back have revealed a landscape that is quite similar to that on Earth—except that the surface is composed of water ice, not rock, and is sculpted by liquid methane, not water. It has some interesting landform features (Lopes et al. 2020), including thousands of linear dunes (Radebaugh et al. 2010), and the largest cover of dune fields in our solar system (Bourke et al. 2010). There are also some stubby drainage networks that may have been generated by methane spring-sapping (Soderblom et al. 2007), tropical endorheic lakes (Tokano 2020), volcanic craters (Keane 2019), and alluvial fans (Birch et al. 2016).

9 Geomorphology and Earth System Science

In the 1980s, Earth System Science (ESS) evolved (see Steffen et al. 2006). It concentrates on modeling, treats the Earth as an integrated system, and seeks a more profound understanding of the physical, chemical, biological and human interactions that determine the past, current, and future states of the Earth's lithosphere, hydrosphere (including the cryosphere), biosphere, and atmosphere. It emerged in response to (i) the realization that biogeochemical systems operate globally and (ii) an increasing appreciation that Earth is a single system. Dadson (2022) provides a good survey of the role of ESS in geomorphology. Geomorphologists have created Earth System Models (Paola et al. 2006; Fan et al. 2019). A prime illustration of the way in which geomorphology contributes to Earth System Science is through understanding the links between silicate weathering in different geomorphological settings (e.g., island areas, mountains, glaciated terrains), the global carbon cycle, and long-term climate changes (Dupré et al. 2003). Examples of the effects of geomorphological change on the Earth System relate to biogeochemical cycling (Viles et al. 2008; Quinton et al. 2010), and silica and carbon budgets (Zhang et al. 2017). Soil erosion by wind may play a significant role in these (Webb et al. 2012; Chappell et al. 2013), but so may water erosion from agricultural fields, and the burning and subsidence of peat.

10 Global Change

In the 1970s, widespread employment of the term 'Global Change' emerged, as seen in the development of the *International Geosphere-Biosphere Programme: A Study of Global Change* (1986). The significance of this for geomorphology is demonstrated in the works of Steffen et al. (2006) and Slaymaker et al. (2009). Global warming, allied with the growth of more local human impacts on the environment, will have major effects on future landscapes. Indeed, they are already doing so.

Climate change is only one of the drivers of landscape change, and it is imperative that we weigh up the relative and/or combined impacts of global climate change and local human impacts. At the regional scale, land cover changes (such as tropical deforestation) may cause climate changes of comparable dimensions to those predicted to arise from global warming (e.g., Deo et al. 2009). Changes in runoff and sediment loads caused by land cover changes or dam construction may surpass those caused by future changes in rainfall quantities (e.g., Xu et al. 2007). Loss of coastal wetlands due to direct human action may be greater than those caused by sea-level rise (Nicholls et al. 1999), and the changing incidence of landslides may owe more to changes in human activity than to climate changes (Crozier 2010).

11 Global Warming

Global warming is one component of global change. Interest in this has developed since the early 1980s and has progressively created considerable interest in its consequences for a range of geomorphological phenomena (Goudie 1990, 2020) (Table 1). Of great importance has been the search for areas that will be particularly sensitive for four reasons: (i) their threshold reliance with respect to particular temperature, precipitation, and vegetation cover conditions, (ii) the compounding effects of climate change on other human actions, (iii) the presence of susceptible, fragile features and (iv) the fact that they are present in zones where climate change will be specially marked (e.g., higher latitudes and the margins of deserts).

Some phenomena that may as a consequence of these characteristics react very substantially to future heating are valley glaciers (especially on tropical mountains), permafrost features, floodplains, relict dune fields, low-lying coasts, areas exposed to tropical storms and hurricanes, and snow-fed rivers (IPCC 2021). Some locations will be subject to very rapid change because of the combined effects of climate change and other anthropogenic pressures, as is the case with many of the world's great deltas (Tessler et al. 2018) and with American rivers (Wan et al. 2017). As recent events in many parts of the world have shown, fire frequencies and severities could change, which would in turn have potentially huge impacts on slope processes (including mudflows) and surface runoff.

Moreover, most of the climatic models from 25 years ago have seemingly been correct in the scenarios they presented. The magnitude of geomorphological changes is becoming more evident by the day. Ongoing monitoring since the mid-1990s has shown that many geomorphological environments are changing rapidly. Equally, active layer thicknesses above permafrost have been increasing in many Arctic regions. Moreover, the World Glacier Monitoring Service has suggested that globally the average annual mass loss of glaciers between 1996 and 2005 was twice that of

Table 1 Some geomorphological consequences of global warming (modified from Goudie and Viles 2016, Table 11.1)

Hydrological
Increased evapotranspiration loss leading to river flow diminution, less soil cohesion, etc.
Overall increase in global precipitation leading to increased flood activity
Increased percentage of precipitation as rainfall at expense of winter snowfall leading to changes in river regimes
Increased precipitation as snowfall in very high latitudes leading to changes in river regimes
Possible increased risk of cyclones (greater latitudinal spread, frequency, and intensity)
Changes in the state of lakes, wetlands, and peatbogs
Less use of water by vegetation because of increased CO_2 effect on stomatal closure
Vegetational Controls
Major changes in latitudinal extent of biomes—reduction in boreal forest, increase in grassland and drylands, etc.
Major changes in altitudinal distribution of vegetation types (i.e., 500 m for 3 °C)
Growth enhancement by CO_2 fertilization
Changes due to increases in fire frequencies
Cryospheric
Permafrost decay, thermokarst, increased thickness of active layer, instability of slopes, degradation of river banks and shorelines
Changes in glacier and ice sheet rates of ablation and accumulation: glacier retreat
Changes in glacier lakes and outburst floods
Removal of glacial buttresses from slopes, leading to slope instability
Sea ice melting increasing wave attack conditions in Arctic regions
Coastal
Inundation of low-lying areas by sea-level rise (including wetlands, deltas, swamps, marshes, reefs, lagoons, etc.)
Increased storm surge activity associated with tropical storms, hurricanes, etc.
Accelerated coast recession (particularly on sandy beaches)
Changes in rates of reef growth and coral bleaching
Spread of mangrove swamps into higher latitudes
Aeolian
Increased dust storm activity in areas of moisture deficit, but reduced activity in areas of global stilling
Dune reactivation in areas of moisture deficit
Soil Erosion
Changes in response to changes in land use, fires, natural vegetation cover, rainfall erosivity, etc.
Changes resulting from soil erodibility modification (e.g., sodium and organic contents)

(continued)

Table 1 (continued)

Subsidence
Desiccation of clays under conditions of increased summer drought Thermokarst as a result of permafrost melting
Weathering
Reduction in number of frosts Salt weathering changes in response to groundwater levels and temperature and humidity cycles

the previous decade (1986–1995) and over four times that from 1976–1985. Retreat rates are unprecedented (Zemp et al. 2015).

Some selected studies from 2020/2021 are listed in Table 2. The list is not comprehensive but gives a taste of the huge increase in studies that have taken place in the

Table 2 Select studies of the geomorphological effects of global warming undertaken in 2020/2021

Phenomenon	Source
Coastal erosion	Masselink et al. (2020)
Coastal plain submergence	Antinioli et al. (2020)
Coral bleaching	Goreau and Hayes (2021)
Coral reefs—turbid situations	Morgan et al. (2020)
Coral reefs—accreting situations	Masselink et al. (2021)
Cryosphere melting	Ding et al. (2020)
Fire-induced erosion	Moran-Ordonez et al. (2020)
Glacial lake formation	Shugar et al. (2020)
Glacier outburst floods	Zheng et al. (2021)
Glacier retreat	Sommer et al. (2020)
Ice cap retreat	Wood et al. (2020)
Mangrove swamps	Bozi et al. (2021)
Peat bog degradation	Lin et al. (2021)
River floods	Di Sante et al. 2021
Salt marshes	Cahoon et al. (2021)
Sedimentary conditions	East and Sankey (2020)
Siberian discharges	Wang et al. (2021a, b)
Slope instability	Savi et al. (2021)
Small island submergence	Lin et al. (2020)
Soil erosion and desertification	Ma et al. (2021)
Storm surges	Chen et al. (2020)
Thermokarst	Turetsky et al. (2020)
Wave attack	Morim et al. (2021)

twenty-first century. They demonstrate both the range and the importance of global warming for geomorphological processes and forms.

The complexity of future changes in the environment creates severe problems for prediction and modeling (Blum and Törnqvist 2000). As Bogaart et al. (2003) pointed out, landscape response to climate change is (i) highly non-linear, and (ii) characterized by numerous feedbacks between different variables and by lead-lag phenomena. An example they cite is a precipitation increase in an initially semi-arid area. This would initiate hillslope erosion and increased sediment transport capacity. However, over time, soil and vegetation conditions would adjust to the new moisture conditions, resulting in an improved soil structure and greater vegetation cover. As a result, after a time lag, slope erosion and sediment yields might diminish.

Interest has also arisen in the role that global warming might play in accentuating or triggering geohazards (McGuire 2010). For example, accelerated thawing of submarine permafrost and the release of gas hydrates therefrom might promote submarine slope failure within turn might lead to tsunamis (Day and Maslin 2010). Equally, changes in the extent of ice sheets would modify the amount of loading on Earth's crust and so might have an influence on seismic and volcanic activity.

12 The Human Impact and the Anthropocene

Particularly over the last few centuries of the Anthropocene, and over the 'Great Acceleration' since the 1950s (Steffen et al. 2010), humans have become major agents of landscape change (Goudie and Viles 2016; Goudie 2018; Hudson et al. 2015), not least in Brazil (Junior et al. 2018) The Anthropocene concept has arisen (Ellis 2018). This was introduced by Crutzen (2002) as a name for a new epoch in Earth's history—an epoch when human activities have 'become so profound and pervasive that they rival, or exceed the great forces of Nature in influencing the functioning of the Earth System' (Steffen 2010, p. 443).

Anthropogeomorphology studies both the nature of deliberate land-forming processes (Szabo et al. 2010; da Luz and Rodrigues 2015), such as the creation of sea defenses, artificial islands, embankments, levees, spoil heaps, agricultural terraces, mines, quarries, canals and reservoirs, and the less deliberate changes in the operation of processes. Deforestation, grazing, plowing, city growth, atmospheric pollution, construction, and hydrological manipulation, have a wide range of impacts. They may accelerate a number of hazards, including mass movements, ground subsidence, soil erosion, rock weathering, and even seismic activity caused by fracking (Table 3).

Direct human interventions can have linked unforeseen and unwanted indirect impacts on landscapes. For instance, there are many examples of attempts to reduce coastal erosion which exacerbated it rather than solved it. Protecting one piece of coast, which comprises a component of a natural sediment circulation system, without realizing its larger setting, can lead to unanticipated knock-on effects elsewhere. For example, groyne construction to stop beach erosion, by reducing sediment transport downdrift can deplete beaches and lead to accelerated cliff retreat.

Table 3 Some major anthropogeomorphic processes (based on Goudie 2018, Table 6.2)

Direct processes
Constructional
Tipping, molding, plowing, terracing, reclamation
Excavational
Digging, cutting, mining, blasting of cohesive or non-cohesive materials Trampling, churning
Hydrological
Flooding, damming, canal construction, dredging, channel modification, draining, coastal protection
Indirect processes
Acceleration of erosion and sedimentation
Agricultural activity and clearance of vegetation Engineering, especially road construction and urbanization Modifications of hydrological regime by dams, etc.
Subsidence: collapse, settling
Mining (e.g., of coal and salt) Hydraulic (e.g., groundwater and hydrocarbon pumping) Thermokarst (melting of permafrost) Draining and desiccation of organic soils
Slope failure: landslides, flows, accelerated creep
Loading by spoil, buildings etc. Undercutting by road construction, etc. Shaking Lubrication by irrigation water, broken sewers, etc.
Seismic activity
Loading by reservoirs Lubrication along fault planes Fracking
Weathering
Acidification of precipitation by sulfate emissions Accelerated salinization following changes in groundwater levels Lateritization following vegetation removal

Anthropogenic modifications of erosion and sedimentation rates have been a major concern. Various studies (e.g., Hooke 1994; Douglas and Lawson 2001; Walling 2006) suggest that the amount of material moved by humans is somewhat greater than that moved by the world's rivers to the oceans. As technology evolves, this ability grows still further (Haff 2010). Furthermore, land-use changes, and in particular developments in farming, have led to a leap in erosion rates (Wilkinson and McElroy 2007). Conversely, Syvitski et al. (2005) calculated that sediment retention behind dams has led to a reduction in the annual net flux of sediment reaching the world's coasts by around 1.4 billion tonnes, with a total of more than 100 billion tonnes being trapped within the last 50 years. Syvitski and Milliman (2007) estimated that reservoirs behind dams now trap around 26% of the global sediment delivery to the oceans. Data on increasing sediment accumulation rates in eastern USA are presented in Rodriguez et al. (2020). Cooper et al. (2018, p. 222) argued that 'the annual direct anthropogenic contribution to the global production of sediment in 2015 was conservatively some 316 Gt (150 km^3), a figure more than 24 times greater than the sediment supplied annually by the world's major rivers to the oceans.'

It is now appreciated that human impacts on geomorphology go back a long way into prehistory (Braje 2015). Smith and Zeder (2013) argued that the Anthropocene commenced around 10,000 years ago at the Holocene/Pleistocene boundary, with the first domestication of plants and animals and the development of agriculture and pastoralism. In antiquity, huge changes in land cover in Europe took place (Kaplan et al. 2009), and there is increasing evidence to suggest that Bronze and Iron Age valley fill resulted from accelerated slope erosion produced by the activities of early farmers. Macklin et al. (2014) employed the term 'Anthropocene Alluvium' to describe human-generated floodplain sediments. Indeed, in recent years, studies in Britain have shown the importance of changes in sedimentation rate caused by humans at different times in the Holocene (e.g., Foster et al. 2009). In some parts of the world, more landscape change may have been achieved in prehistoric times than has been achieved by humans since. For example, in the circum-Mediterranean lands and the Levant, huge tracts of land are characterized by terraces, check dams, rainwater harvesting structures, and the like, while in Central America there are raised fields, drainage channels, reservoirs, and other structures produced in what Beach et al. (2015) described as the 'Mayacene'.

13 Geomorphological Hazards

As Latrubesse (2009) and Alcántara-Ayala and Goudie (2010) have shown, geomorphologists have become more and more concerned with geomorphological hazards. Although high magnitude, low frequency, catastrophic events, such as hurricanes or earthquakes with their concomitant geomorphic hazards, gain attention because of the casualties and financial losses they lead to, there are many more pervasive and less spectacular changes that are also highly significant for the welfare and livelihoods of human populations. These may have slower speeds of onset, longer durations, wider spatial extents, and a higher frequency. Examples include weathering phenomena (Goudie and Viles 1997), which can threaten a wide range of engineering structures (Goudie and Viles 2010a, b), and soil erosion (Boardman and Poesen 2006), which causes soil loss and the incision of adlands.

Geomorphological hazards are very diverse. Mass movements are one major category (Crozier 2010). There is also a range of fluvial hazards, such as flooding and changes in channels. In areas with volcanic activity, disasters are caused by eruptions, lava flows, ash falls, and lahars (Thouret 2010). In coastal regions inundation caused by storm surges, rapid coastal erosion and siltation, dune encroachment, and sea-level rise are all significant (Walker and McCraw 2010). In glacial areas surging glaciers, outwash floods, pro-glacial lake formation, and impedance of drainage are severe hazards. Permafrost regions are hazardous because of ground heaving, thermokarst development, slumping of slopes and banks, icings, etc. There is also a wide range of ground subsidence hazards caused, *inter alia*, by solution of limestone, dolomites, and evaporites (Gutierrez 2010), degradation of organic soils and peats, sediment hydrocompaction, and mining of groundwater, brines, and hydrocarbons. In drylands, wind erosion (Shao 2008), flash floods, deflation of susceptible surfaces, dust storm generation (Goudie and Middleton 2006), and dune migration, pose hazards. Large modern cities are not immune from these sorts of hazards (Garcia-Soriano et al. 2020), and urbanization may increase their incidence.

14 Applied Geomorphology

For many years, geomorphologists, collaborating with engineers and engineering geologists (Fookes et al. 2005), have used their skills to mitigate problems facing humanity, including hazards of the type mentioned above (Cooke and Doornkamp 1990; Hooke 2020). Indeed, applied geomorphology is a developing field (Keller et al. 2020; Griffiths and Lee 2022).

Notable examples of recent work in this area include mapping geomorphological phenomena for terrain evaluation (Smith et al. 2011); assessing the effects of river restoration following dam removal (e.g., Foley et al. 2017; Wohl 2020); developing means of forest management to control erosion (Phillips et al. 2018); managing of coasts to reduce erosion (Lazarus et al. 2016); establishing the flood histories of rivers in the Holocene by surveying and dating slack-water deposits laid down by earlier floods (Harden et al. 2010); and management of the effects of water and sediment control structures on river flows (Nichols et al. 2018). Geomorphologists are no longer simply spectators of geomorphological change but have become active in promoting it. Slope stabilization and river channelization, for example, clearly manifest the role of engineering geomorphology in modifying the landscape. Recognition of negative and persistent human impacts has encouraged research and applications in river restoration, including large-scale dam removal (Wohl 2014).

15 Geoconservation and Education

Geomorphologists have taken an increasing interest in how they can make an impact in terms of landscape conservation. There are major contributions that geomorphologists can make to landscape conservation and the preservation of Geodiversity (Gray 2013; Singh et al. 2021). The 'Convention Concerning the Protection of the World Cultural and Natural Heritage' was adopted by UNESCO in November 1972, and came into force in December 1975. This created a burgeoning interest in landscape conservation and interpretation. UNESCO has now established an annual International Geodiversity Day. It has also commissioned reports on the need to designate areas of karst and caves (Williams 2008), volcanic landforms (Wood 2009), and deserts (Goudie and Seely 2011). There are already a large number of essentially geomorphological World Heritage Sites in the natural category, as well as some cultural sites that may also have geomorphological value (https://whc.unesco.org/en/list/) (accessed 22nd September 2021). Related to this is the establishment of Geoparks and Geomorphosites (Joyce 2010; Santos et al. 2019). At present, there are 169 UNESCO Global Geoparks in 44 countries (https://en.unesco.org/global-geoparks) (accessed 22nd September 2021). Individual countries, such as the USA, have national parks and State Parks that may exist primarily because of their beautiful landforms. Geotourism is a developing field that shows the need for geomorphological education and explanation (e.g., Wang et al. 2019). The series edited by Piotr Migoń, *World Geomorphological Landscapes* (published by Springer), is an immensely valuable source of information on geomorphological diversity and contains 25 volumes (https://www.springer.com/series/10852) (accessed 22nd September 2021).

16 Conclusions

In the early twenty-first century, Geomorphology has become a discipline that is both wide-ranging and speedily evolving. This is because of the development of a wide spectrum of techniques, by the arising of the plate tectonics paradigm, by the ability to explore both extra-terrestrial and submarine landscapes, by the continued success of Quaternary studies, by an appreciation of the growing role of human and biological activities, by its engagement with research on the newly developed and controversial concept of the Anthropocene, by the application of the discipline to solving and managing various issues of concern to humans, including hazards, and geoconservation and stewardship of landscapes. However, geomorphology is also engaged, though perhaps not yet sufficiently, with issues raised by both Earth System Science, and by global environmental change associated with land cover changes and with global warming.

References

Alcántara-Ayala I, Goudie AS (2010) Geomorphological hazards and disaster prevention. Cambridge University Press, Cambridge, p. 304

Algeo TJ, Scheckler SE (1998) Terrestrial-marine teleconnections in the Devonian: links between the evolution of land plants, weathering processes and marine anoxic events. Phil Trans Roy Soc London B 353:113–130

Allchin B, Goudie A, Hegde K (1978) Prehistory and Palaeogeography of the Great Indian Desert. Academic Press, London, p. 389

Anderson DE (2022) Revolution in techniques: temporal. In: Burt, TP, Goudie AS, Viles HA (eds.) 2022. The History of the study of landforms or the development of geomorphology. Volume 5: Geomorphology in the second half of the twentieth century. Geological Society, London, Memoirs, 58, in press
Anderson DE, Goudie AS, Parker AG (2013) Global environments through the Quaternary. Oxford University Press, Oxford, p. 424
Anka Z, Séranne M, di Primio R (2010) Evidence of a large upper-Cretaceous depocentre across the Continent-Ocean boundary of the Congo-Angola basin. Implications for palaeo-drainage and potential ultra-deep source rocks. Mar Petrol Geol 27: 601–611
Antonioli F, De Falco G, Lo Presti V, Moretti L, Scardino G, Anzidei M, Bonaldo D, Carniel S, Leoni G, Furlani S, Marsico A (2020) Relative sea-Level rise and potential submersion risk for 2100 on 16 coastal plains of the Mediterranean Sea. Water 12(8): p. 2173
Baker VR 2008. Planetary landscape systems: a limitless frontier. Earth Surf Proc Landf 33:1341–1353
Beach T, Luzzadder-Beach S, Cook D, Dunning N, Kennett DJ, Krause S, Terry R, Trein D, Valdez F (2015) Ancient Maya impacts on the Earth's surface: An Early Anthropocene analog? Quat Sci Rev 124:1–30
Benn DI, Evans DJA (2010) Glaciers and glaciations (2nd edition). Hodder, London, p. 816
Bennett SJ, Simon A (2004) Riparian vegetation and fluvial geomorphology. American Geophysical Union, Washington D.C., p. 282
Bhardwaj A, Sam L, Gharehchahi S (2021) Four decades of understanding Martian geomorphology: Revisiting Baker's 'The geomorphology of Mars'. Progr Phys Geogr: Earth Environ: p. 03091333211026215
Birch SPD, Hayes AG, Howard AD, Moore JM, Radebaugh J (2016) Alluvial fan morphology, distribution and formation on Titan. Icarus 270:238–247
Bishop, P. (2007) Long-term landscape evolution: linking tectonics and surface processes. Earth Surf Proc Landf 32:329–365
Blum MD, Törnqvist TE (2000) Fluvial responses to climate and sea-level change: a review and look forward. Sedimentology 47 (Suppl. 1):2–48
Boardman J, Poesen J (eds.) (2006) Soil erosion in Europe. Wiley, Chichester, p. 878
Bogaart PW, Van Balen RT, Kasse C, Vandenberghe J (2003) Process-based modeling of fluvial system response to rapid climate change—I: model formulation and generic applications. Quat Sci Rev 22:2077–2095
Bourke MC, Goudie AS (2009) Varieties of barchan form in the Namib Desert and on Mars. Aeolian Res 1:45–54
Bourke MC, Lancaster N, Fenton LK, Parteli EJR, Zimbelman JR, Radebaugh J (2010) Extraterrestrial dunes: an introduction to the special issue on planetary dune systems. Geomorphology 121:1–14
Bozi BS, Figueiredo BL, Rodrigues E, Cohen MC, Pessenda LC, Alves EE, de Souza, AV, Bendassolli JA, Macario K, Azevedo P, Culligan N (2021) Impacts of sea-level changes on mangroves from southeastern Brazil during the Holocene and Anthropocene using a multi-proxy approach. Geomorphology 390: p. 107860
Braje TJ (2015) Earth systems, human agency, and the Anthropocene: Planet Earth in the human age. J Archaeol Res 23:369–396
Büdel J (1982) Climatic geomorphology (translated by L. Fischer and D. Busche). Princeton University Press, Princeton, New Jersey, p. 444
Burbank DW, Anderson RS (2011) Tectonic geomorphology: A frontier in Earth science (2nd edition). Blackwell Science, Malden, Mass., p. 460
Burt TP, Goudie AS, Viles HA (eds.) (2022) The history of the study of landforms or the development of geomorphology. Volume 5: Geomorphology in the second half of the twentieth century. Geological Society, London, Memoirs, 58, in press
Burt TP, Chorley RJ, Brunsden D, Cox NJ, Goudie AS (eds.) (2008) The history of the study of landforms or the development of geomorphology volume 4: Quaternary and recent processes and

forms (1890–1965) and the mid-century revolutions. Geological Society of London, London, p. 1056
Butler DR (1995) Zoogeomorphology: Animals as geomorphic agents. Cambridge University Press, Cambridge, p. 240
Cahoon DR, McKee KL, Morris JT (2021) How plants influence resilience of salt marsh and mangrove wetlands to sea-level rise. Estuaries Coasts 44(4):883–898
Carson MA, MJ Kirkby (1972) Hillslope Form and Process. Cambridge University Press, Cambridge
Carter NEA, Viles HA (2005) Bioprotection explored: the story of a little known earth surface process. Geomorphology 67:273–281
Chan MA, Nicoll K, Ormö J, Okubo C, Komatsu G (2011) Utah's geologic and geomorphic analogs to Mars—an overview for planetary exploration. Geol Soc Am Special Papers 483:349–375
Chappell A, Webb NP, Butler HJ, Strong CL, McTainsh GH, Leys JF, Viscarra Rossel, RA (2013) Soil organic carbon dust emission: an omitted global source of atmospheric CO_2. Global Change Biol 19:3238–3244
Chen J, Wang Z, Tam CY, Lau NC, Lau DSD, Mok HY (2020) Impacts of climate change on tropical cyclones and induced storm surges in the Pearl River Delta region using pseudo-global-warming method. Sci Rep 10(1):1–10
Church M (2010) The trajectory of geomorphology. Progr Phys Geogr 34:265–286
Church M (2022) Physical experiments in geomorphology. In: Burt TP, Goudie AS & Viles HA (eds.) 2022. The history of the study of landforms or the development of geomorphology. Volume 5: Geomorphology in the second half of the twentieth century. Geological Society, London, Memoirs, 58, in press
Clare M, Lintern DG, Rosenberger K, Clarke JEH, Paull C, Gwiazda R, Cartigny MJ, Talling PJ, Perara D, Xu J, Parsons D (2020) Lessons learned from the monitoring of turbidity currents and guidance for future platform designs. Geol Soc, London, Special Pub 500(1):605–634
Conway S (2022) Planetary geomorphology. In Burt TP, Goudie AS & Viles HA (eds) The history of the study of landforms or the development of geomorphology. Volume 5: Geomorphology in the second half of the twentieth century. Geological Society, London, Memoirs, 58, in press
Coombes MA (2016a) Biogeomorphology. International encyclopedia of geography: People, the Earth, environment and technology. John Wiley and Sons, Chichester, 1–9
Coombes MA (2016b) Biogeomorphology: diverse, integrative and useful. Earth Surf Proc Landf 41(15):2296–2300
Cooke RU, Doornkamp JC (1990) Geomorphology in environmental management. (2nd edition). Oxford University Press, Oxford, p. 434
Cooper AH, Brown TJ, Price SJ, Ford J, Waters CN 2018. Humans are the most significant global geomorphological driving force of the 21st century. The Anthropocene Review, 5(3):222–229
Coppo NP, Schnegg P-A, Falco P, Costa R (2009) A deep scar in the flank of Tenerife (Canary Islands): Geophysical contribution to tsunami hazard assessment. Earth Planetary Sci Letters 282:65–68
Crozier MJ (2010) Deciphering the effect of climate change on landslide activity: a review. Geomorphology 124:260–267
Crutzen PJ (2002) The "anthropocene". J Phys IV 12:1–5
da Luz RA, Rodrigues C (2015) Anthropogenic changes in urbanised hydromorphological systems in a humid tropical environment: River Pinheiros, Sao Paulo, Brazil. Z Geomorph Supplementband 59:109–135
Dadson S (2010) Geomorphology and Earth system science. Progr Phys Geogr 34:385–398
Dadson S (2020) Geomorphology and Earth system science. In Burt TP, Goudie AS & Viles HA (eds.) (2022) The history of the study of landforms or the development of geomorphology. Volume 5: Geomorphology in the second half of the twentieth century. Geological Society, London, Memoirs, 58, in press
Day SJ, Maslin M (2010) Gas hydrates: a hazard for the twenty-first century? Phil Trans Roy Soc A 368:2579–2583

Deo RC, Syktus JI, McAlpine CA, Lawrence PJ, McGowan HA, Phinn SR (2009) Impact of historical land cover change on daily indices of climate extremes including droughts in eastern Australia. Geophys Res Letters 36: L08705. doi:https://doi.org/10.1029/2009GL037666

de Paula Barros LF, Junior AP (2020) Late Quaternary landscape evolution in the Atlantic Plateau (Brazilian highlands): Tectonic and climatic implications of fluvial archives. Earth-Sci Rev 207:103228

Di Sante F, Coppola E, Giorg, F (2021) Projections of river floods in Europe using EURO-CORDEX, CMIP5 and CMIP6 simulations. Int J Climatol 41(5):3203–3221

Diniega S, Bramson AM, Buratti B, Buhler P, Burr DM, Chojnacki M, Conway SJ, Dundas CM, Hansen CJ, McEwen AS, Lapôtre MG (2021) Modern Mars' geomorphological activity, driven by wind, frost, and gravity. Geomorphology, p. 107627

Ding Y, Mu C, Wu T, Hu G, Zou D, Wang D, Li W, Wu X (2020) Increasing cryospheric hazards in a warming climate. Earth-Sci Rev: p. 103500

Douglas I, Lawson N (2001) The human dimensions of geomorphological work in Britain. J Industr Ecol 4:9–33

Drewry D (1986) Glacial geologic processes. Arnold, London

Dupré B, Dessert C, Oliva P, Goddéris Y, Viers J, François L, Millot R, Gaillardet J (2003) Rivers, chemical weathering and Earth's climate. CR Geosci 335:1141–1160

Duszyński F, Ford DC, Goudie A, Migoń P. 2022. Rock properties and rock-controlled landforms. In Burt TP, Goudie AS & Viles HA (eds) The history of the study of landforms or the development of geomorphology. Volume 5: Geomorphology in the second half of the twentieth century. Geological Society, London, Memoirs, 58, in press

East AE, Sankey J.B (2020) Geomorphic and sedimentary effects of modern climate change: current and anticipated future conditions in the western United States. Rev Geophys 58(4): p. e2019RG000692

Eckardt, F. 2022. Geomorphology from Earth Orbit 1957–2000. In Burt TP, Goudie AS & Viles HA (eds.) (2022) The history of the study of landforms or the development of geomorphology. Volume 5: Geomorphology in the second half of the twentieth century. Geological Society, London, Memoirs, 58, in press

El-Baz F, Maxwell TA (eds.) (1982) Desert landforms of Southwest Egypt: a basis for comparison with Mars. NASA, Washington D.C. p. 385

Ellis EC (2018) Anthropocene: a very short introduction. Oxford University Press, Oxford, p. 208

Fan W, McGuire JJ, Shearer PM (2020) Abundant spontaneous and dynamically triggered submarine landslides in the Gulf of Mexico. Geophys Res Letters 47(12): p. e2020GL087213

Fan Y, Clark M, Lawrence DM, Swenson S, Band LE, Brantley SL, Brooks PD, Dietrich WE, Flores A, Grant G, Kirchner JW (2019) Hillslope hydrology in global change research and Earth system modeling. Water Resources Research, 55(2):1737–1772

Ferguson R, Hardy R, Lewin J (2022). Fluvial processes and forms. In Burt TP, Goudie AS & Viles HA (eds.) (2022) The history of the study of landforms or the development of geomorphology. Volume 5: Geomorphology in the second half of the twentieth century. Geological Society, London, Memoirs, 58, in press

Flemming NC (1999) Archaeological evidence for vertical movement on the continental shelf during the Palaeolithic, Neolithic and Bronze Age periods. Geol Soc London Special Pub 146(1):129–146

Foley MM, Magilligan FJ, Torgersen CE, Major JJ, Anderson CW, Connolly PJ, Wieferich D, Shafroth PB, Evans JE, Infante D, Craig LS (2017) Landscape context and the biophysical response of rivers to dam removal in the United States. PloS one 12(7): p. e0180107

Fookes PG, Lee EM, Milligan G (eds.) (2005) Geomorphology for engineers. Whittles Publishing, Dunbeath, p. 874

Ford DC, Williams PW (2007) Karst geomorphology and hydrology (2nd Edition). Wiley, Chichester, p. 576

Foster GC, Chiverrell RC, Thomas GSP et al. (2009) Fluvial development and the sediment regime of the lower Calder, Ribble catchment, northwest England. Catena 77:81–95

French HM (2017) The periglacial environment (4th edition). John Wiley and Sons, Chichester, p. 544
Gardner J (2020) How water, wind, waves and ice shape landscapes and landforms: Historical contributions to geomorphic science. Geomorphology, 366, p. 106687
García-Soriano D, Quesada-Román A, Zamorano-Orozco JJ (2020) Geomorphological hazards susceptibility in high-density urban areas: A case study of Mexico City. J South Am Earth Sci 102:102667
Gillette DA (1977) Fine particulate emissions due to wind erosion. Transactions of the ASAE 20(5):890–897
Goreau TJ, Hayes RL (2021) Global warming triggers coral reef bleaching tipping point. Ambio, 50(6):1137–1140
Goudie AS (ed.) (1990) Geomorphological techniques (2nd edition). Routledge, London and New York, p. 592
Goudie AS (2005) The drainage of Africa since the Cretaceous. Geomorphology 67: 437–456
Goudie AS (2006) The Schmidt Hammer in geomorphological research. Progress in Physical Geography 30:703–718
Goudie AS (2013) Arid and semi-arid geomorphology. Cambridge University Press, Cambridge, p. 461
Goudie AS (2016a) Geomorphology: History. International Encyclopedia of Geography: People, the Earth, Environment and Technology. Wiley, Chichester, 1–8
Goudie AS (2016b) Quantification of rock control in geomorphology. Earth-Sci Rev 159: 374–387
Goudie AS (2018) The human impact on the natural environment (8th edition). Wiley-Blackwell, Oxford, p. 472
Goudie AS (2020) The human impact in geomorphology–50 years of change. Geomorphology 366: p. 106601
Goudie AS, Middleton NJ (2006) Desert dust in the global system. Springer, Heidelberg, p. 298
Goudie AS, Seely MK (2011) World Heritage Desert Landscapes: Potential Priorities for the Recognition of Desert Landscapes and Geomorphological Sites on the World Heritage List, IUCN World Heritage Studies N° 9
Goudie AS, Viles HA (1997) Salt weathering hazards. John Wiley and Sons, Chichester, p. 256
Goudie AS, Viles HA (2010a) Landscapes and geomorphology. Oxford University Press, Oxford, p. 152
Goudie AS, Viles HA (2010b) Weathering hazards. In Alcántara-Ayala & Goudie AS, (eds.). Geomorphological hazards and disaster prevention. Cambridge University Press, Cambridge, 145–159
Goudie AS, Viles HA (2012) Weathering and the global carbon cycle: Geomorphological perspectives. Earth-Sci Rev 113(1–2):59–71
Goudie AS, Viles HA (2016) Geomorphology in the Anthropocene. Cambridge University Press, Cambridge, p. 380
Graf WL (2001) Damage control: restoring the Physical integrity of America's rivers. Ann Ass Am Geogr 91:1–27
Gray M (2013) Geodiversity: valuing and conserving abiotic nature (2nd edition). Wiley, Chichester, p. 512
Greeley R, Iversen JD (1985) Wind as a Geological Process on Earth, Mars, Venus and Titan. Cambridge University Press, Cambridge
Griffiths JS, Lee EM (2022) Geomorphology in environmental management 1965–2000. In Burt TP, Goudie AS & Viles HA (eds) The history of the study of landforms or the development of geomorphology. Volume 5: Geomorphology in the second half of the twentieth century. Geological Society, London, Memoirs, 58, in press
Gutierrez F (2010) Hazards associated with karst. In Alcántara-Ayala I & Goudie AS, (eds.), Geomorphological hazards and disaster prevention. Cambridge University Press, Cambridge, 161–175

Haff PK, (2010) Hillslopes, rivers, plows and trucks: mass transport on Earth's surface by natural and technological processes. Earth Surf Proc Landf 35:1157–1166
Harden T, Macklin MG, Baker VR (2010) Holocene flood histories in south-western USA. Earth Surf Proc Landf 35:707–716
Hooke JM (2020) Changing landscapes: Five decades of applied geomorphology. Geomorphology 366:106793
Hooke RL (1994) On the efficacy of humans as geomorphic agents. USA Today, 4, 217:224–225
Hudson PF, Butzer KW, Beach T (2008) Fluvial deposits and environmental history: Synthesis. Geomorphology 101(1):xvii
Hudson P, Goudie A, Asrat A. (2015) Human Impacts on landscapes: Sustainability and the role of Geomorphology. Z Geomorph 59, Suppl 2: 1–5
Ielpi A, Lapôtre MG (2020) A tenfold slowdown in river meander migration driven by plant life. Nature Geoscience 13(1):82–6
Innocenti C, Battaglini L, D'Angelo S, Fiorentino A (2021) Submarine landslides: mapping the susceptibility in European seas. Q J Eng Geol Hydrogeol 54, no. 1
IPCC (2021) Climate change 2021: The Physical Science Basis. Contribution of Working Group I to the Sixth Assessment Report of the Intergovernmental Panel on Climate Change. Cambridge University Press, Cambridge (in press)
Joyce EB (2010) Australia's geoheritage: history of study, a new inventory of geosites and applications to geotourism and geoparks. Geoheritage 2(1–2):39–56
Junior PF, Korb CC, Brannstrom C (2018) Research on Technogene/Anthropocene in Brazil. Quat Environ Geosci 9(1):1–10
Kaplan JO, Krumhardt KM, Zimmermann N (2009) The prehistoric and preindustrial deforestation of Europe. Quat Sci Rev 28:3016–34
Keller E, Adamaitis C, Alessio P, Anderson S, Goto E, Gray S, Gurrola L, Morell K (2020) Applications in geomorphology. Geomorphology 366: p.106729
Keane JT (2019) Maar on Titan. Nature Geosci 12(10):789
Kirkby M (2022) Hillslope form and process: History 1960–2000+. In Burt TP, Goudie AS & Viles HA (eds.) (2022) The history of the study of landforms or the development of geomorphology. Volume 5: Geomorphology in the second half of the twentieth century. Geological Society, London, Memoirs, 58, in press
Lane SN, Bradbrook KF, Richards KS, Biron PA, Roy AG (1999) The application of computational fluid dynamics to natural river channels: three-dimensional versus two-dimensional approaches. Geomorphology 29(1–2):1–20
Latrubesse E (ed.) (2009) Natural hazards and human-exacerbated disasters in Latin America. Elsevier, Amsterdam, p. 550
Lazarus ED, Ellis MA, Murray AB, Hall DM, (2016). An evolving research agenda for human–coastal systems. Geomorphology 256:81–90
Leopold LB, Wolman MG, Miller JP (1964) Fluvial processes and geomorphology. W.H. Freeman, San Francisco, p. 544
Li P, Kneller B, Hansen L (2021) Anatomy of a gas-bearing submarine channel-lobe system on a topographically complex slope (offshore Nile Delta, Egypt). Marine Geol 437:p. 106496
Lin S, Liu Y, Huang X (2021) Climate-induced Arctic-boreal peatland fire and carbon loss in the 21st century. Sci Total Environ 796:p. 148924
Lin W, Sun Y, Nijhuis S, Wang Z (2020) Scenario-based flood risk assessment for urbanizing deltas using future landuse simulation (FLUS): Guangzhou Metropolitan Area as a case study. Sci Total Environ 15:139899
Lopes RM, Malaska MJ, Schoenfeld AM, Solomonidou A, Birch SPD, Florence M, Hayes AG, Williams DA, Radebaugh J, Verlander T, Turtle EP (2020) A global geomorphologic map of Saturn's moon Titan. Nature Astron 4(3):228–233
Løvholt F, Bondevik S, Laberg JS, Kim J, Boylan N (2017) Some giant submarine landslides do not produce large tsunamis. Geophys Res Letters, 44(16):8463–8472

Ma X, Zhao C, Zhu J, (2021) Aggravated risk of soil erosion with global warming—A global meta-analysis. Catena 200:p. 105129
Macklin MG, Lewin J, Jones AF, (2014) Anthropogenic alluvium: an evidence-based meta-analysis for the UK Holocene. Anthropocene 6:26–38
Mann GA, Clarke JDA, Gostin VA (2004) Surveying for Mars analogue research sites in the central Australian deserts. Austral Geogr Studies 42:116–124
Martin Y (2022) Modelling. In Burt TP, Goudie AS & Viles HA (eds) The history of the study of landforms or the development of geomorphology. Volume 5: Geomorphology in the second half of the twentieth century. Geological Society, London, Memoirs, 58, in press
Masselink G, Hughes MG (2003) An introduction to coastal processes and geomorphology. Routledge, London
Masselink G, Russell P, Rennie A, Brooks S, Spencer T (2020) Impacts of climate change on coastal geomorphology and coastal erosion relevant to the coastal and marine environment around the UK. MCCIP Science Review 2020: 158–189. doi: https://doi.org/10.14465/2020.arc08.cgm
Masselink G, McCall R, Beetham E, Kench P, Storlazzi C (2021) Role of future reef growth on morphological response of coral reef islands to sea-level rise. J Geophys Res: Earth Surface: 126(2):e2020JF005749
McGuire B (2010) Climate forcing of geological and geomorphological hazards. Phil Trans Roy Soc A, 368:2311–2315
Mescolotti PC, do Nascimento Pupim F, Ladeira FS, Sawakuchi AO, Santa Catharina A, Assine ML (2021) Fluvial aggradation and incision in the Brazilian tropical semi-arid: Climate-controlled landscape evolution of the São Francisco River. Quat Sci Rev 263:106977
Micallef A, Berndt C, Masson DG, Stow DAV (2007) A technique for the morphological characterization of submarine landscapes as exemplified by debris flows of the Storegga Slide. J Geophys Res 112: F02001. doi:https://doi.org/10.1029/2006JF000505
Micallef A, Masson DG, Berndt C, Stow DAV (2009) Development and mass movement processes of the north-eastern Storegga Slide. Quat Sci Rev 28:433–448
Micallef A, Krastel S, Savini A (eds.) (2018) Submarine geomorphology. Springer, Cham, p. 571
Micallef A, Krastel S, Savini A (2022) Submarine geomorphology. In Burt TP, Goudie AS, Viles HA (eds.), The history of the study of landforms or the development of geomorphology. Volume 5: Geomorphology in the second half of the twentieth century. Geological Society, London, Memoirs, 58, in press
Migoń P 2007 Granite landscapes of the world. Oxford University Press, Oxford, p. 412
Moore, R., Davis, G. and Dabson, O., 2018. Applied geomorphology and geohazard assessment for deepwater development. In Micallef A, Krastel S, Savini A (eds.) Submarine geomorphology. Springer, Cham, 459–479
Morán-Ordóñez A, Duane A, Gil-Tena A, De Cáceres M, Aquilué N, Guerra CA, Geijzendorffer IR, Fortin MJ, Brotons L (2020) Future impact of climate extremes in the Mediterranean: Soil erosion projections when fire and extreme rainfall meet. Land Degrad Devel, 31(18):3040–3054
Morgan KM, Perry CT, Arthur R, Williams HT, Smithers SG (2020) Projections of coral cover and habitat change on turbid reefs under future sea-level rise. Proc Roy Soc B, 287(1929):p. 20200541
Morim J, Vitousek S, Hemer M, Reguero B, Erikson L, Casas-Prat M, Wang XL, Semedo A, Mori N, Shimura T, Mentaschi L (2021) Global-scale changes to extreme ocean wave events due to anthropogenic warming. Environ Res Letters 16(7):p. 074056
Nichols MH, Magirl C, Sayre NF, Shaw JR (2018) The geomorphic legacy of water and erosion control structures in a semiarid rangeland watershed. Earth Surf Proc Landf 43:909–918
Nicholls RJ, Hoozemans FMJ, Marchand M (1999) Increasing flood risk and wetland losses due to global sea level rise: regional and global analyses. Global Environ Change 9:S69–87
Ottesen D, Dowdeswell JA (2009) An inter–ice-stream glaciated margin: Submarine landforms and a geomorphic model based on marine-geophysical data from Svalbard. Geol Soc Am Bull 121:1647–1665
Ouchi S (1996) Studies of process geomorphology in Japan in the first half of the 1990s. Geographical Review of Japan, Series B 69(2):126–133

Paola C, Foufoula-Georgiou E, Dietrich WE, Hondzo M, Mohrig D, Parker G, Power ME, Rodriguez-Iturbe I, Voller V, Wilcock, P (2006) Toward a unified science of the Earth's surface: Opportunities for synthesis among hydrology, geomorphology, geochemistry, and ecology. Water Resources Res 42(3). doi:https://doi.org/10.1029/2005WR004336

Phillips C, Marden M, Basher LR (2018) Geomorphology and forest management in New Zealand's erodible steeplands: An overview. Geomorphology 307:107–121

Quinton JN, Govers G, Van Oost K, Bardgett RD (2010) The impact of agricultural soil erosion on biogeochemical cycling. Nature Geosci 3:311–314

Radebaugh J, Lorenz R, Farr T, Paillou P, Savage C, Spencer C (2010) Linear dunes on Titan and Earth: initial remote sensing comparisons. Geomorphology 121:122–132

Rodriguez AB, McKee BA, Miller CB, Bost MC, Atencio AN (2020). Coastal sedimentation across North America doubled in the 20th century despite river dams. Nature Comms 11(1):1–9

Santos DS, Reynard E, Mansur KL, Seoane JC (2019) The specificities of geomorphosites and their influence on assessment procedures: A methodological comparison. Geoheritage, 11(4):2045–2064

Savi S, Comiti F, Strecker MR (2021) Pronounced increase in slope instability linked to global warming: A case study from the eastern European Alps. Earth Surf Proc Landf. doi:https://doi.org/10.1002/esp.5100

Selby MJ (1980) A rock mass strength classification for geomorphic purposes: with tests from Antarctica and New Zealand. Z Geomorph 24:31–51

Shao Y (2008) Physics and modelling of wind erosion (2nd edition). Springer, Berlin and Heidelberg, p. 472

Shugar DH, Burr A, Haritashya UK, Kargel JS, Watson CS, Kennedy MC, Bevington AR, Betts RA, Harrison S, Strattman K (2020) Rapid worldwide growth of glacial lakes since 1990. Nature Climate Change 10(10):939–945

Singh RB, Wei D, Anand S (eds.) (2021) Global geographical heritage, geoparks and geotourism: Geoconservation and development. Springer Nature, Singapore, p. 498

Slaymaker O, Spencer T, Embleton-Hamann C (eds.) (2009) Geomorphology and environmental change. Cambridge University Press, Cambridge, p. 450

Smith BD, Zeder MA (2013) The onset of the Anthropocene. Anthropocene 4:8–13

Smith M, Griffiths J, Paron P (eds.) (2011) Applied geomorphological mapping. Elsevier, Amsterdam, p. 610

Soderblom LA and 18 others (2007) Topography and geomorphology of the Huygens landing site on Titan. Planetary Space Sci 55:2015–2024

Sommer C, Malz P, Seehaus TC, Lippl S, Zemp M, Braun MH (2020) Rapid glacier retreat and downwasting throughout the European Alps in the early 21st century. Nature Comm 11(1):1–10

Spencer T, French JR (2022) Coastal processes and landforms. In Burt TP, Goudie AS & Viles HA (eds.) (2022) The history of the study of landforms or the development of geomorphology. Volume 5: Geomorphology in the second half of the twentieth century. Geological Society, London, Memoirs, 58, in press

Steffen W (2010) Observed trends in Earth system behaviour. Interdis Rev Climate Change 1:428–449

Steffen W, Sanderson RA, Tyson PD, Jäger J, Matson PA, Moore III B, Oldfield F, Richardson K, Schellnhuber HJ, Turner BL, Wasson RJ (2006) Global change and the earth system: a planet under pressure. Springer Science & Business Media

Summerfield MA (1991) Global geomorphology: An introduction to the study of landforms. Longman, Harlow, p. 560

Summerfield MA (ed.), (2000) Geomorphology and global tectonics. John Wiley and Sons, Chichester, p. 386

Summerfield MA (2005) A tale of two scales, or the two geomorphologies. Trans Inst Brit Geogr NS 30:402–415

Summerfield MA (2022) Plate tectonics and macrogeomorphology. In Burt TP, Goudie AS & Viles HA (eds) The history of the study of landforms or the development of geomorphology. Volume

5: Geomorphology in the second half of the twentieth century. Geological Society, London, Memoirs, 58, in press
Suzuki T, Matsukura Y, Ehlen J, Tanaka Y (2000) (eds.) Rock Control in Geomorphological Processes. Trans Jap Geomorph Union 23(2):1–386
Syvitski JPM, Milliman JD (2007) Geology, geography and humans battle for dominance over the delivery of fluvial sediment to the coastal ocean. J Geol 115:1–19
Syvitski JPM, Vörösmarty CJ, Kettner AJ, Green P (2005) Impact of humans on the flux of terrestrial sediment to the global coastal ocean. Science 308:376–380
Szabó J, Dávid L, Lóczy D (eds.) (2010) Anthropogenic geomorphology. Springer, Dordrecht, p. 260
Tessler ZD, Vörösmarty CJ, Overeem I, Syvitski JP (2018) A model of water and sediment balance as determinants of relative sea level rise in contemporary and future deltas. Geomorphology 305:209–220
Thomas MF (1994) Geomorphology in the tropics: A study of weathering and denudation in low latitudes. John Wiley & Sons, Chichester, p. 482
Thouret J-C (2010) Volcanic hazards and risks: a geomorphological perspective. In Alcántara-Ayala I & Goudie AS (eds.), Geomorphological hazards and disaster prevention. Cambridge University Press, Cambridge, 13–32
Tokano T (2020) Stable existence of tropical endorheic lakes on Titan. Geophys Res Letters 47(5):p. e2019GL086166
Turetsky MR, Abbott BW, Jones MC, Anthony KW, Olefeldt D, Schuur EA Grosse, G., Kuhry, P., Hugelius, G., Koven, C., Lawrence, D.M., (2020) Carbon release through abrupt permafrost thaw. Nature Geosci 13(2):138–143
Viles HA (ed.) (1988) Biogeomorphology. Blackwell, Oxford, p. 376
Viles HA (2004) Biogeomorphology. In Goudie AS (ed.) Encylopedia of geomorphology. Routledge, London, 83–86
Viles HA (2020) Biogeomorphology: Past, present and future. Geomorphology 366: p.106809
Viles HA, Goudie AS, Goudie A (2021) Ants as geomorphological agents: A global assessment. *Earth-Science Rev* 213:p. 103469
Viles HA, Goudie AS, Grab S, Lalley J, (2011) The use of the Schmidt Hammer and Equotip for rock hardness assessment in geomorphology and heritage science: A comparative analysis. Earth Surf Proc Landf 36:320–333
Viles HA, Naylor LA, Carter NEA Chaput, D., (2008) Biogeomorphological disturbance regimes: progress in linking ecological and geomorphological systems. Earth Surf Proc Landf 33:1419–1435
Vita-Finzi C (1967) The Mediterranean valleys. Cambridge University Press, Cambridge, p. 150
Walker HJ, Grabau WE (eds.) (1993) The evolution of geomorphology. A nation-by-nation summary of development. Wiley, Chichester, p. 556
Walker HJ, McGraw M (2010) Geomorphology and coastal hazards. In Alcántara-Ayala I & Goudie AS (eds.), Geomorphological hazards and disaster prevention. Cambridge University Press, Cambridge, 129–144
Walling DE, (2006) Human impact on land-ocean sediment transfer by the world's rivers. Geomorphology 79:192–216
Wan W, Zhao J, Li HY, Mishra A, Ruby Leung L, Hejazi M, Wang W, Lu H, Deng Z, Demissisie Y, Wang H, (2017) Hydrological drought in the Anthropocene: Impacts of local water extraction and reservoir regulation in the US. J Geophys Res: Atmosph 122(21):11–313
Wang B, Lyu F, Li S, Li J, Yang Z, Li L, Wang X, Lu Y, Yang T, Wu J, Sun G (2021a) A buried submarine canyon in the northwest South China Sea: architecture, development processes and implications for hydrocarbon exploration. Acta Oceanologica Sinica, 40(3):84–93
Wang P, Huang Q, Pozdniakov SP, Liu S, Ma N, Wang T, Zhang Y, Yu J, Xie J, Fu G, Frolova NL (2021b) Potential role of permafrost thaw on increasing Siberian river discharge. Environ Res Letters 16(3):p. 034046

Wang Y, Wu F, Li X, Chen L (2019) Geotourism, geoconservation, and geodiversity along the belt and road: A case study of Dunhuang UNESCO Global Geopark in China. Proc Geol Ass 130(2):232-241
Washburn AL (1979) Geocryology. Arnold, London
Webb NP, Chappell A, Strong CL, Marx SK, McTainsh GH (2012) The significance of carbon-enriched dust for global carbon accounting. Glob Change Biol 18(11):3275–3278
Whalley WB (ed.) (1978) Scanning electron microscopy in the study of sediments. GeoAbstracts, Norwich, p. 414
Whipple KX (2009) The influence of climate on the tectonic evolution of mountain belts. Nature Geosci 2:97–104
Whipple KX, Meade BJ (2006) Orogen response to changes in climatic and tectonic forcing. Earth Planetary Sci Letters 243:218–228
Whittaker AC, Cowie PA, Attal M, Tucker GE, Roberts GP (2007) Bedrock channel adjustment to tectonic forcing: implications for predicting river incision rates. Geology 35:103–106
Wilkinson BH, McElroy BJ (2007) The impact of humans on continental erosion and sedimentation. Bull Geol Soc Am 119:140–156
Williams PW (2008) World heritage caves and karst, a thematic study: global review of karst world heritage properties: present situation, future prospects and management, requirements. IUCN World Heritage Studies N° 2
Wohl E (2014) Time and the rivers flowing: Fluvial geomorphology since 1960. Geomorphology 216:263–282
Wohl E (2020) Rivers in the Anthropocene: the US perspective. Geomorphology 366:106600
Wood C (2009) World heritage volcanoes: a thematic study: a global review of volcanic world heritage properties: Present situation, future prospects and management requirements. IUCN World Heritage Studies N° 8
Wood LR, Neumann K, Nicholson KN, Bird BW, Dowling CB, Sharma S (2020) Melting Himalayan glaciers threaten domestic water resources in the Mount Everest Region, Nepal. Frontiers Earth Sci. 29:128.
Xiao L, Wang J, Dang Y, Cheng Z, Huang T, Zhao J, Xu Y, Huang J, Xiao Z, Komatsu G (2017) A new terrestrial analogue site for Mars research: the Qaidam Basin, Tibetan Plateau (NW China). Earth-Sci Rev 164:84–101
Xu K, Milliman JD, Yang Z, Xu H (2007) Climatic and anthropogenic impacts on water and sediment discharges from the Yangtze River (Changjiang), 1950–2005. In A. Gupta (ed.) Large rivers: Geomorphology and management. Wiley, Chichester, 609–626
Yatsu E (1966) Rock control in geomorphology. Sozosha, Tokyo, p. 135
Yatsu E (1988) The nature of weathering: an introduction. Sozosha, Tokyo, p. 624
Young RW, Wray AL, Young ARM (2009) Sandstone landforms. Cambridge University Press, Cambridge, p. 314
Yu S, Oguchi CT (2009) Complex relationships between salt type and rock properties in a durability experiment of multiple-rock treatments. Earth Surf Proc Landf 34:2096–2110
Zemp M, Frey H, Gärtner-Roer I, Nussbaumer SU, Hoelzle M, Paul F, Haeberli W, Denzinger F, Ahlstrøm AP, Anderson B, Bajracharya S (2015) Historically unprecedented global glacier decline in the early 21st century. J Glaciol 61:745–762
Zhang Q, Tao Z, Ma Z, Tang W, Gao Q, Xu P, Ding J, Liu Z, Lin Y, Su D, Zheng W (2017) Influences of anthropogenic activities on dissolved silica migration in a granite-hosted basin, Hainan Island, China. Quat Int 440:99–110
Zheng G, Allen SK, Bao A, Ballesteros-Cánovas JA, Huss M, Zhang G, Li J, Yuan Y, Jiang L, Yu T, Chen W (2021) Increasing risk of glacial lake outburst floods from future third pole deglaciation. Nature Climate Change 11(5):411–417

Geomorphological Units of Brazil: A Review in the Context of Brazilian Spatial Planning

Jurandyr Luciano Sanches Ross, Marciel Lohmann, and Camila Cunico

Abstract This paper aimed to elaborate a new division of Geomorphological Units of Brazil based on the division of Relief Units of IBGE (Mapa das Unidades de Relevo do Brasil. Brazilian Institute of Geography and Statistics, Rio de Janeiro, 1993) and on the Geomorphological Map of Brazil of Ross (Ecogeografia do Brasil: subsídios para planejamento ambiental. São Paulo, Oficina de Textos, 2006). The derived product at the scale of 1:5,000,000 was used in the research "Brazilian Territorial Planning: Natural Potentialities and Social Vulnerabilities, to characterize the physical-natural environment of Brazil, to identify the Natural Environmental Units, and finally, to integrate the analyzed variables in land units for the territorial planning of the mentioned geographic clipping. The technical procedures involved the georeferencing and digitalization of the Geomorphological Map of Ross (Ecogeografia do Brasil: subsídios para planejamento ambiental. São Paulo, Oficina de Textos, 2006) since it was in analogical format. The relief unit map of IBGE (Mapa das Unidades de Relevo do Brasil. Brazilian Institute of Geography and Statistics, Rio de Janeiro, 1993) was obtained from the official website of the institution, already in format compatible with Geographic Information System environment. In possession of the bases in digital format, a rereading of the maps was elaborated, resulting in a new proposal of geomorphological units for Brazil. The results provided the identification of a greater number of Geomorphological Units, especially those associated with sedimentary basins and orogenic belts.

J. L. S. Ross (✉)
University of São Paulo—FFLCH, São Paulo, Brazil
e-mail: juraross@usp.br

M. Lohmann
Department of Geosciences, Londrina State University, Londrina, Brazil
e-mail: marciel@uel.br

C. Cunico
Department of Geosciences, Federal University of Paraíba, João Pessoa, Brazil

G. Barbosa dos Santos et al. (eds.), *Geomorphology of Brazil: Complexity, Interscale and Landscape*, Springer Proceedings in Earth and Environmental Sciences,
https://doi.org/10.1007/978-3-031-05178-4_2

1 Introduction

The Brazilian relief results from morphogenetic processes related to the Meso Cenozoic epigenesis resulting from the drift of continents and from the Tertiary and Quaternary denudational processes, acting over ancient structures represented by cratons, ancient orogenetic belts, and Phanerozoic sedimentary basins. The altimetric variations result from the combination of Meso Cenozoic tectonics promoting unequal uplift of pre-existing macrostructures and denudational downdrafts with also unequal speeds, produced by climatic variations (dry and humid climate) in the Tertiary and Quaternary.

For the application of the Brazilian relief map in land use planning, it was necessary to integrate two distinct cartographic products, i.e., the IBGE map, published in 1993, and the Ross map, from 2006, which presented different levels of detail: the first more detailed and the second more generic, producing a map of intermediate resolution. Thus, the objective of this work was to generate the division of the geomorphological units of Brazil in order to meet the Brazilian Territorial Planning research prepared at the scale of 1:5,000,000.

The mentioned research was structured from the perspective of ensuring environmental quality, sustainable development, and improvement of living conditions of the Brazilian population, having Brazil as the adopted geographic cutout. Based on the premise that the territorial planning needs to be grounded on the foundations of sustainable development, it is essential to consider in its implementation the environmental and ecological, social, and economic approaches.

Many possibilities arise in parallel to the initiative of territorial planning, such as: integrated socio-environmental diagnosis, involving society-nature relations, the implementation of integrated public policies guided by the principles of environmental conservation and social and economic development and the possibility of identifying the potentials, vulnerabilities, and fragilities of the territory. Given these perspectives, the understanding of geomorphology, as well as its characterization through relief units, becomes a basic principle.

The new division of the geomorphological units of Brazil was necessary for the physical-natural characterization of the territory, from the perspective of an integrated socio-environmental diagnosis; for the identification of the Natural Environmental Units, which consider the geology, geomorphology, pedology, as well as information on the climate and natural vegetation cover; and finally, for the identification of the Land Units for the territorial planning of the mentioned geographic clipping.

2 Study Area

The major relief compartments found in Brazil and South America are associated with macrostructures, such as cratons or platforms, ancient and recent orogenic belts, large paleo-mesozoic and cenozoic sedimentary basins.

The oldest terrains are the cratons dating from the middle to lower Precambrian, with ages of hundreds of millions of years. They are composed of a diversity of lithologies and structures, among which dominate the very old metamorphic rocks of the Middle to Lower Precambrian (Archeozoic) and the old intrusive rocks of the Middle to Upper Precambrian (Proterozoic) overlain by sedimentary rocks dated to the Upper Precambrian (Proterozoic), which residually cover more restricted areas of the cratons or platforms. To represent these ancient terrains, the Amazon, São Francisco, and Uruguaio-sul-rio-grandense cratons or platforms are cited as examples.

It corresponds to the extensive strips of folded structures located in the Brazilian territory, the so-called orogenetic belts. Such structures are also very old and were generated during the Precambrian Superior (Proterozoic). They are characterized for portraying bands of suture between cratons, founding in South America, the lands of the Precambrian. Recognized examples of these fold belts in Brazil are the Atlantic, Tocantins, and Paraguayan belts.

The large Paleo-Mesozoic sedimentary basins correspond to the third type of macrostructures, which confer the Oriental Amazon, the Parnaíba-SanFranciscana, the Paraná, and the Parecis basins. These large basins were formed in lower altimetric conditions and underwent tectonic processes of syneclizes and amphiclizes, mainly from the Jura-Cretaceous. The sedimentary strata formed by marine, continental, glacial and desert deposits constitute the great South American basins. It is relevant to note that at the end of the Mesozoic (Cretaceous) there was an interruption of extensive sedimentation in these basins.

The Andes mountain range and the cenozoic sedimentary basins formed simultaneously. Tectonic processes have reactivated ancient faults and promoted the formation of escarpments with epeirogenic uplifts and archegations in the central and eastern parts of the continent. In parallel, an intensification of erosional upheaval took place, which was decisive for the lowering of the plateaus and mountain ranges in the central and eastern parts of the continent. The Cretaceous is a very important temporal divisor to unravel the enigmas of the morphogenesis of the relief of Brazil and South America in general, as emphasized by Ross (2016).

3 Materials and Methods

The methodological support of the relief map presented here follows the theoretical assumptions defined by the orientation of the Russian geomorphologist I. P. Guerassimov along the 1960s and Mescerjakov (1968), which were the basis for the taxonomic proposal of Ross (1992) concerning the geomorphological cartography. In this context, Ross (1992) presented a taxonomic sequence to be applied in geomorphological mappings in Brazil in six taxa.

The first and second taxons correspond to the structural and sculptural macro-influences in the genesis of the Brazilian relief, following what was established by the Russian authors, that is, the morphostructures and the morphoscultures.

The third taxon corresponds to the smaller units, contained in the morphostructures and morphoscultures, but being determined by more recent sculptural processes, being defined by the relief dissection modulations, also more recently called topographic rugosity.

The fourth taxon represents individual forms such as hills, hills, river plains, sea plains, escarpments, structural terraces, erosional terraces, and others.

The fifth taxon is represented by the slope typologies, such as concave, convex, rectilinear slopes, convex tops, flat tops, sharp tops, and the sixth taxon is represented by the forms generated through current processes, such as furrows, ravines, gullies, embankments, slope cuts, opening of channels produced by anthropogenic processes. In this work, due to the scale, the first two taxa of Ross (1992) were used, i.e., the Morphostructural Units, and, in the context of each of these, the Morphocultural Units, in which the differentiation in the topographic rugosity can be identified, conditioning the territorial planning and signaling for the investigation of the potential and natural environmental fragilities in the face of human interventions.

The technical procedures involved, at first, the georeferencing and digitization of the Geomorphological Map of Ross (2006), because it was in analog format. To do so, it was used a geoprocessing *software* in which one can make the screen scanning of each polygon that represented a geomorphological unit. After this procedure, a file was obtained in. SHP (*shapefile) format was obtained* with all the geomorphological units.

The relief unit map of IBGE (1993) was obtained from the official website of the institution, already in a format compatible with the Geographic Information System (GIS) environment.

In possession of the two databases, the vectors were architected in a GIS environment with the final purpose of elaborating a rereading of the maps, resulting in a new division of geomorphological units for Brazil. Each of the defined units was characterized and integrated into the final map containing 34 geomorphological units.

4 Brazilian Geomorphological Units: General Characterization Applied to Land Use Planning

The results resulted in the identification of a greater number of relief units, especially those associated with sedimentary basins and orogenic belts, becoming more compatible with the interests and needs of the Brazilian Territorial Planning research at the 1:5,000,000 scale. To illustrate the geomorphological units, Figs. 1 and 2 follow, with the divisions of the first and second taxa. In the first taxon are found the morphostructures of the Amazon Craton, the Ancient Orogenic Belt, the Paleo-Mesozoic Sedimentary Basins, and the Cenozoic Sedimentary Basins. In the second taxon, the morphosculptural divisions are recognized, i.e., plateaus, depressions, mountains,

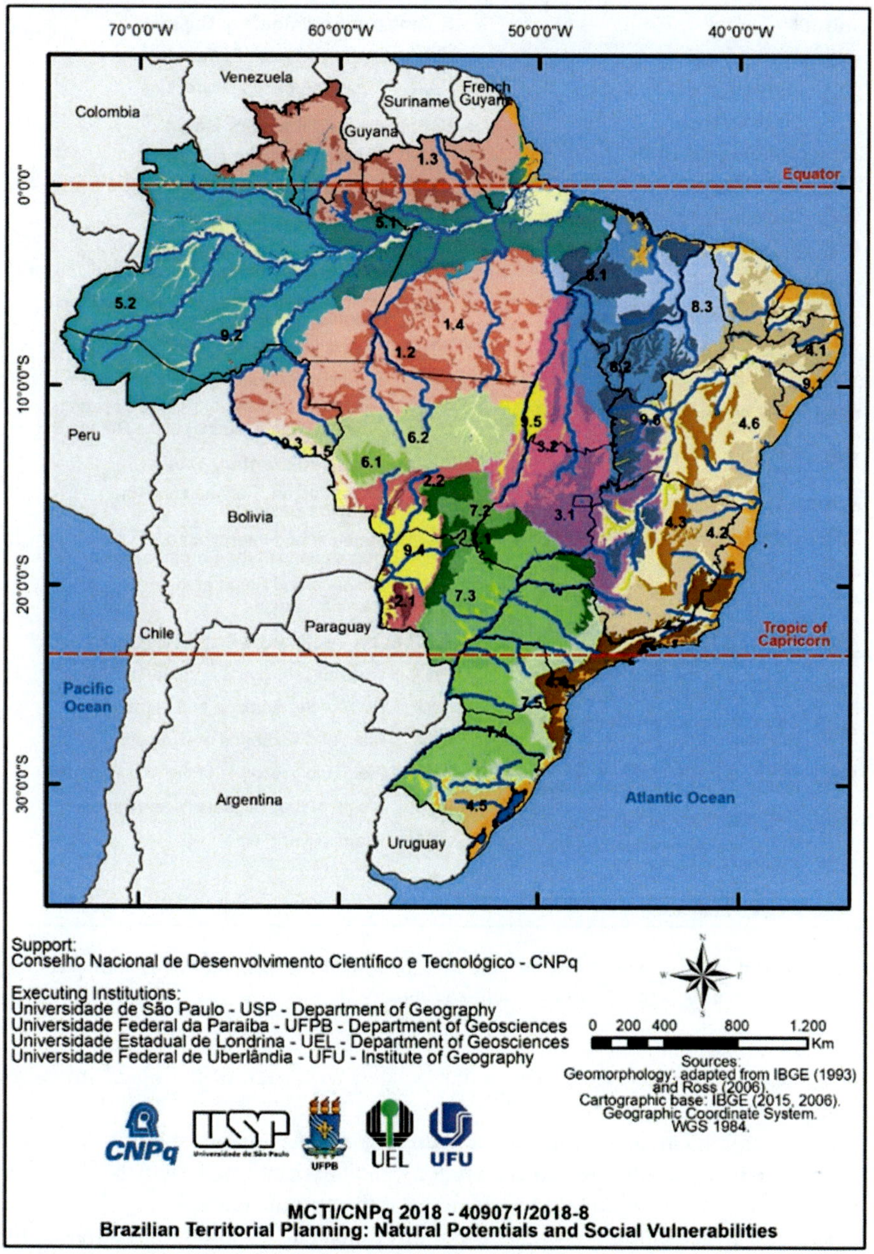

Fig. 1 Map of the division of geomorphological units of Brazil. *Source* of data: adapted from IBGE (1993); and Ross (2006). Elaboration: The authors

Fig. 2 Legend of the geomorphological units division map of Brazil. *Source* of data: adapted from IBGE (1993); and Ross (2006). Elaboration: The authors

and plains. The morphosculptures are associated with each of the mega-structures presented in the first taxon.

The Brazilian landforms predate the present configuration of the South American continent. The effects of the Andean orogeny and the opening of the Atlantic Ocean since the Jurassic (130 Ma) are responsible for the current shape.

The global geotectonic processes that are reflected in the conformation of the South American continent undergo a significant change in behavior from the Jurassic, extending through the Cenozoic. Consequently, this change is also reflected in the geological structure and in the genesis of the relief of what would become, in the Tertiary and Quaternary, the South American continent. The continuous weathering progressively sculpts the plateaus, mountain ranges, mountains as well as the relative depressions that surround the Brazilian sedimentary basins.

In the South American Central Depression and the basins of the Western Amazon (Solimões), Orinoco, Paraguay-Paraná (Pantanal of Matogrosso and Guaporé, as shown in Fig. 3), Araguaia, and Atlantic rift, Cenozoic deposits compose sedimentary basins in *rift valleys* and platform coverings.

In Brazil and Venezuela, the oldest terrains are related to the Amazon Craton, identified on the map as the Amazon Craton Morphostructure (Fig. 4). The largest

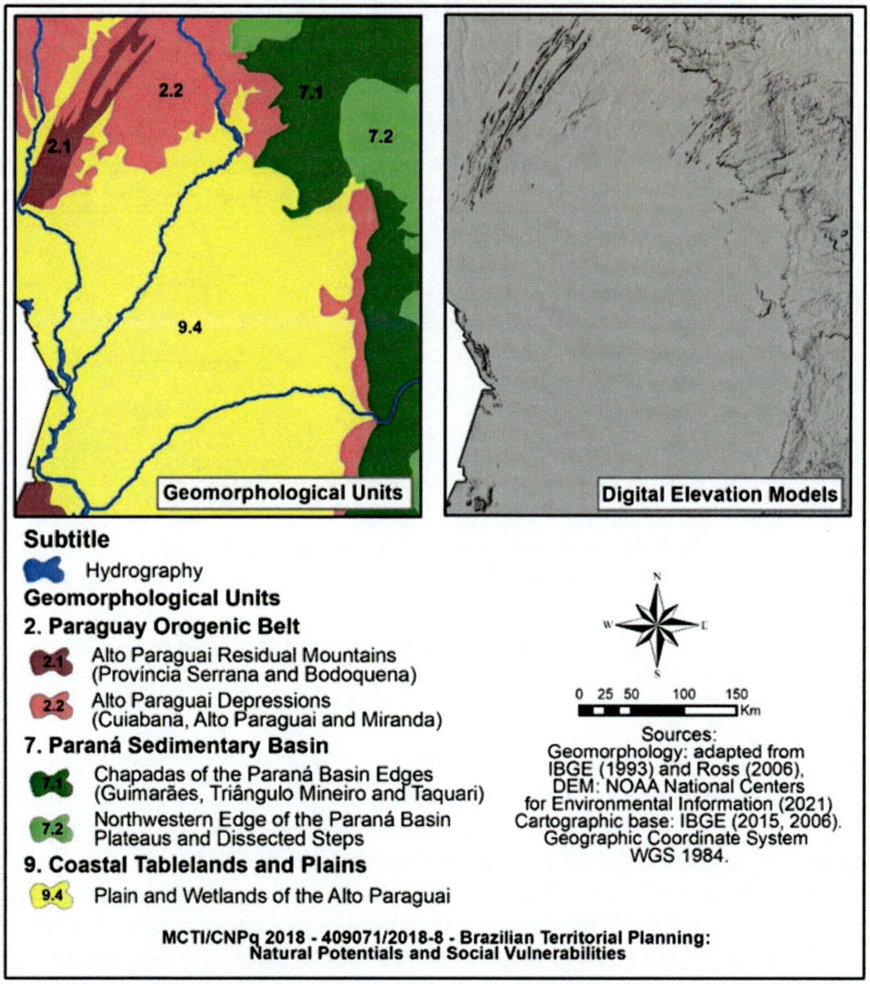

Fig. 3 Detailing of the geomorphological unit plain and wetlands of the Alto Paraguay and its surroundings. *Sources* adapted from IBGE (1993); Ross (2006). Elaboration: The authors

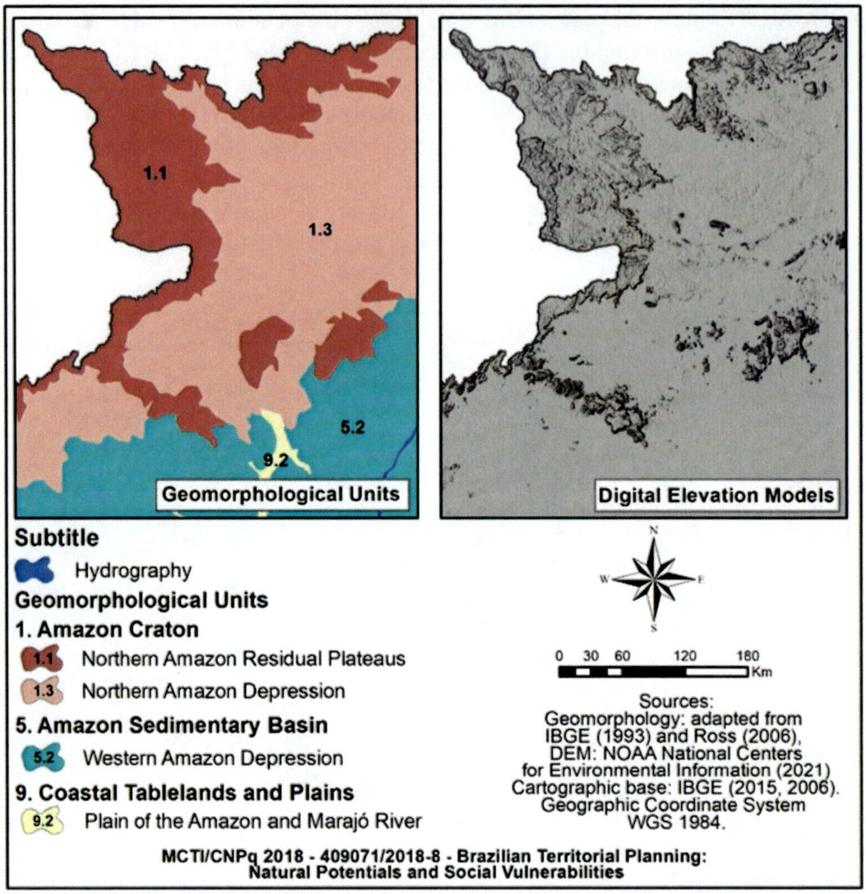

Fig. 4 Detailing of the geomorphological unit Northern Amazon Residual Plateaus and its surroundings. *Sources* adapted from IBGE (1993); Ross (2006). Elaboration: The authors

territorial extension in this morphostructure, according to Ross (1990), is represented by the flattened and lowered surface with altitudes that range between 100 and 300 m. The metamorphic rocks of the Middle Precambrian (1.8 to 2.5 billion years) give support to this surface, denominated by Ab'Saber (1972) as North and South Amazonian Marginal Depression in relation to the presence of the Eastern Amazon Sedimentary Basin. The northern and southern edges of this basin reveal cuestiform reliefs, containing higher parts between 350 and 400 m. Between the tops of the Cuesta reverses and their base, there are 200 to 300 m of unevenness which Ab'Saber (1972) interpreted as an "eversion surface", due to the erosive processes that would have occurred during the Tertiary-Quaternary period. This surface to the north of the Sedimentary Basin of the Oriental Amazon receives the name Depression or North Amazonian Surface and its correspondent to the south, Depression or South Amazonian Surface.

Throughout this low and dissected surface in hills and low hills, with accentuated topographic rugosity, higher reliefs, with altitudes above 600, reaching in restricted areas, more than 1000 m. The igneous rocks of the granite family, the presence of acid volcanic rocks such as rhyolites, and the residual platform coverings composed of silicified sandstones, support these higher reliefs and are examples of the great lithological diversity present.

There is also the presence of high reliefs, where the tops are preserved by iron and manganese formations, products of supergene deposits from the Precambrian, very resistant to erosive processes. Examples are the Carajás mountain range complex, located in the state of Pará, and Urucum, located in the state of Mato Grosso do Sul. It is pertinent to note that these higher elevations are represented in the geomorphological units by the North and South Amazonian Residual Plateaus.

The fold belts of the Brazilian cycle are represented by the Paraguay (Fig. 5), Tocantins, and Atlantic Orogenetic Belts. From a morphogenetic perspective, they correspond to residual reliefs resulting from erosive processes that developed during the Phanerozoic, lowering the ancient mountain chains to current topographic levels.

In general, these macrostructures support reliefs marked by excavated anticlines, raised synclinals, fault scarps, and tectonic pits that define complex mountain systems composed of aligned mountains roughly parallel to each other. They are present in these orogenetic belts underlying intrusive masses, represented by igneous with prominence for the granites, that are exposed on surface as a result of the erosive lowering that eroded the highest parts of these old mountain chains. Examples are the Caparaó, Cantareira, São Francisco, Itaqui and countless others in the Atlantic Belt, or still the Dourada and Mesa Mountains, in the Tocantins-Brasilia Belt, and Serra de São Vicente, in the Paraguay Belt.

However, what most highlights these mountainous reliefs of Brazil are the mountain ranges held by metamorphic rocks, such as quartzite or sedimentary, and by ancient rocks, such as silicified sandstones that support the edges of excavated anticlines and uplifted synclinals. These aligned mountain ranges are observed mainly in the Tocantins Orogenetic Belt, such as the Serra da Canastra; in the Paraguayan Belt, especially the Serra do Espinhaço; and in the Atlantic Orogenic Belt, such as the Chapada Diamantina. Among the mountains that configure the relief of these morphostructures, there are surfaces lowered by erosion, which interpenetrate forming the Intermontane Depressions or those that accompany parallel mountain ranges (Ross 2016).

The sedimentary basins of the Oriental Amazon and Parecis partially cover the Amazon Craton. The Parecis basin, located in the watershed of the rivers that flow into the Amazon River, to the north, and the Paraguay/Paraná River, to the south, is composed of Cretaceous sandstones. These geological formations correspond to the superior extracts of these basins, being that the highest parts in the Amazon Basin, are between 350 and 400 m, while the Parecis Plateau, oscillate between 400 and

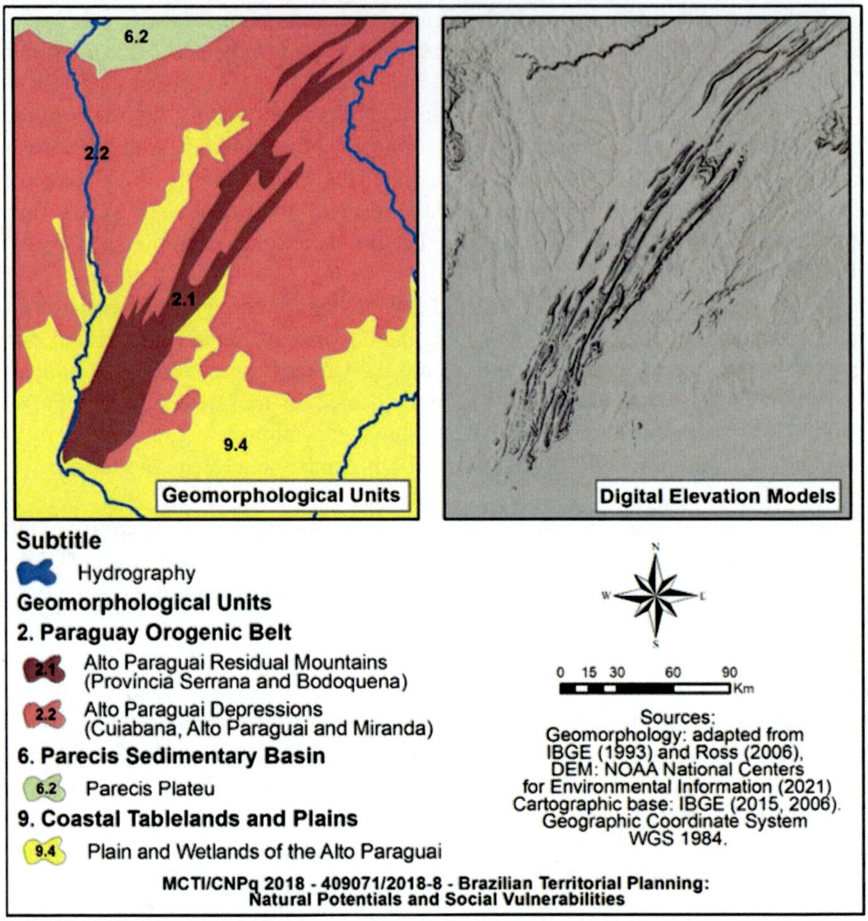

Fig. 5 Detailing of the geomorphological unit Alto Paraguay Residual Mountains (Serrana and Bodoquena Province) and its surroundings. *Sources* adapted from IBGE (1993); Ross (2006). Elaboration: The authors

800 m. These altimetric levels indicate that the Amazon Craton was also affected by the Cenozoic epigenetic processes in an uneven manner.

The sedimentary basins of Paraná and Parnaíba-Sanfranciscana are configured in Plateaus and Chapadas and have their edges uplifted by the effects of the Meso Cenozoic tectonics with the highest altitudes oscillating between 800 and 1200 m. However, they reach 1500 m in the Planalto de Vacaria, in the region known as Aparados da Serra, in the northeast of the State of Rio Grande do Sul and southeast of Santa Catarina (Fig. 6).

The craggy edges of these large sedimentary basins of Paraná and Parnaíba-Sanfranciscana present, in general, cuestiform reliefs that are accompanied by Peripheral and Marginal Depressions, with regional differentiation among them. The

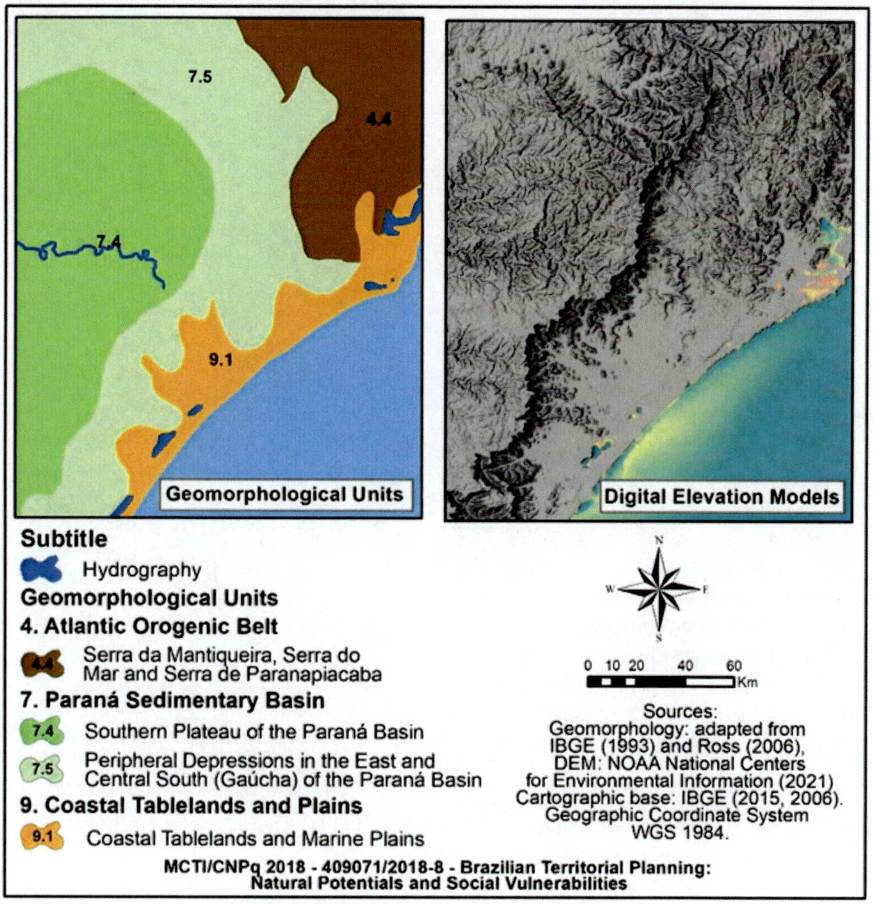

Fig. 6 Detailing of the geomorphological unit Southern Plateau of the Paraná Basin and its surroundings. *Sources* adapted from IBGE (1993); Ross (2006). Elaboration: The authors

altitudes of these depressions, which are characterized by surfaces eroded during the Cenozoic, are very varied and depend on the greater or lesser effects of the post-Cretaceous epeirogeny (Fig. 7).

The Sertaneja and São Francisco Depressions are between a few tens of meters as in the State of Ceará and reach around 400 m in the São Francisco basin in the State of Minas Gerais. The Peripheral Depression of the eastern edge of the Paraná Basin, in the state of São Paulo known as Peripheral Depression Paulista, has altitudes between 550 and 700 m, but changes its morphological aspect in the east of the states of Paraná and Santa Catarina, when it is called Second Plateau because it defines a physiognomy of a wide and well-marked structural plateau supported by Devonian

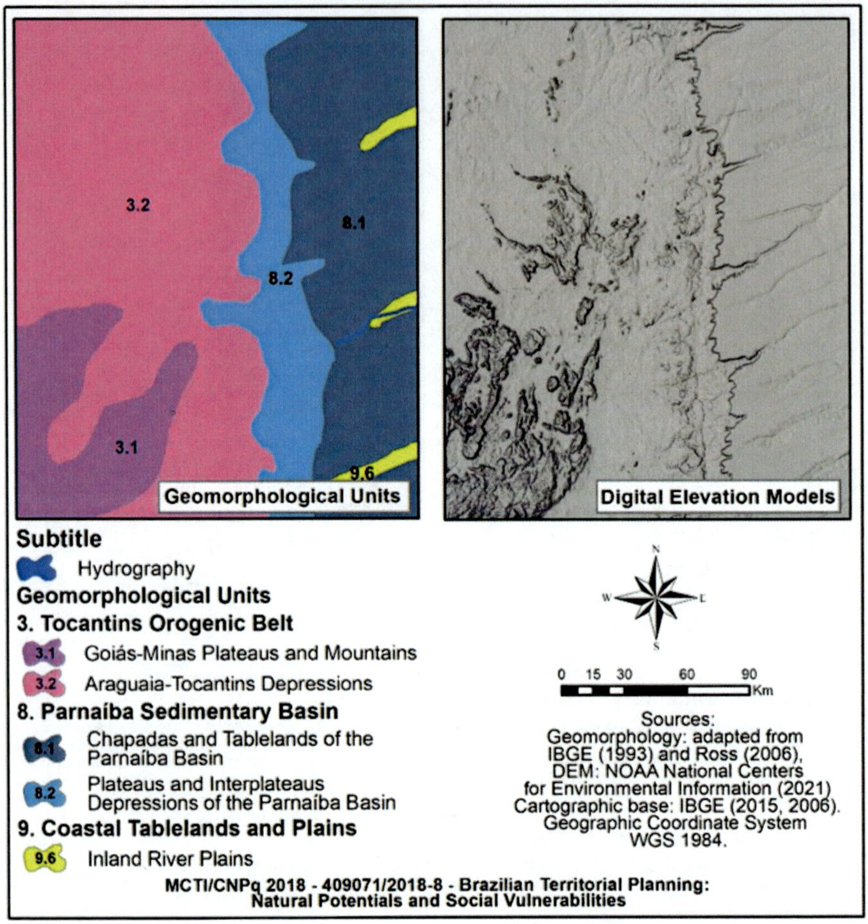

Fig. 7 Detailing of the geomorphological unit Chapadas and Tablelands of the Paranı́ba Basin and its surroundings. *Sources* adapted from IBGE (1993); Ross (2006). Elaboration: The authors

rocks. In the state of Rio Grande do Sul, the Central Depression has altitudes not much higher than 200 m in the highest parts.

A similar fact also occurs with the Upper Paraguay River Depressions in the states of Mato Grosso and Mato Grosso do Sul, whose altimetric values are around 100 to 300 m.

Ross (2014) considers that Chapada dos Guimarães, to the northwest of the Paraná sedimentary basin, as well as Chapada dos Parecis, further to the northwest, have their genesis associated with the combination of geotectonic processes such as crustal movement, resulting from the opening of the Atlantic, the Andean orogeny and the generalized uplift of the South American platform from the Jura-Cretaceous to the Cenozoic. These movements promoted dome arcing along major structural alignments, such as the one that occurs along the fold belts of the Paraguay Orogenetic Belt, positioned between the above-mentioned plateaus and known in the geological literature as the São Vicente Arch (Ross 2014), has genetic relationships with these movements.

On the flat tops of these chapadas, which are composed of rocks of the Cretaceous (Bauru and Parecis Groups), there is a thick clay and ferruginous pedological cover constituting deep soils. Surrounding these are escarpments articulated to the depressions of the upper Paraguay.

These facts aligned to interpret the morphogenesis of Chapada dos Guimarães also apply to Chapada dos Parecis and the other plateaus that are located on the edges of the sedimentary basins of Paraná and Parnaíba-Sanfranciscana, considering, obviously, the specificities of each one of them (Fig. 8). The mentioned plateaus and

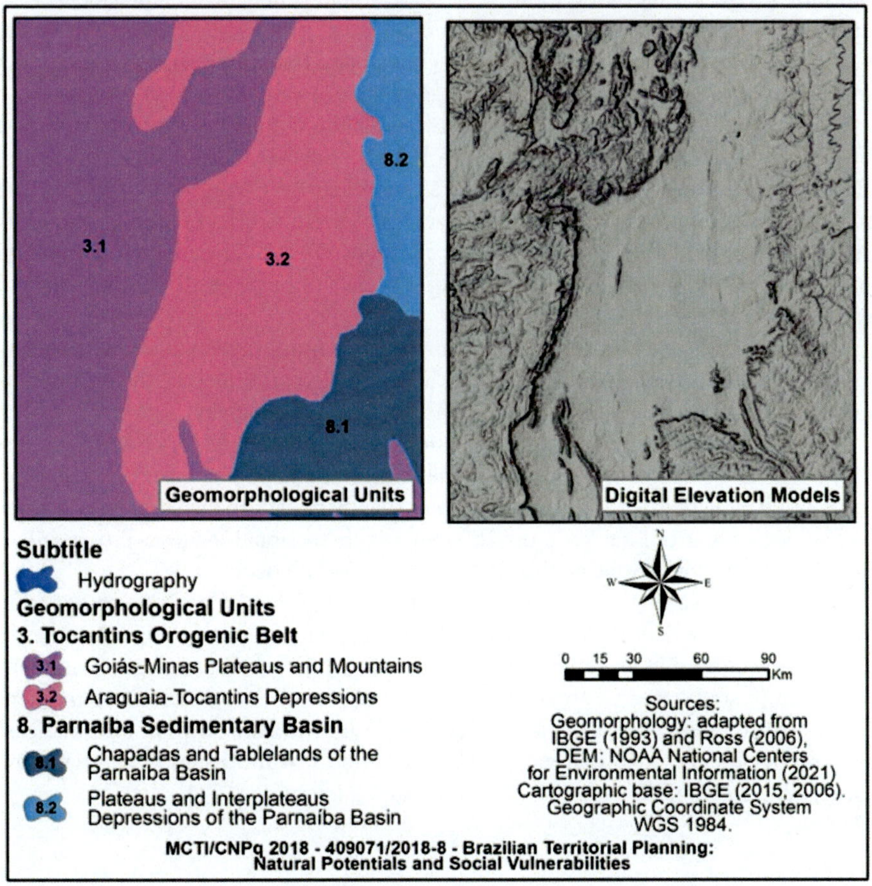

Fig. 8 Detailing of the geomorphological unit Araguaia-Tocantins Depressions and its surroundings. *Sources* adapted from IBGE (1993); Ross (2006). Elaboration: The authors

the depressions that surround them, to the south as well as to the north and west, were generated by simultaneous processes of tectonics and erosional lowering for more than 60 Ma, according to Ross (2014). This interpretation is in contrast to that of Ab'Saber (1972), who considered them to have resulted from denudational processes from the Plio-Pleistocene onwards.

The surfaces lowered by erosion gave rise to the Peripheral and Marginal Depressions, according to Ab'Saber (1949, 1972). According to the interpretation, these were sculpted by more or less intense phases, combined with the intermittent impulses of tectonic activities (neotectonics) and greater or lesser erosive vigor, depending on whether the climate conditions were drier or more humid and hot.

The *rift valleys* marked by transcurrent faults, scarps, generation of grabens/horsts, and formation in these tectonic depressions of syntectonic sedimentary basins (Fig. 9) both continental and marine are also associated with the tectonics from the Jurassic extending to the Cenozoic. These traphogenetic sedimentary basins in emersed lands correspond to the Cenozoic basins of São Paulo, Taubaté, Rezende, Curitiba, Pariquera-Açu, Volta Redonda, Guanabara, Itaboraí, generated, according to Riccomini (1989), from the Oligocene–Miocene.

The submerged basins composed of marine and continental sediments are part of this *rift* system. Among them are the Pelotas, Santos, Campos Espírito Santo, Bahia-Sul and Sergipe-Alagoas basins, as recorded by Chang et al. (1992). The materials that compose the sedimentary packages are from different phases of marine sedimentation in shallow or deeper sea conditions. These variations are a function of the movement processes of the structural blocks throughout the Cretaceous and especially in the Tertiary.

In the *rift* concentration areas according to Riccomini (1989) and reaffirmed by Gontijo (1999), the tectogenetic activities are still active, revealed by the testimonies observed in the Rezende, Taubaté, and São Paulo basin deposits.

The application of dating techniques using fission trace analysis (TFA) on apatites and U-Th/He ages in vertical profiles in the central-southern portion of Serra do Mar, as reported by Ribeiro et al. (2011), indicates that there was, in the Upper Cretaceous to Paleocene, a strong tectonic uplift, from which erosional cycles were installed, and that this probably repeated itself in the Eocene–Oligocene.

The results revealed by the data of 2013 and 2014 of the Brazilian Seismographic Network and revealed by Assumpção et al. (2015), strengthen the interpretation of the active tectonics in current times. The publication shows us, through a summary map, that in some axes or areas the frequency of seismic tremors is more frequent in the Brazilian territory. This is very evident in the Atlantic strip, in the southeast region, and for the strip that enters the center north of the country following the Orogenetic Belts of the Atlantic, Tocantins, and Paraguay.

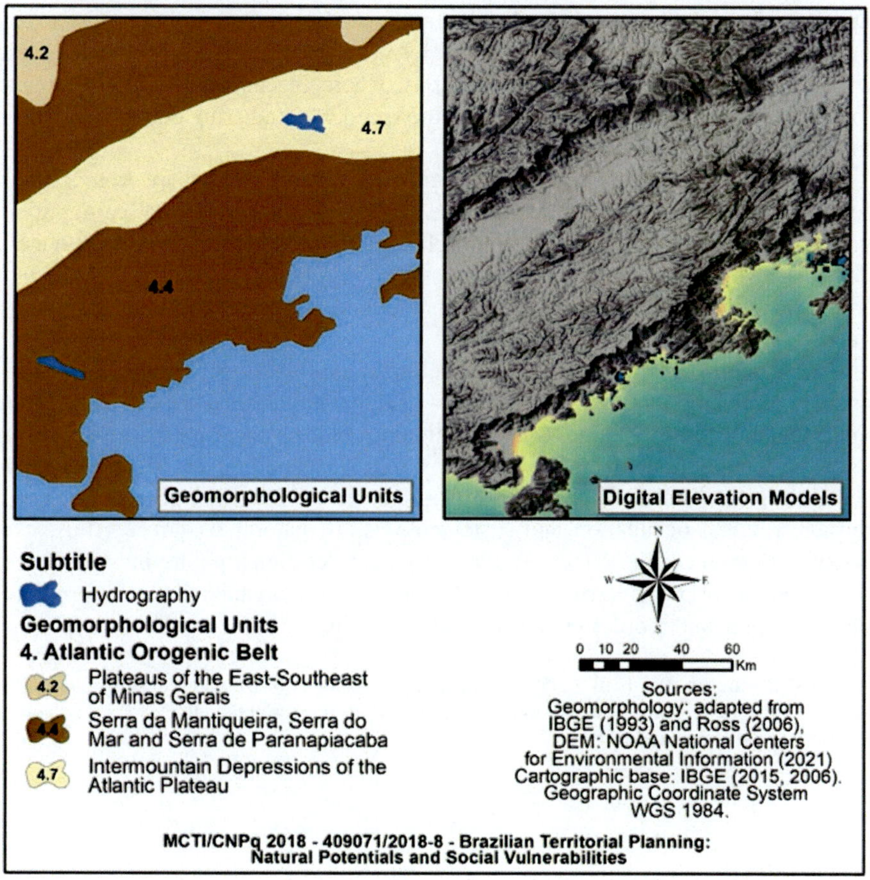

Fig. 9 Detailing of the geomorphological unit intermountain depressions of the Atlantic Plateau. *Sources* adapted from IBGE (1993); Ross (2006). Elaboration: The authors

5 Final Considerations

Some regional generalizations made here allow us to highlight two conclusions: the first is that flattened, or at least rectilinear, levels of the tops of fold belts (belts) bear witness to Cretaceous or earlier erosional phases. They are manifestations observed in the extensive leveled and low surfaces that are carved on the Amazon and São Francisco Cratons. Another fact considered scientifically accepted is that the inter-planaltic depressions and surfaces embedded in the edges of large sedimentary basins are of more recent age (Tertiary and Quaternary).

It is possible to affirm that the local and regional morphological or topographic levels are associated with diverse origins and genesis, contemplating the combined relationship of the Meso Cenozoic tectonics, which promoted uplift/arcs/basculations, and the erosive processes that acted and continue to act in the lowering and deposition, whether by processes triggered by drier climatic conditions or in hot and humid conditions, where mechanical and geochemical lowering are determinant to differentiate the speed of weathering due to differences in mineralogical resistance of the rocks.

From this morphogenetic analysis, presented in more detail by Ross (2016), followed by the technical work to merge the two geomorphological cartographic products, a new division of the geomorphological units of the relief is presented, considering the general context of South America. The relief division was thought from the geological macrostructures that define the mega-units of relief and geodiversity of Brazil.

The rereading of the geomorphological units applied to territorial planning has enabled more emphatic integrated studies of the national territory, considering the understanding of the dynamics of the functioning of the physical-natural environment with the intervention of human actions. Land use planning, based on technical-scientific principles, optimizes and orders the use of Brazilian geographic space through rational occupation and sustainable use of natural resources. Thus, it is possible to increase the effectiveness of political decisions, public intervention in the management of the territory, as well as an effective organization of information needed to plan and reorder the territory of the country.

Acknowledgements We thank CNPq for the partial funding of the research through the project registered under No. 409071/2018-8 and entitled "Brazilian Territorial Planning: Natural Potentialities and Social Vulnerabilities".

References

Ab'Saber AN (1949) Regions of Post-Cretaceous Circundesnudation in the Brazilian Plateau. Boletim Paulista de Geografia 1:1–21

Ab'Sáber AN (1972) Participation of peripheral depressions and flattened surfaces in the compartmentalization of the Brazilian Plateau. Geomorphology 28:1–38

Assumpção M, Barbosa JR, Bianchi M, Rouet S, Rocha MP, Ferreira JM (2015) Regional Earthquakes Detected by the Brazilian Seismographic Network-RSBR in 2013–2014. In: XV National Symposium of Tectonic Studies

Chang HK, Kowsmann RO, Figueiredo AMF, Bender AA (1992) Tectonics and stratigraphy of the east Brazil Rift system: an overview. Tectonophyscs 213:97–138

Gontijo AHF (1999) Morphotectonics of the Middle Paraíba do Sul River Valley: Serra da Bocaina Region, SP-RJ States. PhD Thesis: Institute of Geosciences and Exact Sciences, UNESP Rio Claro

IBGE (1993) Mapa das Unidades de Relevo do Brasil. Rio de Janeiro. Brazilian Institute of Geography and Statistics

Mescerjakov JP (1968) Les concepts de morphostructure et de morphosculpture: un nouvel instrument de l'analyse géomorphologique. Annales de Geographie, 423:539–552

Ribeiro MCS, Hackspacher PC, Ribeiro LFB, Neto JCH (2011) Tectonic and Denudational Evolution of the Serra do Mar-Southeastern Brazil at the Boundary between the Upper Cretaceous

and Paleocene Using Fission Trace Analysis and U-TH/HE in Apatites. Brazilian Journal of Geomorphology. 12(3):03–14. https://doi.org/10.20502/rbg.v12i0.254
Riccomini C (1989) The Continental *Rift of* Southeastern Brazil. PhD Thesis: Institute of Geosciences, USP
Ross JLS (1990) Relevo Brasileiro: Uma nova proposta de classificação. Journal of the Department of Geography, 4:25–39. https://doi.org/10.7154/RDG.1985.0004.0004
Ross JLS (1992) The cartographic record of geomorphic facts and the question of relief taxonomy. Journal of the Department of Geography, 6:17–30. https://doi.org/10.7154/RDG.1992.0006.0002
Ross JLS (2006) Ecogeografia do Brasil: subsídios para planejamento ambiental. São Paulo, Oficina de Textos
Ross JLS (2014) Chapada dos Guimaraes: Edge of the Paraná Basin. Journal of the Department of Geography, 28:180–197. https://doi.org/10.11606/rdg.v28i0.525
Ross JLS (2016) The Brazilian Relief in the Context of South America. Revista Brasileira de Geografia, IBGE 61(1): 21–58. https://doi.org/10.21579/issn.2526-0375_2016_n1_art_2

Geomorphological Structuring and Tectonic Control in the Southeastern Brazilian Stepped Reliefs: Relation with the Evolution of Crystalline Scarps

Roberto Marques Neto, Felipe Pacheco da Silva, Juliana Alves Moreira, Juliana Costa Baptista Barreto, and Matheus da Silva Frauches

Abstract The strong tectonic control acting on the relief evolution in the context of the southeastern Brazilian rift margin has been treated by several methodological prisms. The great escarpments of the Brazilian passive margin present important crustal elements acting in its geomorphological evolution and structuring, being the present work in charge of discussing the relations between the structural and tectonic control in the evolution of such geomorphological systems, focusing on specific compartments of Serra do Mar and Serra da Mantiqueira, the two main orographic steps of the Brazilian Atlantic margin. Through the elaboration of paleotopographic models and structural and geochronological analysis, the results pointed out strong neotectonic deformational effects superimposed on Precambrian structures reactivated in the Cretaceous-Paleocene, reinforcing the action of a post-Miocene tectonics influencing the maintenance of scarps and enhancing local and regional amplitudes.

1 Introduction

The geomorphological structuring of the crystalline terrains of southeastern Brazil conforms to relief systems characterized by well-marked regional scaling, controlled by different base levels that contain distinct organizations in terms of morphogenetic and surface processes. From the base level defined in the Atlantic Ocean to the more receded interfluves of the basins that flow toward the oceanic level, there stands out a graded relief marked by high ridges in continuous *horsts*, structural hills, interplanaltic erosive depressions, tectonic depressions (grábens), and other geomorphic facts carved in ancient structures and in great measure influenced by neotectonic efforts.

R. M. Neto (✉) · J. A. Moreira · J. Costa Baptista Barreto · M. da Silva Frauches
Federal University of Juiz de Fora, Juiz de Fora, Brazil
e-mail: roberto.marques@ufjf.edu.br

F. P. da Silva
Federal University of Rio de Janeiro, Rio de Janeiro, Brazil

G. Barbosa dos Santos et al. (eds.), *Geomorphology of Brazil: Complexity, Interscale and Landscape*, Springer Proceedings in Earth and Environmental Sciences,
https://doi.org/10.1007/978-3-031-05178-4_3

The distensional tectonics that engendered the separation of the Afro-Brazilian paleoplate from the Mesozoic remobilized the Proterozoic structures inherited from the Brazilian-Pan African orogeny through epeirogenetic efforts that defined the plateau levels of the Atlantic margin, including the taphrogenic tectonics responsible for generating the Serra da Mantiqueira and Serra do Mar, as well as the Paraíba do Sul river tectonic depression. Such geodynamic event has been defined as the southeast Brazilian continental rift (Riccomini 1989). With the frank opening of the Atlantic Ocean, the evolution of the hydrographic basins that drain toward the oceanic level took course, and that in the context of the rift system under consideration has in the Paraíba do Sul river its main regional collector, whose base level is defined since the Eocene.

Despite the existence of a regional Paleogene base level, the base levels have not been stable in the region due to the effect of neotectonic deformational efforts (Hasui 1990; Saadi, 1991), with well-documented evidence in the context of the rift margin (Santos 1999; Chiessi 2004; Zalan and Oliveira 2005; Silva and Mello 2011; Marques Neto et al. 2019). Such structuring derives from a strong erosive imperative defined by the proximity to the Atlantic Ocean, partly differential due to lithostructural contrasts, but augmented by a morphotectonic control responsible for the reactivation of headwaters, re-hierarchical drainage, lateral migration of channels by block basculation, colluvialization on the slopes, among other processes.

Among the main results of this entanglement of controls, some of them are worth mentioning: (1) The Atlantic crystalline margin of southeastern Brazil is geomorphologically characterized by staggered plateau levels defined in successive regional base levels; (2) These base levels are not stable, having been subjected to the influences of post-miocene tectonics; (3) The coastal and sublittoral escarpments have their retraction processes conditioned by successive base levels delimited in different tectonostructural blocks, which repercusses in a distinct regional variation of the erosive dynamics; (4) The upward migration of *knickpoints* tends to be more evident in the rocky segments of the river channels, which has had an impact on the maintenance of slopes along the retreating scarps; (5) In the course of scarp retraction and morphogenesis of the interfluves, it has been common for the interception of channels by transversal drainage that enters the interior of the continent, promoting fluvial captures of various magnitudes as pointed out by authors such as Cherem et al. (2013), Rezende (2018), Salgado et al. (2018), Marent and Valadão (2019), among others.

Focusing on the premises listed above, this chapter discusses aspects related to the morphogenesis and geomorphological structuring of the great escarpments of the Brazilian passive margin, putting the magnifying glass on three regionally significant escarpment extensions positioned in different compartments and subjected to distinct base levels, conforming geomorphological specificities that are very representative of the regional geomorphogenesis: Serra dos Órgãos, in the context of the Serra do Mar, Serra do Ibitipoca, in the context of the Serra da Mantiqueira, and Serra do Relógio, a NNE-SSW alignment corresponding to the festooned structures of the so-called Northern Mantiqueira, cut in deep epigeny by the transverse drainages that trace back to the interplanaltic erosive depressions.

2 Materials and Methods

The selection of the geomorphological compartments collated in the present work took into account (1) the homologous character of the genesis of these structures, (2) the different evolutionary aspects linked to distinct regional base levels and to the conditioning of the Cenozoic tectonics, (3) the regional representativeness in the context of the southeast rift margin. In this sense, the structures analyzed were generated during the reactivations linked to the platform rifting but evolved as a function of specific controls engendered by their respective base levels and particular tectonostructural aspects. The methodological treatment employed coadunited cartographic techniques and paleotopographic, geochronological, structural analysis and field reconstitution.

First, a database was prepared for each area from Landsat images (OLI\TIRS C1 Level 1) and SRTM (*Suttle Radar Topography Mission*) data from the shaded relief, available for download on the website of the USGS (*United States Geological Service*), as well as the planialtimetric bases provided by the Foundation Brazilian Institute of Geography and Statistics (IBGE) at the scale of 1/50,000. The remote sensing products served as a basis for generating baseline information on hydrography, slope, and hypsometry, with their respective cartographic documents processed in ArcGIS software. The planialtimetric bases, together with the base maps that were generated, served for the geomorphological compartmentalization and the generation of paleotopographic maps for the three geomorphological systems studied.

The extraction of the structural lineaments was performed to support the interpretation of the geomorphological structuring of the areas and their regional differences. The relief and drainage lineaments were defined from SRTM radar data in shaded relief. The interpretation took into account four angles of azimuthal illumination (45°, 90°, 315°, and 360°), since each one of these angulations emphasizes more forcefully some lineament orientations to detriment of others, thus allowing a more secure and comprehensive interpretation. The rosette diagrams were generated at 10° angular intervals, using the Georient© software.

The present relief was discussed in relation to the paleotopographic picture obtained from the application of the Sêppomen technique (Motoki et al. 2008), a cartographic approach that estimates the paleorelief from valley fills (Leverington and Teller 2003; Couto et al. 2014; Vargas 2017). The methodology consists of dividing the study area into quadratic cells superimposed on the planialtimetric bases at a scale of 1/50,000, in which the maps are generated. For each cell the highest level curve is considered, which implies the removal of the other curves and the consequent elimination of the effect of fluvial dissection, thus resulting in a geomorphological configuration prior to the most recent phase of drainage incision. The paleotopographic map is generated by interpolation procedures, here performed by the IDW protocol (*Inverse of Distance Weighted*) contained in the toolbox of the GIS software. For the present proposal, we opted for the modulation in an intermediate grid of 1 km × 1 km, which presumably indicates a more recent temporality of the relief, emphatically neoquaternary, but that reveals to satisfaction past situations of altimetry, density, and depth of dissection.

The comparison between the current and past topographic pictures was also subsidized by the interpretation of the existing fluvial captures along the three compared interfluvial segments, gauged from the evidence expressed in the topographic sheets at the scale of 1/50,000, in Google Earth Pro images, and from field control. The following criteria were considered: (1) presence of anomalous curve (*elbow of capture*) linked or not to a transcurrent displacement; (2) occurrence of blind valley (*wind gap*) visible in the planialtimetric bases e\ou in the field; (3) same altimetric position along the surface where the capture occurred, between the captured basin and the capturing basin.

For the Serra do Ibitipoca context, the presence of terraces generated by the post-capture fluvial notch still preserved enabled the dating of the deposit by Optically Stimulated Luminescence (OLE), thus allowing to bring to light geochronological aspects of the local and regional Quaternary morphogenesis. The material collected in the field followed the usual protocols (Jacobs and Roberts 2007; Sallun 2007; Anderson and Anderson 2010; Bierman and Montgomery 2014), removing the sediment mass by incising a dark-colored, properly sealed PVC plastic tube. In this case, the collection was performed at a depth of 1.05 m. Then, the collected material was sent to the laboratory Datações, Comércio e Prestação de Serviços Ltda., where the dating test was performed from the SAR protocol (*Single Aliquot Regenerative Doses*) in fifteen aliquots.

3 Study Area

The region focused on in the present study comprises the area influenced by the southeast Brazilian continental rift (Riccomini 1989). The geomorphological structuring of the southeastern rift margin is driven by successively reactivated Precambrian structures of general NNE-SSW orientation, the dominant one in the continental rift system. Such structures intersect different geomorphic compartments that evolved throughout the Cenozoic and control regional base levels arranged at distinct altitudinal positions (Fig. 1). Indefectibly, they are the main structural conditioning factors in the current scalings as the main alignments correspond to the major regional geomorphological compartments defined in the *horsts of* Serra do Mar and Serra da Mantiqueira and the tectonic depression of the Paraíba do Sul river. Secondarily, a tectonic control has favored the reaffirmation of the scarps and the generation of stepped landforms and, finally, a mainly erosive control is related to the retraction of the scarps and the consequent opening of interplanaltic depressions in the E-W orientation.

The selected geomorphological compartments comprise terrains inserted in different hydrographic basins of regional importance. The Piabanha river basin, fully attached to the Rio de Janeiro state, covers a large part of the stepped reverse of the Serra dos Órgãos, whose slopes facing the Baixada Fluminense are dissected

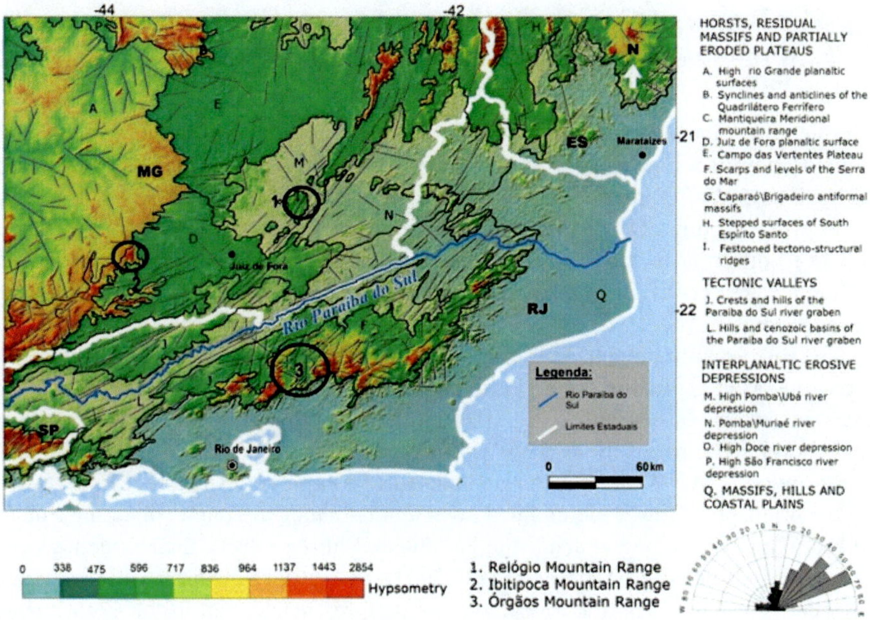

Fig. 1 Regional geomorphological structure and compartmentalization

by smaller channels. Serra do Ibitipoca is positioned in an important interfluvial segment of the Peixe river basin, a direct affluent of Preto river. Regarding Serra do Relógio, its location in the central part of the Pomba river basin stands out, marking the limits between different altimetric levels of the interplanaltic depression linked to the alluded fluvial channel. The three mentioned hydrographic basins, directly or indirectly, imbricate in the Paraíba do Sul river, main regional collecting trunk.

Altimetric variation, slope orientation, lithology, and distance from the Atlantic Ocean are factors that define mesoclimatic differentiations and vegetation cover, dominated by the ombrophilous forests in the Serra do Mar (emphatically in the branch concerning Serra dos Órgãos) and the semideciduous seasonal forests in the interior orographic steps, as occurs in Serra do Relógio in the form of remaining fragments. Serra do Ibitipoca, on the other hand, appears as a higher structure sustained by quartzite, thus showing ombrophilous forests over schist on windward slopes with higher elevations, besides rock fields in the quartzite somites where they vary in different floristic and phytophysiognomic arrangements.

4 The Regional Geomorphological Structuring and the Erosive Control on the Evolution of Scarps

The tectonostructural arrangement in the southeastern Brazilian rift margin, as highlighted, comprises high reliefs of general NE-SW orientation and transversal drainage of copious NW–SE orientation, besides E–W orientations linked to more recent transcurrent regimes. Such structuring, therefore, responds to a complex framework linked to different tectono-erosive controls operating throughout the Cenozoic in different degrees of tropicality. Figure 2 summarizes the main evolutionary events that cover the regional context, serving as a temporal reference for the present discussion.

The three compartments analyzed in a more vertical way replicate the regional typicality, as is explicit in the observation of the structural lineaments extracted for the areas under consideration (Fig. 3), mainly those concerning the relief, pointing to the strong control of the NE–SW shear zones in the regional geomorphological organization. The NW–SE structuring, in its turn, is explored in a more recurrent way by the drainage of lower hierarchical level that demands the main collecting trunks linked to the rift, among which is the Paraíba do Sul river itself. Such structures are dominantly neotectonic (Silva and Mello 2011), and reveal the superposition of post-Miocene stress fields on the Precambrian structures reactivated in the Cretaceous-Paleocene. In Serra dos Órgãos and Serra do Ibitipoca, the relief forms assume a prominent N50E orientation, and in Serra do Relógio N30E.

It is also explicitly noticed that the drainage orientation inherited from the rift, aligned in the NE-SW direction parallel to the great escarpments, diminishes in the direction of the continental interior, being conspicuous in the Serra dos Órgãos and inexpressive in the region of the interfluves bordering the Rio Grande basin where the

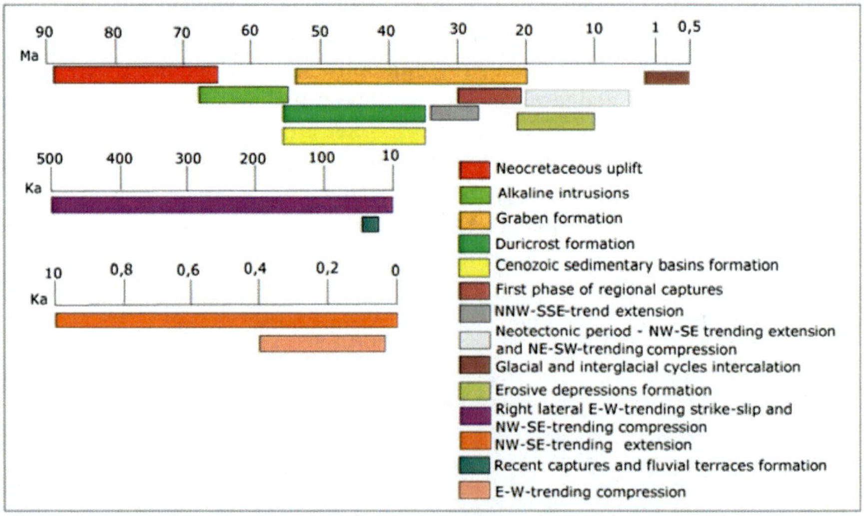

Fig. 2 Main evolutionary events that have affected the region since the Neocretaceous

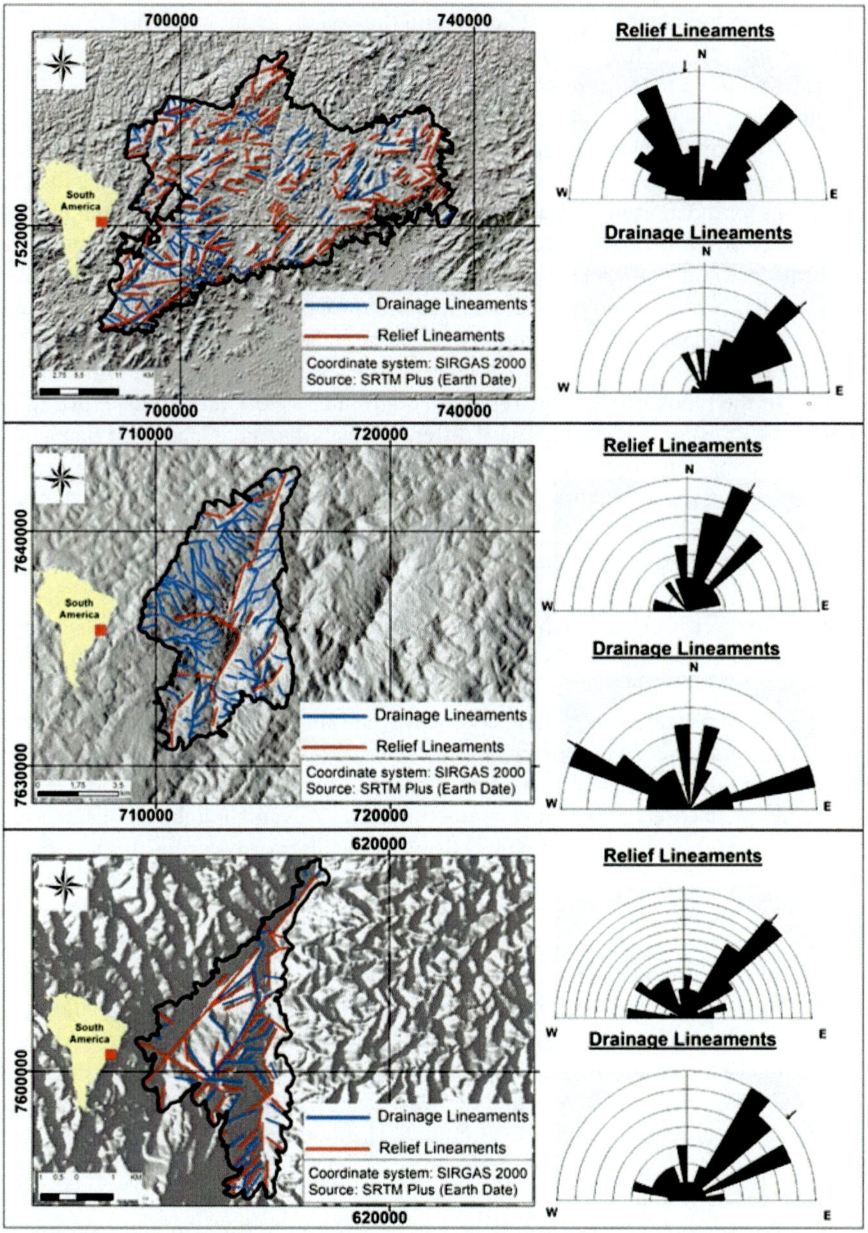

Fig. 3 Topography and drainage lineaments extracted for Serra dos Órgãos (above), Serra do Ibitipoca (below) and Serra do Relógio (in the middle)

rivers are confined to the regional faults until they enter the interplanaltic depression of the Pomba river when the drainage begins to run transversally to the structures genetically linked to the continental rift.

The units presented in greater detail figure as three surfaces of common tectonic genesis given by the platform uplift positioned in different sectors of the continental rift and, therefore, submitted to different base levels that have been imposing distinct conditions for the morphological evolution of these areas. The erosional organization of Serra dos Órgãos is controlled by the general base level of the Atlantic Ocean in its southeast face and by the tectonic depression of the Paraíba do Sul river in the interior plateaus; In Serra do Relógio, the denudational processes are leveled at the second level of the interplanaltic depression of the Pomba river, at an altitude of 300 m; in Serra do Ibitipoca, the regional base level is given by the Peixe river, which in the vicinity of the block dissects the relief at an altitude of 900 m before stabilizing its plain at the threshold of 700 m. These different positions in the landscape denote the regional scaling convergent to the Paraíba do Sul river valley, where the intermingling of the staggered reverses of the Serra do Mar and the Mantiqueira escarpments takes place.

In the central part of the area covered in this study the main regional erosive front opens, which according to Paixão et al. (2020) decreases from NE to SW. This can be well observed in Fig. 1, which also points out the E-W direction erosive wave festooning transversally the NNE-SSW oriented structural highs, to include the Serra do Relógio.

The SW expansion of the alluded erosive entrance is hindered by the tectonic sill of the Paraibuna river, positioned at the southeastern limit of the hemigraben that the aforementioned river dissects in its middle course, as well as by *knick-points* along the Peixe river and other tributaries. Even though, the Peixe river presents a pronounced erosive aggressiveness capable of entering the interfluves of the Mantiqueira mountain range, forming large retractive arcs in gneissic-granitic lithologies that differ from the morphologies of the rectilinear scarps in the quartzite mountains, among which the Ibitipoca mountain range appears as the highest block. In spite of the support propitiated by the resistant quartzite, the dissection is very active on the somite surfaces, promoting the capture of a channel originally belonging to the Grande river basin and that nowadays appears as the main watercourse dissecting the summits in question. The transposition of the immense threshold favored the entrenchment of the channel and the formation of terraces with preserved organic horizons, dated by Optically Stimulated Luminescence (OLE) at 23,000 years (Fig. 4).

The reversal of drainage by fluvial catchments has been common in the evolution of large regional interfluves. The preponderance of catchments draining toward the Atlantic Ocean subtracting areas of basins facing the continental interior has been observed in several works (Cherem et al. 2013; Rezende 2018; Salgado et al. 2018; Marent and Valadão 2019). The capture processes occur, therefore, in the high ridges that mark the interfluves between the Paraíba do Sul river basin and the Rio Grande, Doce, and São Francisco river basins, and there are also important downstream

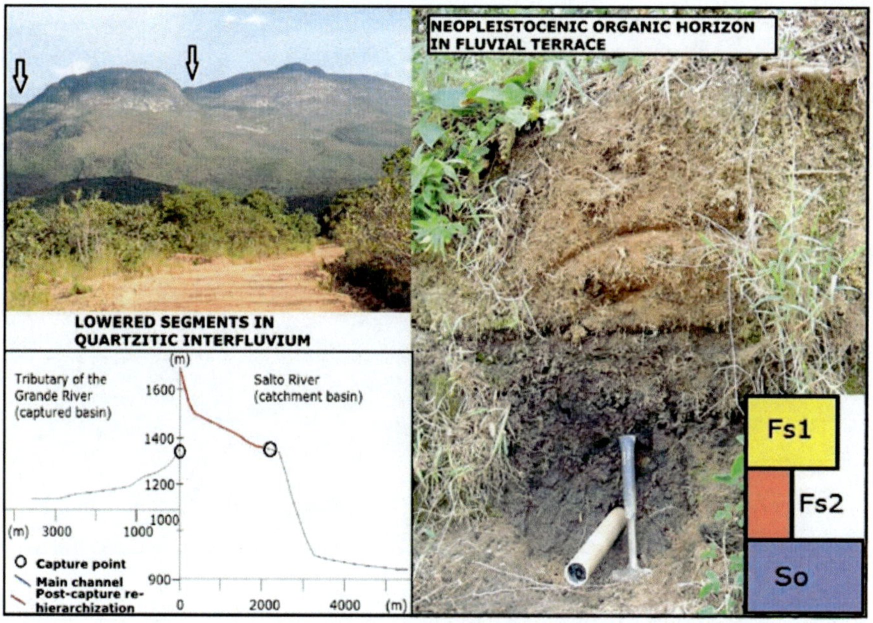

Fig. 4 River catchment in Serra do Ibitipoca with terrace generation in the catchment. Textural distribution. Fs1—Sandy-clay facies referring to the most recent coverings with superficial enrichment of organic matter; Fs2—Sandy-clay facies with greater loss of organic matter; So—sandy facies markedly organic (buried glei horizon)

rearrangements related to the evolution of the relief in the eastern basins directly or indirectly connected to the Atlantic Ocean.

The alluded eastward projection of the capture lines is due to the natural retreat of the mountain escarpments, being the fluvial captures are also quite recurrent in the interfluvial morphogenesis of Serra dos Órgãos, a compartment of Serra do Mar that shows abundant evidences of active tectonic control. The capture front follows the interfluvial lines that separate the basins of the rivers Piabanha and Paquequer (tributaries of the Paraíba do Sul river) from the small channels that drain directly toward the Baixada Fluminense. The two cited rivers dissect deeply the relief, accompanied by the strong incision of its tributaries and of diverse rearrangements for capture in the interior of these basins, that assume thus a strong aggressive condition capable to capture segments of small opposing channels directed directly toward the Atlantic. Figure 5 is representative of the catchment patterns in Serra dos Órgãos, sampling a situation in which the catchment turns toward the Paraíba do Sul river and another in which the catcher condition is from the opposite basin facing the Atlantic base level.

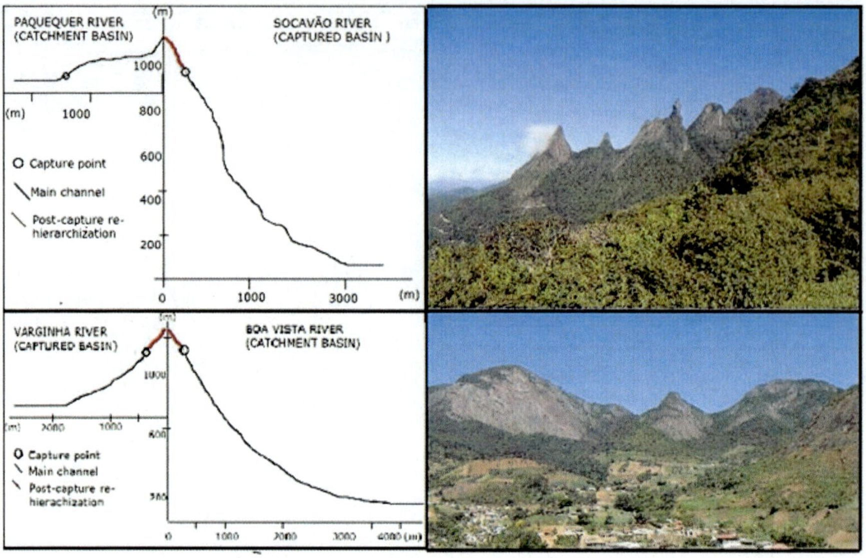

Fig. 5 River catches in the interfluvial strip of Serra dos Órgãos

5 Neotectonic Conditioning

Besides erosional and structural aspects, evidence of morphotectonic control also abounds in the region, especially in Serra do Mar (Riccomini 1989; Gontijo 1999; Hartwig 2006; Nascimento et al. 2013) and Serra da Mantiqueira (Saadi 1991; Santos 1999; Morales 2005; Marques Neto 2017). The presence of facets on different surfaces, both on high escarpments and on structural hills of intermontane surfaces, is associated with other geomorphological evidence of tectonic control, such as *shutter ridges,* outcrops on terraces and low slopes, lateral migration of channels by block basculation, strangulation of fluvial plains, hanging valleys disarticulated from local base levels, misalignment of interfluves, etc.

The retractive morphogenesis of scarps and interfluves, associated with tectonic uplift and reactivation of headlands, have been shared in the Cenozoic evolution of the regional relief. Structural and geochronological interpretations have been associated with geochemical denudational approaches (Rezende et al. 2013) for a more comprehensive understanding of the regional morphological evolution. Paleotopographic reconstructions are a cartographic technique that spatializes a previous altimetric organization and dissection, providing important morphometric and morphographic elements to support the interpretation of the relief evolution. The paleotopographic reconstructions performed in the three interfluvial alignments under consideration were accompanied by the elaboration of current topographic profiles superimposed on the profiles generated from the paleo-model (Fig. 6) to aid the visualization of the current compartments and their past projection from transects.

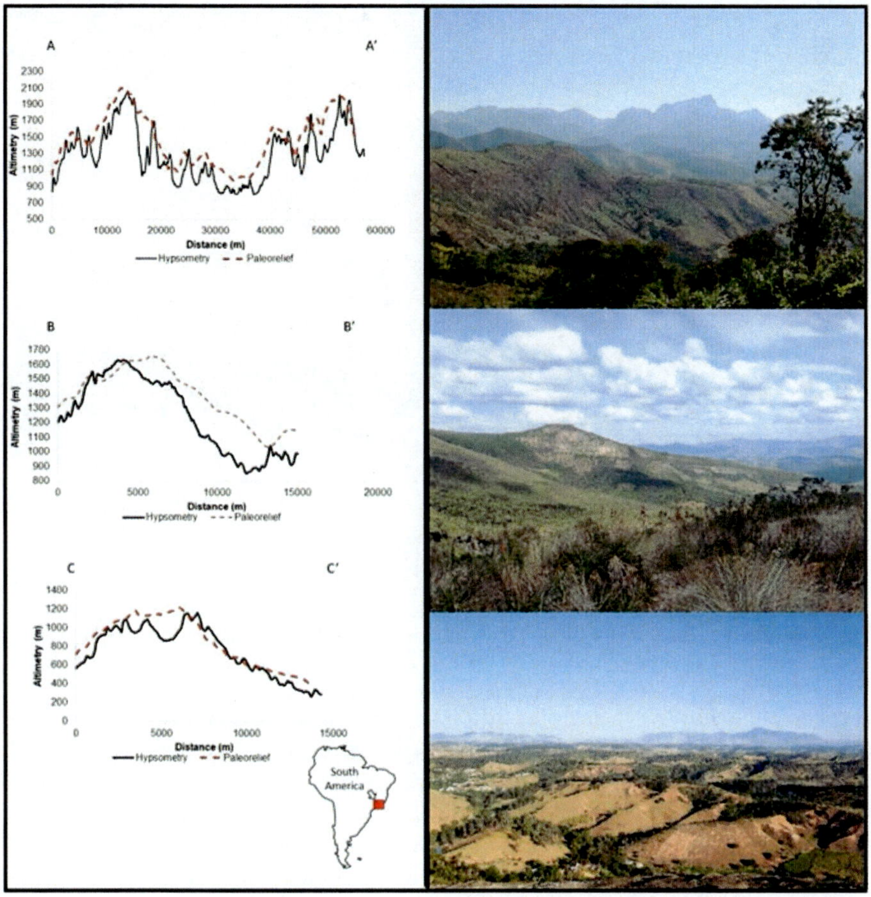

Fig. 6 Overlapping topographic and palaeotopographic profiles for the three interfluvial alignments compared. A-A′—Serra dos Órgãos, B-B′—Serra do Ibitipoca, C–C′—Serra do Relógio

Neotectonic processes visibly affect the geomorphological structure of Serra dos Órgãos, which shows morphological evidence of such effects in the form of deep and embedded valleys, sills, and high-mountain valleys, scarps with trapezoidal facets, etc. The paleotopographic map generated for the area (Fig. 7) denotes the maintenance of the altimetry in spite of the retractive processes of the scarps, which have presented a marked conservation of the slopes and a tendency of remontant migration of the *knickpoints*, with dominance of rocky bed rivers in the interior of the alluded compartment.

Currently, the altimetric levels of Serra dos Órgãos exceed 2100 m in the culminating plateaus, figuring as one of the highest compartments of the entire Serra do Mar. The paleotopographic map suggests a pattern in which the evolution of the valleys tends to maintain and even increase the slopes, due to the predominance of *bedrock rivers* within this compartment. Synchronously, the sustaining of the slopes

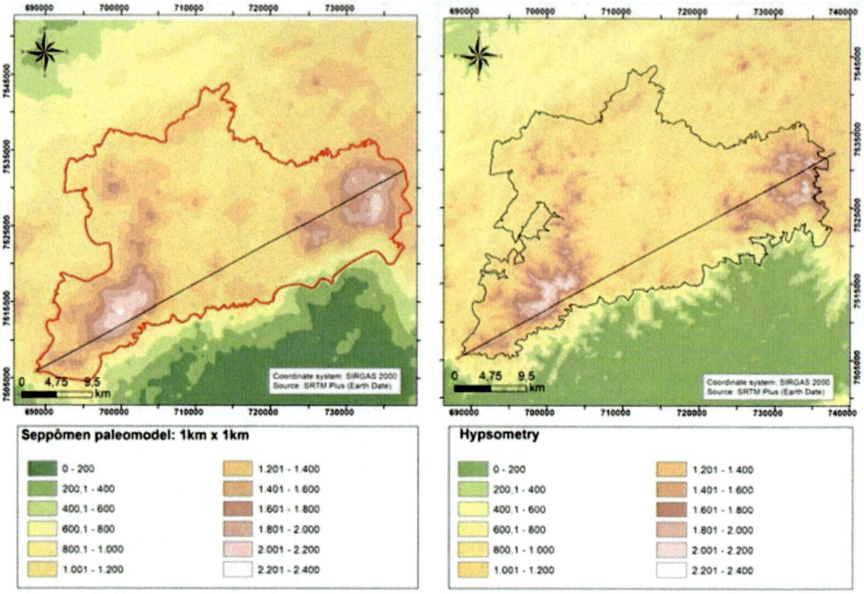

Fig. 7 Paleotopographic reconstitution of Serra dos Órgãos (RJ) compared to present altimetry

is accompanied by the increase of the dissection density, revealing intense rehierarchization and vertical notching in higher-order channels due to the regional *uplift*, a configuration that converges with other research that points to regional morphotectonic control, both in Serra dos Órgãos (Hartwig 2006) and in the Paraíba do Sul valley (Silva and Melo 2011). The very lineaments linked to the Piabanha river and its main formers, well marked on the current hypsometric map, do not appear well defined on the paleotopographic map, suggesting a relatively recent intensification of the vertical notch that favored the formation of deep low-order valleys, many of them suspended.

The maintenance of the interfluvial lines associated with the high slopes suggests vertical evolution of the landscape commanded by the fluvial carving, and the reworking of the headwaters by the main channels has partially reworked the interfluves sustained by uplift. On the other hand, the staggering levels reveal local controls exerted by several thresholds arranged in steps, controls that overlap the regional *trend* and influence the interfluvial erosive remodeling not only from the slopes facing the Guanabara Bay but also from those facing the Paraíba do Sul valley. Such geomorphological organization is reflected, for example, in the superimposition of Paquequer River on a NW–SE structure well marked in the relief that lines up in the interior extensions of the scarp ridges of the reverse of Serra dos Órgãos, showing the existence of a higher plateau in the upstream confining an altimontane depression limited by a local sill that was obliterated by the epigenetic cut of superimposition. In the current topographic configuration, the referred process resulted in a greater leveling of the valley bottom and reaffixed the crest spur in local hills.

In spite of the strong tectonic conditioning of Serra dos Órgãos, the comparison of the present topography with the paleomodel shows that the main slope breaks that delimitate the compartment did not suffer significant alterations, differently from what occurred in the interior blocks that climb in the direction of the tectonic valley of the Paraíba do Sul river, where a strongly controlled drainage has been aggressively dissecting the relief. Possibly, this is due to the differential resistance of the lithotypes that compose the granite-gneissic batholith of Serra dos Órgãos.

Once the Serra do Mar is transposed, the *horst* of Serra da Mantiqueira stands out as the second orographic step of southeast Brazil. Serra do Ibitipoca is positioned in the northeastern extremity of this important regional geomorphological compartment, one of the continuously higher structures of the whole Brazilian territory. Its limit is given by a structural sill that represents a strong erosive obstacle for the drainage that has been causing the retreat of this sector of the Mantiqueira mountain range controlled by the base levels imposed on the rivers captured by the basins that drain into the Paraíba do Sul river. The high altimetries are sustained by quartzite, which flanked by biotite-gnaisses and schists establish local amplitudes above 700 m. The Salto River is the main channel that dissects the block, crossing stepped sills until it flows into the Peixe river, the first base level that rises in the regional scaling until the Atlantic Ocean. Together with the Conceição stream, the upper course of the channel in question integrated an organized drainage toward the Grande river, sharing the hydrographic basins that drain inland, having been captured by the erosive wave that has been forcing the retraction of the escarpments of the Mantiqueira mountain range, as reported in the previous section.

Although the quartzite landscapes are markedly resilient, thus presenting a strong structural significance, evidence of neotectonic diastrophic effects are present in Serra do Ibitipoca as in the entire regional context. As occurs in Serra dos Órgãos, ridges with preserved facets incarcerating hanging valleys and disarticulated local and regional base levels are common, associated with vertical notches that deepen the structural canyons. The paleotopographic reconstitution of the block under consideration (Fig. 8) indicates sectors in which the present altimetry is higher than its previous correspondents, suggesting a preferential uplift of this block limited by craggy fronts. On the other hand, it is clear the erosive entrance of the Salto River, whose incision in the escarpment forms an inverted "V" that is not observed in the pale-model, suggesting an accentuated post-capture resurfacing.

In this case, the importance of the capture processes can be endorsed by the appreciable resurfacing unleashed on a relief sustained by quartzite, even if the underlying shale levels favor preferential alteration fronts and more forceful erosive inputs. Notoriously, the southern and eastern faces of the Serra do Ibitipoca facing the catchment basin present indentations and sinuosities that are much more prominent compared to its western side, facing the Rio Grande basin.

The block corresponding to the Serra do Ibitipoca clearly marks the limits of continuity of the elevated relief of the Serra da Mantiqueira and the passage to the hills of the lower plateau levels that climb up to the Paraíba do Sul river plateau. On the geomorphic surface in question, the regional NE-SW orientation is maintained, but elevated ridges are much more discontinuous, missing in the Paraibuna river basin

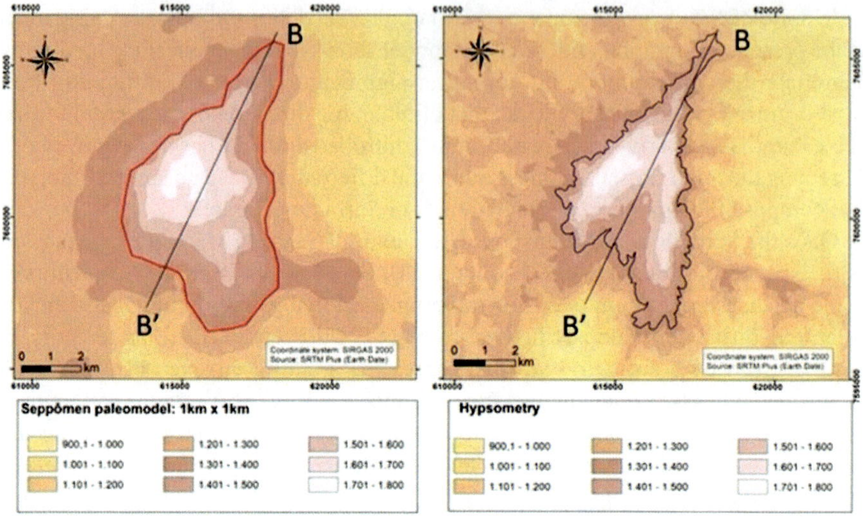

Fig. 8 Paleotopographic reconstitution of the Serra do Ibitipoca (MG) and surroundings compared to the present altimetry

and standing out in the medium Pomba river basin, where they are festooned by the passage of the Pomba and Novo rivers. In this context is the Serra do Relógio, a gneissic-granitic massif that appears as a geomorphological heritage of more continuous alignments quite denuded by the hydrographic network that was arranged in the interplanaltic depressions.

The Serra do Relógio marks the contact of two geomorphological plateaus with approximately 200 m of unevenness, which concern the two main levels of the interplanaltic depression of Pomba river. The higher one is elevated at about 500 m of altitude in the backyard of the structural highlands and the lower one at medium altitudes of 300 m, closer to the Paraíba do Sul river and figuring as a functional base level for the higher compartment positioned upstream of the epigenetic cut.

Convergently to what was verified in the other two structures analyzed, the paleotopographic reconstitution suggested that the dissection of Serra do Relógio did not imply altimetric loss, despite the advance of the lowered surfaces in its surroundings (Fig. 9), suggesting tectonic influence in the sustaining of the structure in question. This behavior is in revelation of its position in the main regional erosive fronts verified in the Pomba river depression compartments, being sustained, inclusively, by the same lithotypes that frame the hills of the depression compartment to west (Fig. 10), given by the enderbitic gneisses of the Juiz de Fora Complex (Soares et al. 2002). In contrast, in the southernmost branch of the Mantiqueira, although the Peixe river itself presents a transversal behavior and a position very close to the interfluvial line of the Grande river basin, the great sills were not yet cut by superimpositions in this sector, leaving the relief more preserved in comparison with the festonations

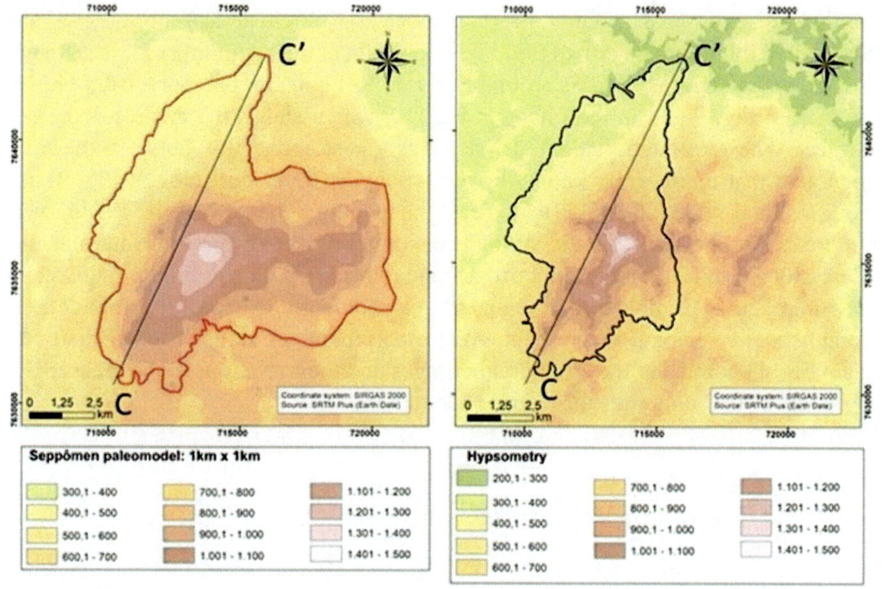

Fig. 9 Paleotopographic reconstitution of the Serra do Relógio and surroundings compared to the present altimetry

Fig. 10 View of the Serra do Relógio. Around it, the originally forested hills of the Pomba river Depression, dominant form pattern in the Zona da Mata region of Minas Gerais

verified in the northeast, context in which Serra do Relógio and a set of NE-SW and NNE-SSW ridges are inscribed (see Fig. 1, brought at the beginning of this chapter).

The erosive aggressiveness around Serra do Relógio can be appreciated both in the map in Fig. 9 and in the image in Fig. 10, which shows the erosive depressed surfaces. The generated paleomodel suggests a past connection between the main alignment that is presently known by the toponymy Serra do Relógio (Fig. 9) and more residual hills and ridges to the east, connections that reveal the existence of a more continuous past somite surface consumed by the opening of interplanaltic depressions, which obliterated surfaces underlain by both the biotite-gnaisses of the Andrelândia Complex and the enderbites and charnockites of the Juiz de Fora Complex. The regional view of the great escarpments of southeastern Brazil (see again Fig. 1) points out the structural heritages and their relations to the present-day morphology that underlies these more residual dividers.

6 Some Trends in Regional Evolution

Considering what has been presented, the region of the great escarpments of the southeastern Brazilian rift margin houses geomorphological systems of genetic linkage, both erosional and tectonic. The compartments analyzed more specifically share a tectonic genesis that is still influential in their respective evolutions, although erosional forces have acted differently in the present and in the past in each one of them, as well as bequeathing profound regional geomorphological differentiations.

In a summarized way, it can be considered that the morphogenesis of the interfluves and the intermontane areas in the domain of the great escarpments of southeastern Brazil is markedly controlled by tectonic uplift associated with the retreat of the escarpments, indicating that it is not only the reequilibrium by mass balance that influences the maintenance of high topographies but also distensive, compressive and transpressive crustal efforts well marked in the relief forms, in the drainage arrangement and in the structures present in the outcrops.

The results show a regional differentiation in the tectono-erosive processes, with a lowered area within the interplate depressions that surround Serra do Relógio contrasted with the relief of the great regional tectonic pillars—Serra do Mar and Mantiqueira, characterized by the recurrence of fault scarps with deep, confined and markedly rectilinear valleys. The major downdip in the surroundings of Serra do Relógio indicates that the most aggressive denudation zone that is established between the erosional depression of the Pomba river and the tectonic depression of the Paraíba do Sul river can be relatively old and refers to superimposition processes that engendered epigenetic cuts in the NNE-SSW structures, in agreement with the age of the alteration coverages estimated in this region by Antonioli et al. (2005) as Miocene to Pliocene age. The remaining interfluves are much more residual, and fluvial catchments currently operate in the intermountain sectors, as the catchment front is already receding into the São Francisco and Doce River basins.

Entering the tributary basins of the Paraíba do Sul river that dissect the Serra da Mantiqueira (Peixe and Preto rivers), the erosive input has not yet reduced the topographic volumetry as in the Pomba river basin. In this way, the interfluvial morphogenetic processes present more recent arrangements, with neoquaternary captures as dated in the quartzite block summits, which signals the erosive imperative of these basins and the advance of denudation on the planar surface of Juiz de Fora toward the interfluves of the Grande river basin. The continuity of erosive regression, in short, can set back the Peixe river to the point of consummating a process of capture of the upper Grande river upstream of its intersection, which is quite probable given the morphogenetic imperative ascertained here. Undoubtedly, this prelude of regional capture, once consummated, would mark a new phase in the geomorphological organization of the great escarpments of the Brazilian rift margin.

References

Almeida FFM, Carneiro CDR (1998) Origin and evolution of the Serra do Mar. Rev Bras de Geociências 28(2): 135–150

Anderson RS, Anderson SP (2010) Geomorphology: the mechanics and chemistry of landscapes. Cambridge University Press, Cambridge

Antonioli L, Tupinambá M, Gracia MJ, Dino R (2005) Palynology and age of Neogene (Upper Miocene/Pliocene) sediments from the Pomba-Muriaé River depression (MG). Symposium on Geology of the Southeast, p. 63

Bierman PR, Montgomery DR (2014) Key concepts in Geomorphology. Freeman and Company, New York

Couto EV, Fortes E, Sordi MV, Marques MJ, Camolezi BA (2012) Seppômens maps for geomorphic developments analysis: the case of Parana Plateau border, Faxinal, State of Parana, Brazil. Acta Scientarium Technology 34(1): 71:81. https://doi.org/10.4025/actascitechnol.v34i1.9944

Chiessi CM (2004) Tectônica Cenozoica no Maciço Alcalino de Passa Quatro (SP-MG-RJ). Dissertação (Mestrado em Geologia). Instituto de Geociências, Universidade de São Paulo

Cherem LF Varajão CA Magalhães Jr A Varajão AFDC Salgado AAR Oliveira LAF Bertolini, WZ (2013) O papel das capturas fluviais na morfodinâmica das bordas planálticas do sudeste do Brasil. Revista Brasileira de Geomorfologia (14)4: 299–308

Gontijo AHF (1999) Morphotectonics of the middle Paraíba do Sul River valley: Serra da Bocaina region, São Paulo and Rio de Janeiro states. D. thesis, Institute of Geosciences and Exact Sciences, Paulista State University

Hartwig ME (2006) Mesozoic-Cenozoic rupile tectonics in the Serra dos Órgãos region, RJ. Dissertation (master's degree), Institute of Geosciences, University of São Paulo

Hartwig ME, Riccomini C (2010) Morphotectonic analysis of the Serra dos Órgãos region, southeastern Brazil. Rev Bras de Geomorfologia 11(1): 11–20

Jacobs Z, Roberts RG (2007) Advances in optically stimulated luminescence dating of individual grains of quartz from archeological deposits. Evolutionary Anthropology, 1(6): 210–223. https://doi.org/10.1002/evan.20150

Leverington DW, Teller JT (2003) Paleotopographic reconstructions of the eastern outlets of glacial Lake Agassiz. Canadian Journal Earth Science 40: 1259–1278. https://doi.org/10.1139/e03-043

Marques Neto R (2017) The southern Mantiqueira *horst*: proposed morphostructural compartmentalization for its mining portion. Revista Brasileira de Geomorfologia 18(3): 561–577

Marent, B Valadão, RC (2019) Contribuição aos estudos da evolução da escarpa entre as bacias hidrográficas dos rios Doce e Paraíba do Sul, na Serra da Mantiqueira-MG-Brasil. Geousp – Espaço e Tempo (26)2: 417–434.

Marques Neto R, Ferraro BV (2018) Regional geomorphological cartography and morphogenesis: methodological contributions. Brazilian Journal of Geomorphology 19(2): 267–281

Marques Neto R, Moreira JA, Silva FP (2019) Evolution of escarpments in rift margins: a discussion on soerguation and denudation in southern Mantiqueira from paleotopographic maps and geomorphometric parameters Revista Brasileira de Geomorfologia. 20(4): 877–890

Morales N (2005) Neotectonics in intraplate environment: examples from southeastern Brazil. Ph. D. thesis, Institute of Geosciences and Exact Sciences, Paulista State University

Motoki AG, Petrakis H, Sichel et al (2008) Origin of the Mendanha syenitic massif reliefs, RJ, based on geomorphological analyses and its relation to the Nova Iguaçu volcano hypothesis Geosciences. 27(1): 97–113

Nascimento ER, Salamuni E, Queiroz GL et al (2013) Evidence of morphotectonic and neotectonic determination in the relief of Serra do Mar in Paraná State. Revista Brasileira de Geomorfologia. 14(3): 287–299

Oliveira, LAF Quaternary fluvial dynamics and landform configuration at the contact between the Rio Pomba Depression and the Campos das Vertentes Plateau (2012) Dissertation (master's degree in Geography) Institute of Geosciences, Federal University of Minas Gerais

Paixão RW, Salgado AAR, Freitas MM et al (2020) Possibility of endorheism and fluvial captures in the morphogenesis of the Paraíba do Sul river basin Revista Brasileira de Geomorfologia 21(4): 821–834

Rezende E, Salgado A, Silva AAR et al (2013) Controlling factors in landform evolution on the NNW flank of the continental rift of southeastern Brazil: an analysis based on the measurement of long-term denudational processes Revista Brasileira de Geomorfologia. 14(2): 221–234. https://doi.org/10.20502/rbg.v14i2.416

Rezende EA (2018) O papel da geodinâmica espaço-temporal da rede hidrográfica na evolução geomorfológica da alta/média bacia do Rio Grande, sudeste brasileiro. Tese (Doutorado em Geologia Ambiental e Recursos Naturais), Escola de Minas, Universidade Federal de Ouro Preto

Riccomini C (1989) The continental rift of southeastern Brazil. Ph.D. thesis, Institute of Geosciences, University of São Paulo

Saadi A (1991) Ensaio sobre a morfotectônica de Minas Gerais: tensões intraplaca, descontinuidades crustais e morfogênese. Ph. D. thesis, Institute of Geosciences, Federal University of Minas Gerais

Sallun AEM (2007) Aloformation Paranavaí: Quaternary colluvial deposits of the upper Paraná River basin (SP, PR and MS). Ph. D. thesis, Institute of Geosciences, University of São Paulo

Salgado, AAR Cherem, LFS Sordi, MV (2018) Grandes capturas fluviais no Brasil: síntese das novas descobertas. Estudos do Quaternário (19): 23–31

Santos M (1999) Serra da Mantiqueira and Planalto do Alto Rio Grande: the Tertiary basin of Aiuruoca and morphotectonic evolution. Ph. D, thesis,- Institute of Geosciences and Exact Sciences, Paulista State University

Silva FP (2019) Role of post-Miocene tectonics in the morphological configuration of hydrographic systems - morphotectonic assessment in the context of rift margins. Dissertation (Master in Geography), Federal University of Rio de Janeiro, Institute of Geosciences

Silva TP, Mello CL (2011) Neotectonic reactivations in the Paraíba do Sul River Shear Zone (southeastern Brazil). Journal of the Institute of Geosciences. 11(1): 95–111. https://doi.org/10.5327/Z1519-874X2011000100006

Soares ACP, Noce CM, Trouw RAJ, et al (2002) Projeto Sul de Minas. COMIG-UFMG-UFRJ-UERJ

Vargas KB (2017) Geomorphological surfaces in the north central Paraná and regional relief evolution. Ph.D. thesis, Centro de Ciências Humanas, Letras e Artes, Universidade Estadual de Maringá

Zalan PV, Oliveira JAB (2005) Origin and structural evolution of the Cenozoic Rift System of Southeastern Brazil Boletim de Geociências da Petrobrás 13(2): 269–300

Pedogeomorphological Compartments of Coastal Tablelands in Amapá, Eastern Amazon

João Santiago Reis, João Carlos Ker, Flávio Rodrigo Lozer de Amorim, Bruno Nery Fernandes Vasconcelos, and Davi Feital Gjorup

Abstract The Barreiras Group sediments extend along the Atlantic coast from northern Rio de Janeiro to Amapá, and soil variability is an intrinsic characteristic. In Amapá State, they occupy a narrow strip of approximately north–south orientation, where soils with high degree of weathering, deep, acidic, dystrophic, cohesive character and low cation exchange capacity predominate, under Cerrado physiognomy. Compartments of Coastal Tablelands with different degrees of dissection are also observed in this region, suggesting the hypothesis of distinct pedogeomorphological dynamics. In face of this, the aim of this study was to characterize distinct relief compartments of the Coastal Tablelands of Barreiras Group in Amapá, through the investigation of soils and their respective attributes in each. For this, 112 soil profiles were observed and collected, being nine profiles selected as representative and submitted to analyses of their sortive complex, granulometry, organic carbon, and mineralogy. In general, the occurrence of similar soil classes was observed in both the Dissected Tableland and the Tableland compartments. However, the results show a clear difference in the granulometry of the soils between the compartments. The aspects related to the distribution of organic carbon, sortive complex, and mineralogy of the soils are similar in the two environments. Thus, the hypothesis of

J. S. Reis (✉)
Department of Geography (CERES), Federal University of Rio Grande do Norte, Natal, Brazil
e-mail: joao.reis@ufrn.br

J. C. Ker · B. N. F. Vasconcelos · D. F. Gjorup
Department of Soils, Federal University of Viçosa, Viçosa, Brazil
e-mail: jcker@ufv.br

B. N. F. Vasconcelos
e-mail: brunonery81@gmail.com

D. F. Gjorup
e-mail: davifeitalgjorup@gmail.com

F. R. L. de Amorim
Department of Geography, Federal University of Viçosa, Viçosa, Brazil
e-mail: flavioramorim@hotmail.com

G. Barbosa dos Santos et al. (eds.), *Geomorphology of Brazil: Complexity, Interscale and Landscape*, Springer Proceedings in Earth and Environmental Sciences,
https://doi.org/10.1007/978-3-031-05178-4_4

soil homogeneity in the Coastal Tablelands of Amapá is discarded and two distinct pedoenvironments are recognized.

1 Introduction

In the Brazilian equatorial zone, between the coastal plains of the mouth of the Amazon River and the hills of the Guiana Shield, in a narrow strip with north–south orientation in Amapá state, we find Coastal Tablelands (more or less dissected) corresponding to the Barreiras Group.

The Barreiras Group corresponds to a sedimentary cover composed of clay, sandy-clay, and sandy sediments, which extends along the Brazilian Atlantic coast from northern Rio de Janeiro to the state of Amapá (Nunes et al. 2011; Moura-Fé 2014).

In the state of Amapá, it is located in the Amazon Lower Plateau, which is constituted by tertiary sediments of the Barreiras Group, presenting smooth fall toward the Amazon River. It is represented by a planing surface that is partially conserved and covered by ferruginous crusts. In this scenario, the high density of the drainage network composed of short and very branched channels is responsible for the forms of dissection of the planing in flattened top hills and tabular interfluves (Brazil 1974).

Namely, the sediments of the Barreiras Group occur, in general, in elevations from 20 to over 200 m altitude (Jacomine 1996), and its composition is varied presenting sequence of detrital, siliciclastic sediments, fluvial and marine origin (Arai 2006), poorly consolidated or not, poorly selected, of variegated colors, ranging from fine to coarse sands, predominantly angular grains, reddish-gray clays, with kaolinitic matrix and scarce occurrence of sedimentary structures (Nunes et al. 2011).

Thus, the soils developed from this substrate are generally characterized by being deep, acidic, acidic, with low cation exchange capacity, little morphological difference between horizons, and frequent presence of cohesive horizons (Nogueira and Nogueira 1996; Oliveira 2000). This landscape unit plays a relevant role in the occurrence of Cerrado biome physiognomies in the middle of the Amazon biome (IEPA 2008).

Furthermore, associated with the relief of Coastal Tablelands, the Barreiras Group in Amapá exhibits some heterogeneity regarding its geomorphological compartments, but even so, it is generally described by Brazil (1974) as associated with the Amazon Lower Plateau, while recent more detailed works subdivide the region of interest in Tablelands and Dissected tablelands (Jorge and Teixeira 2016). Studies on the soils of this region are still scarce (Sena 2016; Ker et al. 2017), lacking more information to better understand their pedogenetic processes, their spatial distribution, and the suitability for use and management.

Thus, the objective of this work was to verify the hypothesis of homogeneity of soils in Coastal Tablelands under the Barreiras Group in Amapá State, aiming to contribute to research related to the understanding of pedogeomorphological compartmentalization of the region.

2 Material and Methods

2.1 Study Area

The study area is located in the municipalities of Porto Grande, Ferreira Gomes, Macapá, and Itaubal, in the state of Amapá (Fig. 1), inserted in the hydrographic divider of the Araguari River watershed to the north and the Pedreira and Matapi River basins to the southwest, the latter two belonging to the Amazon River basin. The distance from the central point of the area to the coast, which coincides with the mouth of the Amazon River, is approximately 70 km. The pluviometric precipitation in the region is between 2750 and 3500 mm and the average temperature is between 26 and 27 °C, reaching maximums around 40 °C between the months of September and November (IEPA 2008).

The region is located on the coastal platform of Amapá, characterized as a monocline with moderate southeast dipping. Under this platform, there is a Cenozoic sedimentary cover of variable width consisting of tertiary sediments, known as Barreiras Group, and Quaternary fluvial and marine sediments, which together characterize the geological province called Cenozoic Coverings of the Amapá Platform (Brasil 1974).

For the accomplishment of the work, a set of 112 soil profiles distributed among the geomorphologic compartments of Dissected tablelands (DT—59 profiles) and Tablelands (T—53 profiles) were described, collected, and observed.

The Dissected tablelands region occupies most of the study area and is characterized by a more rugged terrain, with greater slopes, narrower tops, greater drainage density, and higher altimetric levels (between 100 and 35 m). Furthermore, the Tablelands compartment is characterized by flatter tops and wider slopes, less dissected valleys, and altimetry between 60 and 6 m above sea level (Fig. 2). It is also worth noting that the separation of these two compartments is near the banks of the Pedreira River, which presents a deviation of its course from the west–east to northwest-southeast direction, near the connection between the two environments.

Thus, the collection of 112 sampled profiles represents the variability of soils present in the region, mostly Latossolos Amarelos Distrocoesos—LAdx (Xanthic Ferralsols) of different textural classes, and in smaller proportions Plintossolos Pétricos Concrecionários—FFc (Pisoplinthic Plinthosols), Argissolos Amarelos Distrocoesos (Xanthic Acrisols), Argissolos Acinzentados Distrocoesos (Albic Acrisols), Cambissolos Háplicos Tb Distróficos—CXbd (Xanthic Ferralsols), Plintossolos Argilúvicos Distróficos—FTd (Dystric Plinthosols), and Neossolos Quartzarênicos (Dystric Arenosols).

From this collection, nine representative soil profiles were selected, composing a soil sequence in each geomorphological unit, aiming to extract chemical and mineralogical information for characterization of its respective compartment (Fig. 2).

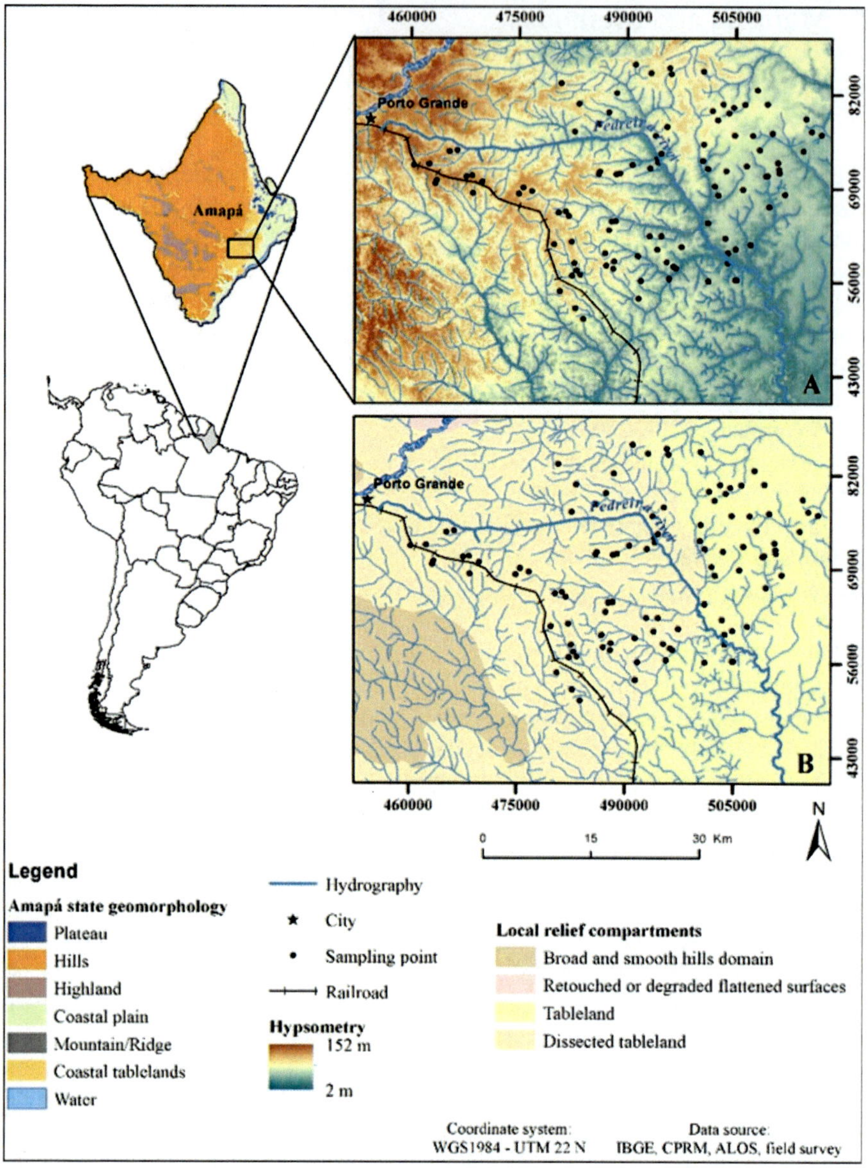

Fig. 1 Location of the study area, showing the sampling points in the different relief compartments

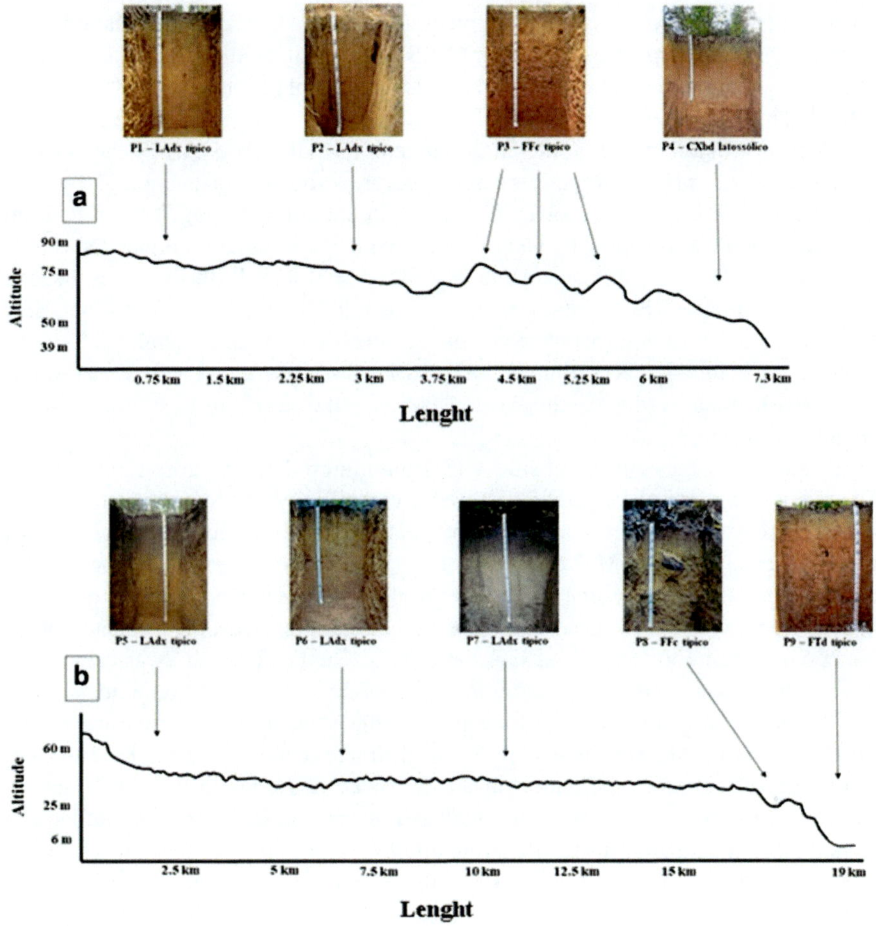

Fig. 2 Sequence of representative soil profiles selected in the Dissected tablelands (**a**) and Tablelands (**b**) compartments

3 Soil Analyses

Firstly, after drying in the shade, samples from representative horizons of all 112 collected profiles were sieved with a 2 mm mesh and subjected to particle size analysis, according to Teixeira et al. (2017). After mechanical and chemical dispersion in a Wagner-type rotary shaker with NaOH solution (0.1 mol L^{-1}), the sand fraction was obtained by sieving (270 mesh), and the silt and clay fractions were separated by sedimentation.

Then, the characterization of the sorptive complex was performed, according to Teixeira et al. (2017), only in the nine profiles selected as representative soils of the geomorphological compartments. pH in water and in 1 mol L^{-1} KCl solution were quantified; available Ca^{2+}, Mg^{2+} and potential acidity (Al^{3+}) contents, quantified by Atomic Absorption Spectrophotometer (AAS) and titration with 0.05 mol L^{-1}(Al^{3+}), after extraction with a solution of KCl 1 mol L^{-1}; available contents of P, K^+, and Na^+ extracted by Mehlich-1 acid solution (HCl 0.05 mol L^{-1} + H_2SO_4 0.0125 mol L^{-1}), and dosage with a flame photometer; and potential acidity (H + Al) by the calcium acetate method (pH 7). Remaining phosphorus (P-rem) was quantified by stirring the soil with a 0.010 mol L^{-1} of $CaCl_2$, containing 60 mg L^{-1} of P. Total organic carbon was obtained by wet oxidation with potassium dichromate ($K_2Cr_2O_7$ 0.167 mol L^{-1}) in acid environment (H_2SO_4, d = 1.84). With these results, the exchangeable bases (EC), cation exchange capacity (CEC), base saturation (BS), and aluminum saturation (m) were calculated according to Santos et al. (2018).

Soils were classified according to the Brazilian Soil Classification System (Santos et al. 2018), and according to the World Reference Base for Soil Resources—WRB (IUSS 2015).

Subsequently, the contents of silicon (Si), aluminum (Al), and iron (Fe) present in Fe oxides, in clay minerals with low degree of crystallinity and complexed to organic matter were extracted by the ammonium oxalate method (0.2 mol L^{-1} and pH 3.0) in the absence of light (McKeague and Day 1966). The Fe and manganese (Mn) contents, present in crystalline and little crystalline oxides and complexed to organic matter, were extracted by dissolution using the dithionite citrate bicarbonate (DCB) method, according to Mehra and Jackson (1960). The Fe_o/Fe_d ratio was calculated, which corresponds to the proportion indicative of crystallinity of iron oxides.

Finally, the mineralogical composition of the sand and clay fractions of the selected horizons was evaluated by X-ray diffractometry. The sand fraction was analyzed with random orientation in an excavated slide, while the clay fraction, besides the excavated slide, was also analyzed in oriented slides (smear method), in its natural state and after desferrification with DCB.

From the diffractograms obtained, the full width at half maximum (FWHM) of the peaks was calculated and the crystallite size (CRS) of the kaolinites was calculated for the iron-free samples using the Scherrer equation (Patterson 1939).

4 Results and Discussions

4.1 Soil Characteristics in the Different Relief Compartments

The data show that there is heterogeneity regarding the physical characteristics of the soils, with significant differences in particle size between soils of the Tablelands (T) and Dissected tablelands (DT) compartments, represented by the mean and median values of sand and clay content (Fig. 3).

That said, it can be seen that the soils, in general, have low silt content, a common characteristic in high weathering degree materials (Ker 1997; Santos et al. 2018). While there is a predominance of loam texture in the Tablelands soils, the very clayey

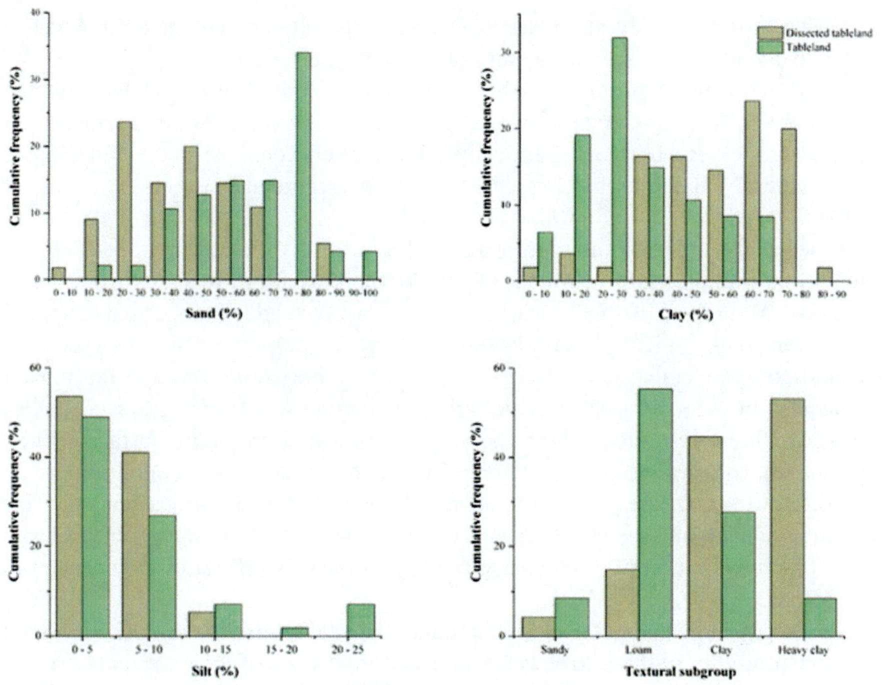

	Sand	Silt	Clay
All profiles (N = 112)			
Mean (%)	50.2	5.8	43.5
Median (%)	49.8	4.6	43.9
Min. (%)	12.6	0.5	7.0
Max. (%)	91.0	28.2	83.2
SD (%)	20.7	4.9	20.0
Dissected tableland (N = 55)			
Mean (%)	40.7	4.8	53.7
Median (%)	41.1	4.4	53.3
Min. (%)	12.6	0.5	9.6
Max. (%)	89.2	12.4	83.2
SD (%)	18.4	3.1	17.5
Tableland (N = 47)			
Mean (%)	61.1	6.9	31.8
Median (%)	68.8	5.3	26.8
Min. (%)	16.0	0.6	7.0
Max. (%)	91.0	28.2	70.0
SD (%)	17.5	6.2	89.7

Fig. 3 Cumulative frequency histograms and descriptive statistics of the grain size of the entire collection of soil profiles sampled. Min.—minimum value; Max.—maximum value; SD—standard deviation

class predominates in the DT. As the silt fraction is always low contents, sand and clay contents are inversely proportional, a behavior that can be seen in Fig. 3.

In the soil profile data of the selected topossequences (Fig. 2), it is observed that in the tabuleiro there is a trend for the fine sand fraction (FS) to increase the lower the altitude. When the CS/FS ratio is analyzed, a significant variation of values between profiles is observed, suggesting lateral heterogeneity of the parent material, which is possible due to the sedimentary nature of the Barreiras Group.

A characteristic of most soils in the region is the increment of clay in depth, which is insufficient to characterize a textural gradient that satisfies the requirements for defining a B textural horizon (argillic horizon). This clay content gradient in depth is more expressive in the soils of Dissected tablelands, recorded by the clayey character in P9.

When analyzing photomicrographs of soil profiles in the study area, Sena (2016) did not find an expressive occurrence of clay illuviationcoating processes in profiles of Latossols Amarelos Distrocoesos argissólicos. This fact, added to the sparse natural vegetation cover typical of cerrado sensu *strictu*, suggests the action of processes of elutriation (selective loss of clay in the surface horizons), through the impact of rainfall (high local rainfall), which triggers the rupture of aggregates in smaller units, which are relocated, filling the surface pores and providing surface sealing and subsequent carriage of fine material by surface runoff (shetflow) (Guerra et al. 1999). However, in areas of cerradão physiognomy there is also a tendency for textural gradient, and the hypothesis of fine particle variation in the parent material cannot be ruled out, even if the coarse sand/fine sand (CS/FS) ratios are constant in all profiles.

As already indicated in the classification of the profiles used in this study, as well as in other profiles of the Barreiras Group in other portions of the Brazilian territory, the local soils present a cohesive character from the transitional AB and/or BA horizons, to depths greater than 1 m. There are different hypotheses to explain the origin of the cohesive horizons, which are based on pore filling by colloids (Oliveira et al. 1968), and on low contents of Fe oxides, causing structural arrangement tending to massive (Achá Panoso 1976; Resende 1982; Bennema and Camargo 1979); degree of crystallinity of kaolinites and face-to-face adjustment (Moreau 2001); geomorphological processes (Anjos 1985; Ribeiro 1998); high values of dispersed clay and fine sand (Souza 1996); weak cementation by amorphous siliceous materials (Silva and Carvalho, 2007); fine clay illuviation (Côrrea et al. 2008), among others.

As in the present work (Fig. 5), analyzing soils of the same region, Sena (2016) also did not find significant levels of silicon extracted by ammonium oxalate, refuting the hypotheses of low degree of crystallinity of kaolinites and cementation by the presence of amorphous silicon compounds. In the same study mentioned above, no significant difference in soil density values is observed in the different horizons at depth, nor between profiles, even varying the levels of Fe oxides (extracted by dithionite citrate bicarbonate) between soils, as verified in this study. The same author, evaluating the micromorphology of soils in the study area, did not observe illuviation features in photomicrographs, but noted differences in the microstructure of profiles with lower degrees of cohesion (microgranular) compared to more cohesive (tending to massive).

This finding suggests that, for the reality studied, the face-to-face adjustment of kaolinites, caused by wetting and drying cycles, may be responsible for cohesion. The constant high temperatures and high rainfall probably have the potential to create an intense dynamic of mechanical clay reorganization in the local soils, due to the large inputs of water and rapid drying time (high evapotranspiration).

As for the sorptive complex, the soils have acidic pH, are dystrophic, acidic, and with low nutrient contents (Table 1). These conditions may explain the permanence of Cerrado physiognomic forms in the middle of the Amazon biome (IEPA 2008). These soil characteristics are consistent with what has been observed in other parts of the country on soils of the Barreiras Group (Oliveira et al. 1968; Anjos 1985; Nogueira and Nogueira 1996; Melo et al. 2002; Corrêa et al. 2008).

The organic carbon contents are low in all profiles, and together with the cation exchange capacity (CEC), they exhibit higher values in the superficial horizon. This correlation explains the great contribution of soil organic matter to the CEC values of tropical soils. The high temperatures and high humidity favor not only intense processes of weathering but also the cycling and more accelerated mineralization of soil organic matter (Kirschbaum 2006), thus maintaining low levels of organic carbon in this system.

There are no significant differences in chemical attributes between the Tablelands and DT profiles. This leads us to believe that the possible heterogeneity of the parent material, indicated by the AG/AF ratio, is more related to the granulometric feature than to the chemical composition in general, or that the advanced process of latossolization/ferralitization, in a humid climate, gave homogeneous chemical characteristics to the local soils.

The X-ray diffractograms of the sand and clay fractions corroborate the previous statements. No significant difference in mineralogical composition was observed between profiles. In the sand fraction, in all the studied profiles, the only mineral identified was quartz. As for the clay fraction, only kaolinite and goethite peaks were identified (except P9).

The iron content extracted by DCB (Fe_d) correlated with clay content (Fig. 4), and it is notable the higher levels in the soils of Tablelands (P5–P9). In this environment, loam-texture soils have the same Fe_d contents as the clayey soils of Dissected tablelands. Both in conditions of poor drainage, past or present, P7 has the lowest values of Fe_d, while P9 has similar Fe_d contents to the topsoils (P5), due to the presence of plintite.

The most clayey profile among the soils studied (P6) has the highest Fe_d contents, but also has one of the highest Fe_o contents among the subsurface horizons, lower only than P10. Even so, it still has a low Fe_o/Fe_d ratio, indicating a predominance of well-crystallized Fe oxides, consistent with its topographic position and similar to the rest of the Latossolos (Ferralsols) profiles.

The Si_o contents were similar in all profiles and between horizons, ruling out the hypothesis of presence of amorphous Si and/or different degrees of crystallinity of kaolinite, contributing to the cohesive character present in the soils.

Another variable that shows a relationship with clay content is Al in low crystallinity clay minerals (Al_o), which, among the elements extracted by ammonium oxalate, had higher contents. A negative correlation between Al_o and remaining phosphorus (P-rem) is also verified. Considering that Al_o comes from kaolinites with low degree of crystallinity, it is possible that this correlation is indicating P

Table 1 Granulometric and exchange complex characterization of soil profiles representative of the Dissected tablelands (P1–P4) and of the Tabuleiros (P5–P9)

Horizon	Depth	Color	Particle size						Exchange complex						OC
			Coarse sand	Fine sand	Sand	Silt	Clay	CS/FS	pH	EB	Al^{3+}	CEC	BS	m	
	cm	wet	dag.kg^{-1}						H_2O	cmol$_c$. kg^{-1}			%		dag.kg^{-1}
P1—Latossolo Amarelo Distrocoeso tfpico/Xanthic Ferralsol (Clayic, Dystric, Densic)—84 m															
A	0–16	10YR 4/2	30.8	11.5	42.3	5.2	52.4	2.7	4.3	0.6	1.3	7.0	8.7	67.4	1.4
Bwl	43–133	10YR 6/6	23.4	7.9	31.3	2.3	66.4	3.0	4.9	0.1	0.0	1.7	5.9	0.0	0.2
Bw2	133–170+	7.5YR 6/6	19.4	8.3	27.7	0.9	71.4	2.3	4.8	0.1	0.0	1.6	6.2	0.0	0.2
P2—Latossolo Amarelo Distrocoeso tfpico/Xanthic Ferralsol (Clayic, Dystric, Densic)—69 m															
A	0–15	10YR 4/3	43.6	20.3	63.9	0.0	36.1	2.1	4.8	0.2	0.6	3.9	5.4	75.0	1.1
Bwl	33–65	10YR 5/6	35.7	18.8	54.5	0.9	44.6	1.9	4.8	0.1	0.3	1.7	4.8	77.1	0.3
Bw2	65–95+	10YR 5/6	33.4	17.6	51.0	0.6	48.4	1.9	4.8	0.1	0.2	1.6	5.7	66.7	0.2
P3—Plintossolo Pétrico Concrecionário êndico/Pisoplinthic Plinthoshol (Geric, Dystric, Ochric, Clayic)—66 m															
A	0–13	10YR 4/4	30.5	16.1	46.6	5.2	48.3	1.9	4.9	0.2	0.7	4.6	3.7	80.9	1.7
Bi	13–40	7.5YR 6/6	23.8	9.6	33.4	4.2	62.4	2.5	4.8	0.1	0.2	2.0	5.0	64.3	0.3
Bc2	90–150+	2.5YR 4/4	23.0	8.9	31.9	11.5	56.6	2.6	4.8	0.1	0.1	1.4	8.5	42.9	0.2
P4—Cambissolo Háplico Tb Distrófico latossólico/Xanthic Ferralsol (Loamic, Dystric, Densic)—63 m															
A	0–19	10YR 5/4	38.6	24.3	62.9	0.2	36.9	1.6	5.5	1.4	0.0	4.0	35.5	0.0	0.8
Bi	19–65	10YR 6/6	34.1	28.4	62.5	3.6	33.9	1.2	4.5	0.3	0.5	2.5	10.6	67.5	0.4
Cl	65–110	5YR 7/6	27.1	20.1	47.2	2.9	49.8	1.3	4.8	0.2	0.3	1.5	12.8	58.7	0.3
P5—Latossolo Amarelo Distrocoeso típico/Xanthic Ferralsol (Loamic, Dystric, Densic)—58 m															
Al	0–15	10YR 3/1	61.2	14.5	75.7	4.4	19.9	4.2	4.2	0.4	1.2	6.2	5.7	77.0	1.0
Bw	58–180+	10YR 5/6	61.2	14.9	76.1	1.0	22.9	4.1	4.9	0.2	0.2	1.1	11.5	58.1	0.0

(continued)

Table 1 (continued)

Horizon	Depth	Color	Particle size						Exchange complex						OC
			Coarse sand	Fine sand	Sand	Silt	Clay	CS/FS	pH	EB	Al^{3+}	CEC	BS	m	
	cm	wet	$dag.kg^{-1}$						H_2O	$cmol_c.kg^{-1}$			%		$dag.kg^{-1}$
P6—Latossolo Amarelo Distrocoeso típico/Xanthic Ferralsol (Clayic, Dystric, Densic)—53 m															
A	0–10	10YR 4/3	25.0	14.0	39.0	5.0	56.0	1.8	4.7	0.6	0.9	5.5	9.1	62.1	2.1
Bwl	45–90	7.5YR 5/8	11.4	7.9	19.3	1.2	79.5	1.4	4.8	0.1	0.0	1.6	7.5	0.0	0.4
Bw2	90–180+	5YR 5/6	9.4	7.0	16.4	2.5	81.1	1.3	4.7	0.2	0.0	1.6	8.6	0.0	0.2
P7—Latossolo Amarelo Distrocoeso típico/Xanthic Ferralsol (Clayic, Dystric, Densic)—45 m															
Al	0–7	10YR 2/1	52.4	10.8	63.2	4.8	32.0	4.9	3.7	0.4	2.2	10.7	3.3	85.7	2.6
Bwl	56–89	10YR 6/3	45.8	9.8	55.6	5.0	39.4	4.7	4.8	0.2	0.7	2.4	8.7	75.8	0.3
Bw2	89–150	10YR 7/4	38.3	10.3	48.6	4.9	46.5	3.7	4.8	0.3	1.0	2.4	9.4	79.8	0.3
P8—Plintossolo Pétrico Concrecionário típico/Pisoplinthic Plinthoshol (Geric, Dystric, Ochric, Loamic)—31 m															
A	0–15	10YR 4/4	41.8	30.6	72.4	3.6	24.0	1.4	4.3	0.4	0.9	5.1	7.5	70.3	1.0
Bi	15–45	7.5YR 5/6	38.6	27.2	65.8	5.6	28.6	1.4	4.9	0.3	0.5	2.5	10.6	63.4	0.3
P9—Plintossolo Argiluvico Distrófico típicoDystric Plinthosol (Acric, Clayic, Ochric, Oxyaquic)—25 m															
A	0–10	10YR 4/2	25.7	46.5	72.2	8.6	19.2	0.6	5.0	2.1	0.5	7.1	29.2	20.8	1.7
Bf2	25–140+	10YR 6/3 (la. pro. ab. 10YR 4/8)	12.8	22.2	35.0	17.6	47.5	0.6	4.7	0.6	4.0	6.5	9.8	86.1	0.2

CS/FS coarse sand and fine sand relation, *EB* exchangeable bases, Al^{3+} exchangeable acidity, *CEC* cation exchange capacity, *BS* bases saturation, *m* aluminum saturation, *OC* organic carbon, *la.* large, *pro.* prominent, *ab.* abundant

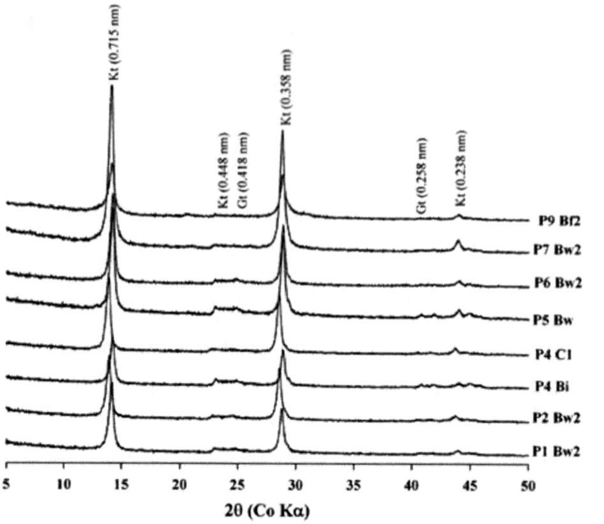

Horizon	Si_o	Al_o	Mn_d	Fe_o	Fe_d	Fe_o/Fe_d	FWHM	CRS	Clay	P-rem
	dag.kg^{-1}							nm	dag.kg^{-1}	mg.dm^{-3}
P1 - Latossolo Amarelo Distrocoeso típico / Xanthic Ferralsol (Clayic, Dystric, Densic)										
A	0.06	1.1	0.03	0.72	10.87	0.07	0.481	20.43	52.4	19.8
Bw1	0.08	0.98	0.01	0.09	9.43	0.01	0.505	19.34	66.4	5.5
Bw2	0.1	0.93	0.02	0.09	10.16	0.01	0.511	19.45	71.4	6.9
P2 - Latossolo Amarelo Distrocoeso típico / Xanthic Ferralsol (Clayic, Dystric, Densic)										
A	0.35	1.01	0.03	0.22	7.37	0.03	0.482	20.23	36.1	30.1
Bw1	0.1	0.69	0.02	0.07	7.37	0.01	0.487	19.84	44.6	13.9
Bw2	0.14	0.89	0.02	0.07	8.86	0.01	0.515	18.89	48.4	8.7
P4 - Cambissolo Háplico Tb Distrófico latossólico / Xanthic Ferralsol (Loamic, Dystric, Densic)										
Bi	0.09	0.64	0.03	0.29	5.62	0.05	0.464	20.89	33.9	17.5
C1	0.06	0.49	0.02	0.05	4.9	0.01	0.443	21.89	49.8	30.1
P5 - Latossolo Amarelo Distrocoeso típico / Xanthic Ferralsol (Loamic, Dystric, Densic)										
A1	0.05	0.52	0.01	0.46	4.17	0.11	0.461	21.39	19.9	25.7
Bw	0.06	0.39	0.02	0.07	3.92	0.02	0.48	20.08	22.9	23.5
P6 - Latossolo Amarelo Distrocoeso típico / Xanthic Ferralsol (Clayic, Dystric, Densic)										
Bw1	0	1.14	0.03	0.15	24.04	0.01	0.529	18.58	79.5	3.1
Bw2	0.14	0.98	0.03	0.13	27.85	0	0.479	20.41	81.1	3.9
P7 - Latossolo Amarelo Distrocoeso típico / Xanthic Ferralsol (Clayic, Dystric, Densic)										
Bw1	0.13	0.59	0.01	0.08	2.39	0.03	0.552	18.11	39.4	14
Bw2	0.13	0.76	0.01	0.09	1.07	0.08	0.571	17.7	46.5	13.3
Bw3	0.14	0.72	0.02	0.07	0.83	0.09	0.576	17.48	50.7	10.6
P8 - Plintossolo Pétrico Concrecionário típico / Pisoplinthic Plinthoshol (Dystric, Ochric, Loamic)										
A	0	0.57	0.03	0.5	12.44	0.04	0.506	19.37	24	17.8
Bi	0.15	0.59	0.03	0.13	14.39	0.01	0.533	18.44	28.6	11
P9 - Plintossolo Argilúvico Distrófico típico / Dystric Plinthosol (Acric, Clayic, Ochric, Oxyaquic)										
A	0.06	0.49	0.03	0.86	2.02	0.43	0.585	16.86	19.2	33.8
Bf1	0.01	0.67	0.02	0.91	3.77	0.24	0.613	15.96	27.4	19
Bf2	0.13	0.47	0.02	0.27	4.06	0.07	0.655	15.07	47.5	11

Fig. 4 X-ray diffractograms of the clay fraction, peak width at half height (FWHM), crystallite size (CRS), remaining phosphorus (P-rem), and contents of silicon (Si), aluminum (Al), manganese (Mn), and iron (Fe) extracted by selective dissolution of selected horizons of the representative soil profiles. Si_o—Si extracted by ammonium oxalate; Al_o—Al extracted by ammonium oxalate; Fe_o—Fe extracted by ammonium oxalate; Mn_d—Mn extracted by dithiocitrate-bicarbonate; Fe_d—Fe extracted by dithiocitrate-bicarbonate

Fig. 5 Latossolos Amarelos Distrocoesos landscape pattern of occurrence, represented by broad tableland tops under natural Cerrado vegetation

adsorption in Al–OH groups of the octahedral layer of these minerals, as suggested by Singh and Gilkes (1992), who observed that the greater the specific surface area and structural disorder of kaolinites, the greater is the P adsorption. This phenomenon was observed by Antonangelo et al. (2020) in a study carried out with Latossolos (Ferralsols) from other Brazilian regions.

In addition to Al_o, the clay content was correlated with Fe_d, Fe_o/Fe_d (negative), and Fe_o (negative), suggesting that the higher the clay content, the higher the content of crystalline Fe oxides. However, P-rem correlated negatively with Fe_d, and positively with Fe_d/Fe_o, suggesting that the greater the proportion of crystalline Fe oxides, the greater the adsorption of P in soils.

In addition, the crystallite size (CRS) of kaolinite show differences between the profiles, i.e., the C horizon analyzed presents the highest values, reflecting, in this way, a lower degree of pedogenetic alteration, as expected. The soil profiles in free drainage conditions (Latossolos, Cambissolos, and Plintossolos Pétricos) show a trend to increase the CRS in depth, which is expected, taking into account that the weathering processes act with more intensity from top to bottom in the profile. On the other hand, the profiles that are in poor drainage conditions, either past or present—represented by P7 and P9, respectively—show a tendency for a decrease in the CRS in depth, which is consistent with the lower stability of the kaolinite in conditions of seasonal or permanent water saturation (Dixon 1989).

4.2 Considerations About Local Soil-Landscape Relationships

From observation of satellite products and extensive fieldwork, it is possible to identify different patterns between the Dissected tablelands (DT) and Tablelands compartments. Figure 1 shows the difference in drainage density and degree of slope notching between the two, which results in different extensions of tops. This reflects in the proportion of the soil types areas, being larger the Latossolos Amarelos Distrocoesos areas in the Tablelands in relation to the DT. There are also more Cambissolos Tb Distróficos in the TD, where the C horizon is closer to the surface and outcrops relatively more frequently.

This distribution of soil types as a result of the degree of slope notching is understandable when considering the different balances between morphogenesis and pedogenesis. In the Tabuleiros, the wider tops of flat relief favor the infiltration of water, and consequently the deepening of the latsolic mantle, resulting in deeper latsolic B horizons (Bw), and the possibility of greater development of pedogenetic processes—also evidenced by the greater occurrence of clay increment in depth in this region. While in the TDs the greater notching of the landscape, with less broad tops and greater drainage density, favors erosive processes, which tend to bring the C horizon and pedogenetically less developed horizons closer to the surface. It can be stated that, to a certain degree, there is a geomorphological control over the distribution of local soil types.

It is common and recurrent in some parts of Coastal Tablelands (Barreiras Group) in other regions of Brazil the occurrence of Espodossolos (Spodosols), as verified by Filizola et al. (2001), Ucha et al. (2002), Moreau et al. (2006), Corrêa et al. (2008), and Oliveira et al. (2010). Unlike the regions where the aforementioned works were developed, in the entire area of this study, the occurrence of Spodosols was not observed. There are small extensions related to Neossolos Quartzarênicos (Arenosols), i.e., there are sandy soils, but which did not undergo the podzolization process for the formation of spodic B horizon. This fact raises questions about the conditions of Espodossolos formation in Coastal Tablelands. Both the climatic regime (temperatures and rainfall) and the type of natural vegetation are different in the study area in relation to the cited articles, raising hypotheses related to the influence of these factors on the podzolization processes, which deserve studies dedicated to this theme.

Figure 2 shows the profiles located in the two compartments and illustrates that the soils in top position are mostly typical LAdx or LAdx argissólicos. Among the LAdx, P7 represents slightly bulging areas with a tendency to water flow direction and presents morphology consistent with past water saturation. Figure 6 shows profiles located in the DT (A, B, and C) and in the Tablelands (D, E, and F), which represent the variability of the soils in top position, mostly occupied by LAdx típicos or LAdx argissólicos—this second type is predominant in the Tablelands. Among the LAdx, P7 (Fig. 6f) represents slightly bulging areas with a trend toward water flow and presents a morphology consistent with past water saturation.

Fig. 6 Latossolos Amarelos Distrocoesos (Xanthic Ferralsols) profiles present in the Dissected tableland (**a–c**) and Tablelands (**d–f**) compartments

As we move toward the edge of the tops, a concretionary horizon (F) becomes closer to the surface (Fig. 7). The Plintossolos Pétricos Concrecionários (FFc—Pisoplithic Plinthosols) occupy areas of isolated hills in rugged topography of bulged and narrow tops, or in steeper gradients of the slopes on the edges of flattened tops (Fig. 8), and are represented by profiles P3 and P8, in their respective geomorphological compartment.

The region studied presents as an outstanding characteristic an extensive high wheatering degree mantle, of yellowish color and variable depth, which presents abrupt contact with concretionary horizons or gradual transition to a horizon more

Fig. 7 Representative profiles of the soil transition toward the edge of the tableland. Latossolo Amarelo Distrocoeso (Xanthic Ferralsol) in top position (**a**) in transition (**b**) to Plintossolo Pétrico Concrecionário (Pisoplinthic Plinthosol) (**c**), in the Dissected tableland compartment. Abrupt contact between the deep wheatered mantle and the concretionary horizon at the top of the landscape is represented in detail **d**

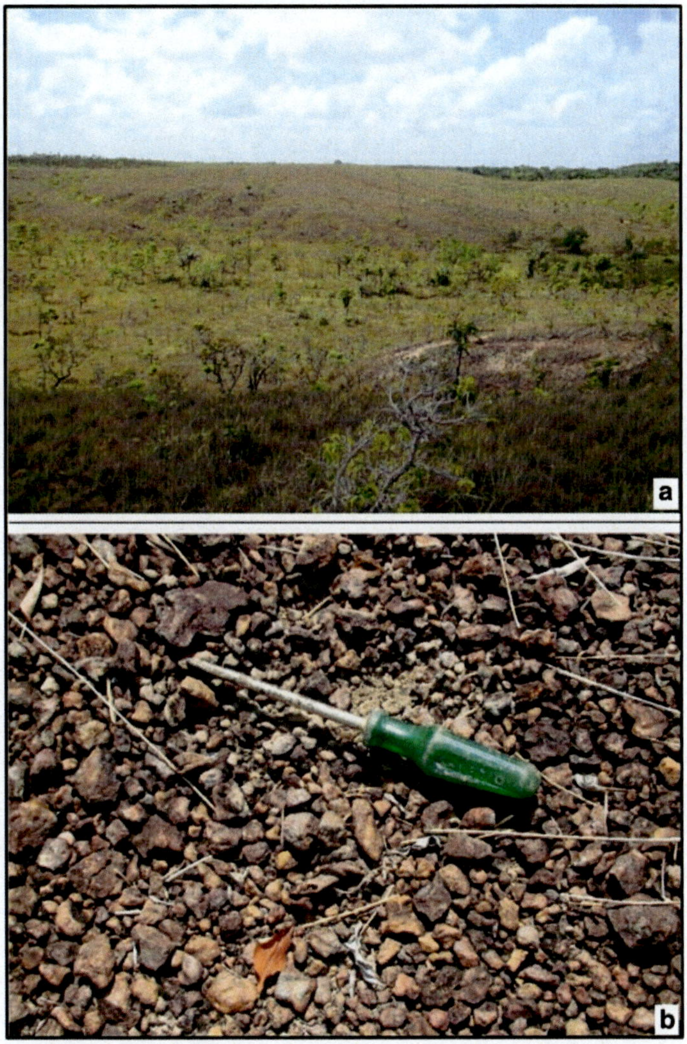

Fig. 8 Typical landscape morphology of occurrence of Plintossolos Pétricos Concrecionários (Pisoplinthic Plinthosols) (**a**), with concretionary horizon from the surface (**b**)

similar to its original material—unconsolidated sandy-clayey sediments composed essentially of quartz and kaolinite.

In situations of sloping slopes, mainly in the Dissected tablelands, the Cambissolos Háplicos Tb Distróficos (P4) occur (Fig. 9), always with a clayey or very clayey texture. The erosion of the wheatered mantle in this slope situation causes the B horizon to have all the chemical characteristics, and some morphological and physical characteristics of Bw horizons, but with insufficient depth to be defined as such, being identified as incipient B horizons (Bi) in these locations.

Fig. 9 Profiles of Cambissolo Háplico Tb Distrófico latossólico, occurring on steep slopes—most common in Tabuleiros Dissecados. Rare outcrop of **c** (**d**) horizon

Also in this context, the soils of the broad grasslands of the Tablelands are represented by the Plintossolos Argilúvicos Distróficos típicos (Dystric Plinthosols) (P9). Furthermore, Gleissolos (Gleysols) and Organosolos (Histosols) occur in the proximities of the river channels, but as the drainage of the region is very embedded, these situations are not representative in terms of area.

The difference in relative altitude (Fig. 2) between the two geomorphological compartments is also evident. The two environments seem to be in different stages of pedological evolution, where it is difficult to affirm if the control over the dynamics of soil genesis is a result of the heterogeneity of the source material, or if it has more influence from the morphology of the landscape. On the other hand, it is possible that the pedological cover itself also exerts structural control over the landforms, as can be observed in the areas of occurrence of FFc, which support sloping slopes on the edges of the tableland or isolated hills with bulged tops.

When observed separately and compared the geomorphological compartments of Dissected tablelands (DT) and Tablelands, it can be seen that in the soils there are differences—more or less subtle—between these two environments, discarding the hypothesis of homogeneity of the Coastal Tabuleiros of the Barreiras Group in Amapá State.

5 Final Considerations

In view of the above, we conclude that the Coastal Tablelands of the Barreiras Group in Amapá present heterogeneous pedological characteristics in their geomorphological compartments of Dissected tablelands and Tablelands. Thus, we can affirm that there are two distinct pedogeomorphological compartments.

Furthermore, it was possible to conclude that the frequency of textural classes between the two environments is inversely proportional. The compartments present clear difference in the predominance of particle size, indicated by the averages and medians of the samples submitted for particle size analysis. While in the Tablelands a loam-texture predominates, in the Dissected tablelands a clayey texture predominates and a very clayey texture occurs with greater frequency.

The soils of the two environments are homogeneous in terms of the characteristics of their sorptive complexes and mineralogy. The nature of the parent material and the process of latossolization/ferralitization in humid climates tend to condition acidic, dystrophic, acidic soils with low cation exchange capacity. The sand fraction is exclusively composed of quartz, while the clay fraction is predominantly composed of kaolinite.

The tabuleiro environment appears to have a greater development of pedogenetic processes. Considering its higher content of Fe oxides, smaller sizes of kaolinite crystals (CRS) that suggest ferralitization processes, and the trend of relative clay accumulation in depth (more expressive than in the Dissected tablelands), it can be affirmed that the soils of Tablelands experienced more intensity pedogenetic processes. In the Dissected tablelands, perhaps due to the higher altimetry, the processes of morphogenesis act in such a way that they inhibit the same rhythm of pedogenesis of the Tabuleiros.

Acknowledgements The authors acknowledge the financial and logistical support offered by Coordenação de Aperfeiçoamento de Pessoal de Nível Superior (CAPES), Conselho Nacional de Desenvolvimento Científico e Tecnológico (CNPq), and Amapá Florestal e Celulose S.A. (AMCEL).

References

Achá Panoso L (1985) Latossolo Vermelho-Amarelo de "Tabuleiro" do Espírito Santo: formation, characteristics and classification. Tese (Doutorado em Agronomia) - Programa de Pós-graduação em Agronomia, Universidade Federal Rural de Pernambuco - UFRPE, Recife

Anjos LHC (1985) Caracterização, gênese, classificação e aptidão agrícola de uma sequência de solos do Terciário na região de Campos-RJ. Dissertation (Master's in Agronomy) - Graduate Program in Soil Science, Federal Rural University of Rio de Janeiro - UFRRJ, Seropédica

Antonangelo JA et al (2020) Phosphorus speciation by P-XANES in an Oxisol under long-term no-till cultivation. Geoderma. https://doi.org/10.1016/j.geoderma.2020.114580

ARAI MA (2006) A Grande elevação eustática do mioceno e sua influência na origem do Grupo Barreiras. Geologia USP Série Científica 6:1–6. https://doi.org/10.5327/S1519-874X2006000300002

Bennema J, Camargo MN (1979) Some remarks on Brazilian latosols in relation to the oxisols of soil taxonomy. In: Internation soil classification workshop, 2 1979, Bangkok. Proceedings. Bangkok, Malaysia. SSD - LDD, 1979. p. 233–235

Bigarella JJ, Andrade GO (1964) Considerations on the stratigraphy of Cenozoic sediments in Pernambuco (Grupo Barreiras). UR, ICT, vol 2, p. 1–14

Brazil (1974) Folha NA/NB 22 Macapá: geologia, geomorfologia, pedologia, vegetação, uso potencial da terra. Ministry of Mines and Energy, Rio de Janeiro

Corrêa MM, Ker JC,; Barrón V, Torrent J, Fontes MPF, Curi N (2008) Crystallographic properties of kaolinites from soils of the Tabuleiros Costeiros, Amazônia and Recôncavo Baiano environments. Rev Bras de Ci do Solo 32:1857–1872. https://doi.org/10.1590/S0100-06832008000500007

Dixon JB (1989) Kaolin and serpentine group minerals. In: Dixon JB, Weed SB (ed) Minerals in soil environments, 2nd ed. Soil Science Society of America, Madison

Filizola HF et al (2001) The fragipans and duripans of the Coastal Tableland depressions of northeastern Brazil: an evolutionary proposal. Rev Bras de Ci do Solo 25:947–963. https://doi.org/10.1590/S0100-06832001000400018

Guerra, AJT, Silva, AS, Botelho RGM (1999) Erosão e conservação dos solos: conceitos, temas e aplicações. Bertrand Brasil, Rio de Janeiro

IEPA (2008) Macrodiagnóstico do estado do Amapá - Primeira Aproximação do ZEE. - Institute for Scientific and Technological Research of the State of Amapá, Macapá

IUSS Working Group WRB (2015) World reference base for soil resources: International soil classification system for naming soils and creating legend for soil maps. FAO, Rome

JACOMINE PKT (1996) Distribuição geográfica, características e classificação dos solos coesos dos Tabuleiros. In: Reunião Técnica Sobre os Solos dos Tabuleiros Costeiros, Anais. Cruz das Almas, Brasil. Embrapa: UFBA, 1996. p. 13–26

Jorge XS, Teixeira SG (2016) Geodiversity of the state of Amapá. CPRM, Belém

Ker JC (1997) Latosols of Brazil: a review. Geonomos 5:17–40. https://doi.org/10.18285/geonomos.v5i1.187

Ker JC, Schaefer CEGR, Romero RE, Corrêa MM (2017) Solos dos Tabuleiros Costeiros. In: Curi N, Ker JC, Novais RF, Vidal-Torrado P, Schaefer CEGR (eds) Pedologia - solos dos biomas brasileiros. Sociedade Brasileira de Ciência do Solo, Viçosa, p. 467–492

Kirschbaum, MUF (2006) The temperature dependence of organic-matter decomposition - still a topic of debate. Soil Biology & Biogeochemistry 38:2510–2518. https://doi.org/10.1016/j.soilbio.2006.01.030

McKeague JA, DAY JH (1966) Dithionite and oxalate extractable Fe and Al as aids in differentiating various classes of soils. Canadian Journal of Soil Science 46:13–32. https://doi.org/10.4141/cjss66-003

Mehra OP, Jackson ML (1960) Iron oxides removal from soils and clays by a dithionite-citrate-bicarbonate system buffered with sodium bicarbonate. Clays and Clay Minerals 7:317–327. https://doi.org/10.1016/B978-0-08-009235-5.50026-7

Melo VF, Schaefer CEGR, Singh B, Novais RF, Fontes MPF (2002) Chemical and crystallographic properties of kaolinite and iron oxides in Barreiras Group sediments in the municipality of Aracruz, State of Espírito Santo. Rev Bras de Ci do Solo 26:53–64. https://doi.org/10.1590/S0100-06832002000100006

Moreau AMSS (2001) Gênese, química e micromorfologia de horizontes coeso, fragipã e duripã em solos do Tabuleiro Costeiro no sul da Bahia. Thesis (Doctorate in Agronomy) - Post-Graduation Program in Soils and Plant Nutrition, Universidade Federal de Viçosa - UFV, Viçosa

Moreau AMSS, Ker JC, Costa LM, Gomes FH (2006) Characterization of soils of two toposse-quences in Tabuleiros Costeiros in southern Bahia. Rev Bras de Ci do Solo 30:1007-1009. https://doi.org/10.1590/S0100-06832006000600010

Moura-Fé, MM (2014) Barriers: series, group or formation? Rev Bras de Geog Fís 7:1055–1061. https://doi.org/10.26848/rbgf.v7.6.p1055-1061

Nogueira LRQ, Nogueira LC (1996) Reunião técnica sobre solos coesosos dos Tabuleiros Costeiros - Pesquisa e desenvolvimento para os tabuleiros costeiros - Anais. EMBRAPA-CPATC/EMBRAPA-CNPMF/EAUFBA/IGUFBA, Cruz das Almas

Nunes FC, Silva EF, Vilas Boas GS (2011) Barriers Group: Characteristics, Genesis and Evidence of Neotectonism. Bol de Pesq e Desenv (Embrapa Solos) 194:1–31 2011

Oliveira LB, Dantas HS, Campelo AB, Galvão SJ, Gomes IF (1968) Characterization of subsoil densification in a tableland area of the Curado Experimental Station, Recife. Pesq Agropec Bras 3:207–214

Oliveira J (2000) Solos coesos dos Tabuleiros Costeiros: limitações agrícolas e manejo. SEAGRI-SPA, Salvador

Oliveira AP et al (2010) Spodosols pedogenesis under Barreiras formation and sandbank environments in the south of Bahia. Rev Bras de Ci do Solo 34:847–860

Patterson AL (1939) The Scherrer Formula for X-Ray Particle Size Determination. Physical Review 56:978–982. https://doi.org/10.1103/PhysRev.56.978

Resende, M. Pedology. (1982) Federal University of Viçosa, Viçosa

Resende M, Curi N, Ker JC, REZENDE SB (2011) Mineralogia de Solos Brasileiros - Interpretações e Aplicações. 2ª ed. Editora UFLA, Lavras

RIBEIRO LP (1998) Os Latossolos Amarelos do Recôncavo Baiano: gênese, evolução e degradação. Seplantec-CADCT, Salvador

Santos HG et al (2018) Sistema Brasileiro de Classificação de Solos. 5th ed. Embrapa, Brasília

Sena NC (2016) Soils developed on sediments of the Formação Barreiras in Amapá state. Dissertation (Master's in Agronomy) - Graduate Program in Soils and Plant Nutrition, Federal University of Viçosa - UFV, Viçosa

Silva AJN, Carvalho FG (2007) Cohesion and shear strength related to physical and chemical attributes of a Yellow Latosol from Tabuleiro Costeiro. Rev Bras de Ci do Solo 31:853-862. https://doi.org/10.1590/S0100-06832007000500003

Singh B, Gilkes RJ (1992) Properties of soil kaolinites from south-western Australia. Journal of Soil Science 43:463–472. https://doi.org/10.1111/j.1365-2389.1992.tb00165.x

Souza LS (1996) Uso e manejo dos solos coesos dos Tabuleiros Costeiros. In: Reunião Técnica Sobre Solos Coesos Dos Tabuleiros Costeiros - Anais. Embrapa Tabuleiros Costeiros, Aracaju

Teixeira PC, Donagemma GK, Fontana A, Teixeira WG (2017) Manual of soil analysis methods. EMBRAPA, 2017

Ucha JM, Botelho M, Vilas Boas GS, Ribeiro LP, Santana OS (2002) Use of ground penetrating radar (GPR) in the investigation of soils of the Tabuleiros Costeiros in the northern coast of Bahia State. Rev Bras de Ci do Solo 26:373–380. https://doi.org/10.1590/S0100-06832002000200011

Vasques GM, Grunwald S, Sickman JO (2008) Comparison of multivariate methods for inferential modeling soil carbon using visible/near-infrared spectra. Geoderma 146:14–25. https://doi.org/10.1016/j.geoderma.2008.04.007

Vidal-Torrado P, Lepsch IF, Castro SS (2005) Concepts and applications of pedology-geomorphology relationships in humid tropical regions. In: Vidal-Torrado, P et al (eds) Tópicos em Ciência do Solo. Sociedade Brasileira de Ciência do Solo, Viçosa, p. 145–192

Geomorphological Evolution of River Forms in Humid and Semi-arid Tropical Environments

Éverton Vinícius Valezio, Kleber Carvalho Lima, and Archimedes Perez Filho

Abstract Fluvial channels are directly affected by changes triggered by natural and anthropic phenomena, adapting to new conditions in different temporal-spatial rhythms. In order to demonstrate how certain techniques can support the interpretation of the geomorphological evolution of tropical river systems, we selected two rivers in different climatic conditions: Jacaré-Guaçu (SP) in humid tropical regimes, and Itapicuru (BA) in semi-arid regime. By means of aerial photographs, orbital images, aerophotogrammetry by remotely piloted vehicle, description of the deposits, and absolute dating by Optically Stimulated Luminescence (OSL), it was verified that, even in distinct systems, the processes of formation of fluvial terraces were active in more humid conditions, however, in different periods, more current for Itapicuru and older for Jacaré-Guaçu, demonstrated by the morphology of the paleochannels and by the absolute dating. Thus, it is believed that the techniques used for the interpretation of the geomorphological evolution of river systems can improve the studies of landscape evolutionary models in line with considerations of the sensitivity of river systems to absorb, resist, or recover from disturbances in their own temporalities.

1 Introduction

Difficulties inherent to the spatialization of relief are recurrent, given the complexity of representation of the terrestrial model. In fluvial systems, sensitive to natural events

É. V. Valezio (✉) · A. Perez Filho
Institute of Geosciences, State University of Campinas (UNICAMP), Campinas, Brazil
e-mail: evertonvalezio@gmail.com

A. Perez Filho
e-mail: archi@ige.unicamp.br

K. C. Lima
Department of Geography, University of Pernambuco—Garanhuns Campus (UPE), São José, Brazil
e-mail: kleber.carvalho@upe.br

G. Barbosa dos Santos et al. (eds.), *Geomorphology of Brazil: Complexity, Interscale and Landscape*, Springer Proceedings in Earth and Environmental Sciences,
https://doi.org/10.1007/978-3-031-05178-4_5

(tectonics and climate) and, more recently, to anthropic factors, the identification of forms and features is crucial for the understanding of their evolution over time.

The interpretation of changes in river channels and the reconstitution of paleoenvironments are linked to the understanding of fluvial dynamics at the most different scales and environmental conditions, being of great importance the association of current and past processes associated with studies about fluvial terraces, paleochannels, and floodplains in hot and humid (Hamilton et al. 2007; Morais et al. 2020) and hot and dry environments (Norton et al. 2016; Larson et al. 2020).

However, the complex analysis and interpretation of fluvial environments in tropical regions are often hindered by the superficial dynamics of the landscape. Hydrological variations, transformations in pedological cover by slope dynamics, and human action end up masking river morphologies that respond to current and past climatic conditions, as well as enhancing the rate of change that took place at geological scales (Sridhar 2007; Hughes et al. 2015; Lima and Lupinacci 2019).

In this sense, the identification, mapping, and characterization of forms and attributes of the river landscape in these environments can be considered complex steps to be developed, constituting real methodological challenges for the researcher. Traditionally, these steps are carried out through systematized cartographic bases, aerial photographs, satellite images, and field procedures, such as translations in toposequences and stratigraphy of alluvial deposits (Straffin et al. 1999; Bisson et al. 2011; Celarino et al. 2013; Piégay et al. 2020; Molliex et al. 2021).

However, new tools have emerged in recent decades as a way to expand the methodological possibilities and improve interpretations regarding the genesis and geomorphological evolution of river systems. New technologies such as remotely piloted vehicles (RPVs), high spatial resolution orbital images, ground-penetrating radar (GPR), and absolute dating have been added to previously used tools in order to obtain more consistent results.

In the Brazilian scalar context, with the different climatic regimes and their structural heterogeneity, we selected two rivers with distinct process characteristics: Jacaré-Guaçu River and Itapicuru River. In the case of the first one, it is disposed of in the Sedimentary Basin of Paraná, latitude 22°, where the hot and humid regime is dominant. The river is a tributary of the right bank of the middle course of the Tietê River, in the state of São Paulo, constituting one of the main rivers of Brazil, both for its capacity for hydroelectric generation and for crossing the most populous state of the country. As a main mark, the Jacaré-Guaçu has meander typology and preserves, in certain sectors of the river plain, forms of its lateral rambling and incision of the bed, while we still encounter hydrodynamic changes that are no longer present in the river plain (Valezio 2016). On the other hand, the Itapicuru River is located in the northeastern region of Brazil, and crosses several structural units, such as the headwaters plateaus, passing through depression and the tablelands in the lower course. Due to its position in low latitude (11°), there are hot and dry air masses with a predominantly semi-arid climate. The longitudinal structural modifications, added to the climatic and anthropic dynamics, are capable of determining patterns and changes in the typology of the river channel throughout its longitudinal extension (Lima et al. 2021).

Thus, we seek to evidence and discuss aspects of fluvial dynamics in these different environments from the identification of forms and processes, as well as the correlation with absolute dating, aiming to understand the functioning of landscapes in the context of geomorphological evolution at different temporal-spatial scales.

2 Study Area

Running 238 km longitudinally, from the headwaters in Itirapina/SP and mouth in Ibitinga/SP, the Jacaré-Guaçu River is part of a humid tropical system, marked by seasonality, with pedogenesis accentuated by the presence of water throughout the year, which quickly decharacterizes the sedimentary deposits from fluvial action (Celarino and Ladeira 2017). The average precipitation of 1402 mm in the upper, 1391 mm in the middle, and 1257 mm in the lower course (IPT 2003), with higher outflows between the months of October and March, links to the higher average annual temperatures in this period 22 °C (Costa 2005). Another fundamental characteristic for understanding the functioning of the river is the lithological heterogeneity along its course and by the transition of geomorphological compartments: from the sandy-basaltic Cuestas to the Paulista Western Plateau, following in the NW direction until the middle course and W until its mouth (Fig. 1). Regarding the geological substratum, the Botucatu and Serra Geral formations are highlighted (Riccomini 1997). The Mesozoic formations vary along the river, alternating meandering alluvial sectors on the sandstone formation of Botucatu and sectors with less lateral migration (in bedrock embedded in the Serra Geral basalts).

The Itapicuru River is 567 km long, whose sources are located in the northern portion of the Diamantina Plateau and crosses the Sertaneja Depression and the Coastal Tablelands until it flows into the Atlantic Ocean (Lima 2017). The topographic variation occurs longitudinally, with low altitudes on the coast, gradually increasing upstream, until reaching more than 1200 m of altitude. Similarly, the spatial variation of the climate is longitudinal (SEI 1999), with humid to sub-humid climate on the coast, sub-humid to dry, semi-arid, and arid climate in the central portion of the drainage basin. In the western sector, in turn, the orographic effect in the transition between the depression and the plateau favors the formation of wet areas in altitudes above 900 m. The lithological and structural variety is significant, whose tectonic domains include the São Francisco Province, composed of syenitic and granitic suites, felsic volcanic rocks, volcanic arcs, orthogneisses associated with granitoids, igneous rocks, among others, dating from the Archean-Proterozoic. The Tucano Central sedimentary basin is of Cretaceous age, composed of the Marizal Formation (associations of sandstones and conglomerates with shales and limestones), São Sebastião Formation (medium to fine sandstones with coarse levels at the base and intercalations of siltstones, argillites, and shales) and Islands Group (medium to coarse sandstones alternately, with intercalations of shales and siltstones) (Kosin et al. 2004).

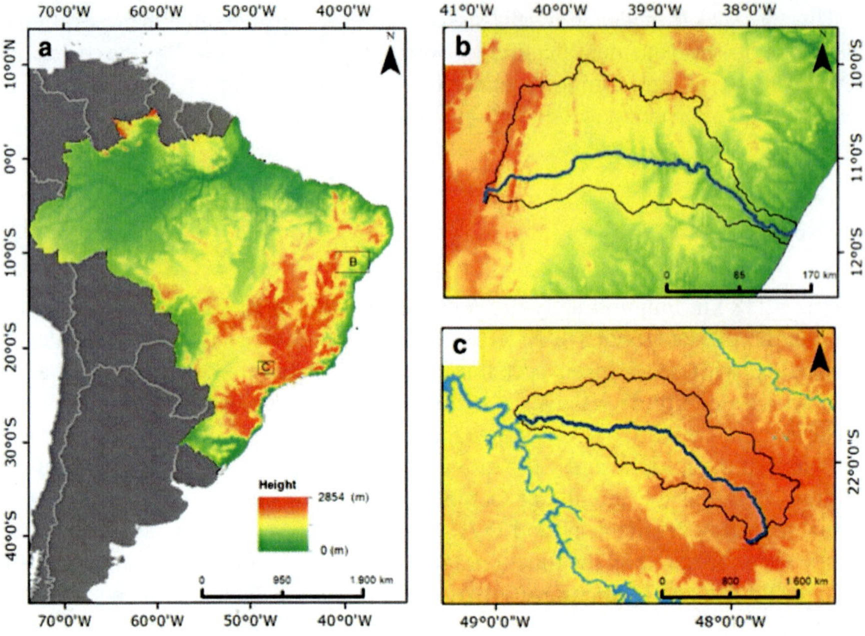

Fig. 1 Location of the study areas. **a** topographic and latitudinal variation of the hydrographic basins in the Brazilian territory. **b** Itapicuru River watershed, Bahia State; **c** Jacaré-Guaçu River watershed, São Paulo State, and its outlet on the Tietê River

3 Materials and Methods

The geomorphological mapping of the fluvial plain of Jacaré-Guaçu and Itapicuru rivers was performed by means of stereoscopic pairs of aerial photographs of the years 1962 and 1975, respectively, both in 1:25,000 scale. From the same, was used for the two cases, orbital images from the Rapideye satellite with spatial resolution of 5 m orthorectified, and images from the CBERS 2B-HRC satellite, with spatial resolution of 2.7 m. In a complementary manner, the SRTM (*Shuttle Radar Topography Mission*) digital elevation model was used, with a spatial resolution of 30 m. For the Jacaré-Guaçu River, an RGB aerophotogrammetric survey was conducted by remotely piloted aircraft (DJI Phantom 4 Advanced), with GSD of 4 cm/pixel and planimetric errors of 11.53 cm and altimetric errors of 29.65 cm, generating orthomosaics, DTM and DEM for an area of approximately 13 km^2 in the middle course of the river.

Fieldwork was carried out for control point inference, translations, and description of sedimentary and pedological profiles at low terrace and floodplain levels. Absolute dating by Optically Stimulated Luminescence (OSL) was used to estimate the formation time of the fluvial plains and embedded forms, correlating them to climatic and/or tectonic factors. Ages were estimated by the SAR (Single Aliquot Regenerative-dose) protocol, as per Murray and Wintle (2000) and Wintle and Murray (2006).

The protocol was used for all samples collected, differing only in the number of calibration curves. Fifteen aliquots were established for the samples of the Itapicuru River and 25 aliquots for the Jacaré-Guaçu River, given the temporal difference of sample collection (years 2014 and 2019, respectively).

Dating by OSL and the use of remotely piloted aircraft have been widely used in Brazilian geomorphological studies in the last ten years. Tools already used for the evolutionary interpretation of the landscape, such as aerial photographs, satellite images (although now with better spatial resolutions), and the descriptions of the forms and their constituents tend to remain, although the results may lose significance when not linked to new technologies and possibilities of analysis. We emphasize that new methodologies should be increasingly incorporated to overcome the limitations imposed by the already consolidated techniques, widely used in geomorphological studies in Brazil.

4 Results

4.1 Jacaré-Guaçu River

On the Jacaré-Guaçu alluvial valley, two levels of low terraces were identified: Level T1, about nine meters above the riverbed, which is preserved in the landscape in the middle and lower course, developed in alluvial environments over sandstone rocks of the Botucatu Formation; and Level T2 elevated about twelve meters above the current level of the fluvial channel. The T2 level, identified by photographs dated 1962, is in a more advanced erosion process, losing its genetic-morphological characteristics. The T1 level, more preserved and easier to access, is characterized by an abrupt transition between sandy material (preponderantly medium sand and fine sand), with millimeter-sized granules and pebbles, and oxirection marks associated with roots, classified, overlaid by fine material (>50% clay and silt compound), oxirreduction marks up to 90 cm deep and medium to large lumps in the first 20 cm (A horizon), clear transition to B1, with light lumpy structure (30–45 cm depth), and plastic and sticky, apparently massive B2 horizon with lumpy structure—associated with modern roots—when manipulated (Fig. 2). Part of level T1 is covered by peat bog, part of which has already been anthropically remobilized. Level T1 was dated by OSL at 140 cm depth, dating 7920 ± 1440 BP (Sample MDT1), in the innermost portion, and 7670 ± 1220 years BP (Sample BT1), in the border portion.

The DEM, DTM, and orthoimagery indicated the presence of upper meander loops in amplitude, width, and length in both low terrace levels of the Jacaré-Guaçu River, differing morphologically and morphometrically from the current and recently abandoned meanders (Figs. 3 and 4). Furthermore, translations confirmed the difference in depth between the meanders sustained on the terraces and the present ones, besides the associated deposits, coarser (coarse sand to blocks) in the older ones, and more sandy-clayey in the modern loops. The scars of the past processes and their presence

Fig. 2 Low terrace level (T1) in the middle course of the Jacaré-Guaçu River. **a** Sedimentary material constituting the low terrace, with subdivision in horizons; **b** Transition from the A to B1 horizon, highlighting the lumpy structure of the superficial part; **c** Location of the sample collection points for dating by OSL. Point in the middle portion of the terrace MDT1 (yellow star) and in the border portion BT1 (red star)

in the landscape allow morphometric comparison between the forms and, consequently, between the patterns established still in the Middle and Lower Holocene in relation to modern characteristics (differences in land use and vegetation) (Fig. 4).

Meander alteration processes typical of deconfined channels in the middle course, with significant cutoffs, reducing both the sinuosity of the sector (2.20–1.85) and the total length (404 m); and in the lower course, especially in the post-confinement sector of the river by the Serra Geral Formation, with four cuts, rotation and enlargement processes, causing the channel to reduce its sinuosity from 2.65 to 2.08 and its length from 5417 to 4259 m, between the years 1962 and 2012. In this second sector, it was possible to identify the presence of settling basins, abandoned meanders, and scrollbars, in addition to, as in the middle course sector, the asymmetric position of the fluvial channel in its fluvial plain (arranged on the right bank).

Another factor addressed in the mappings was the retraction of the riparian vegetation due to the advance of sugar cane and orange plantations, facilitating the erosive process of level T2, heading toward level T1.

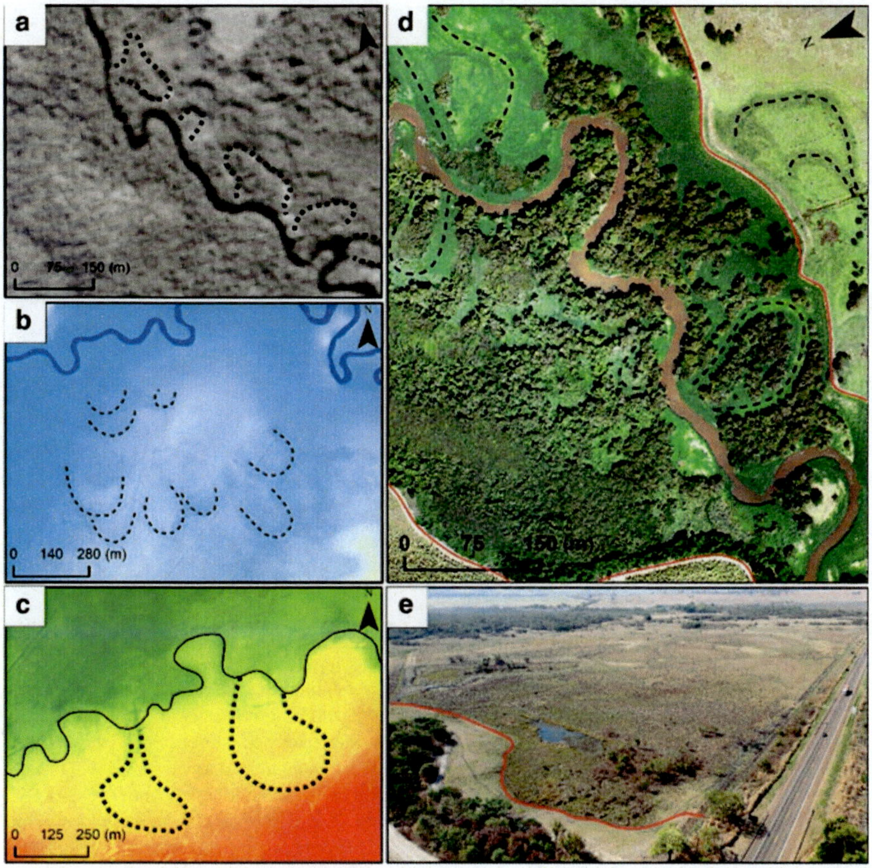

Fig. 3 Current, recently abandoned, and past meanders of the Jacaré-Guaçu River are identified by RPV and CBERS2B images. **a** Jacaré-Guaçu River and centennial-scale abandoned meanders identified by CBERS2B satellite; **b** Intermediate-level floodplain meanders identified by RPV DTM; (C) Lower terrace (T1) paleomeanders identified by RPV DEM; **d** Current river channel, recently abandoned meanders, and subdivision between meander belt and possibly new low terrace level formation made by RPV orthomosaics; **e** Aerial drone image of the division between low terrace (T1) and floodplain/new low terrace level

4.2 Itapicuru River

The orbital products used in the identification and mapping of the terrace levels and floodplain of the Itapicuru River contributed significantly to the understanding of the fluvial dynamics since they allowed the visualization of features indicative of these dynamics. The granulometric analysis of the profiles and the absolute dating by OSL favored the understanding of the dynamic behavior of the river during the Holocene to the current time scale.

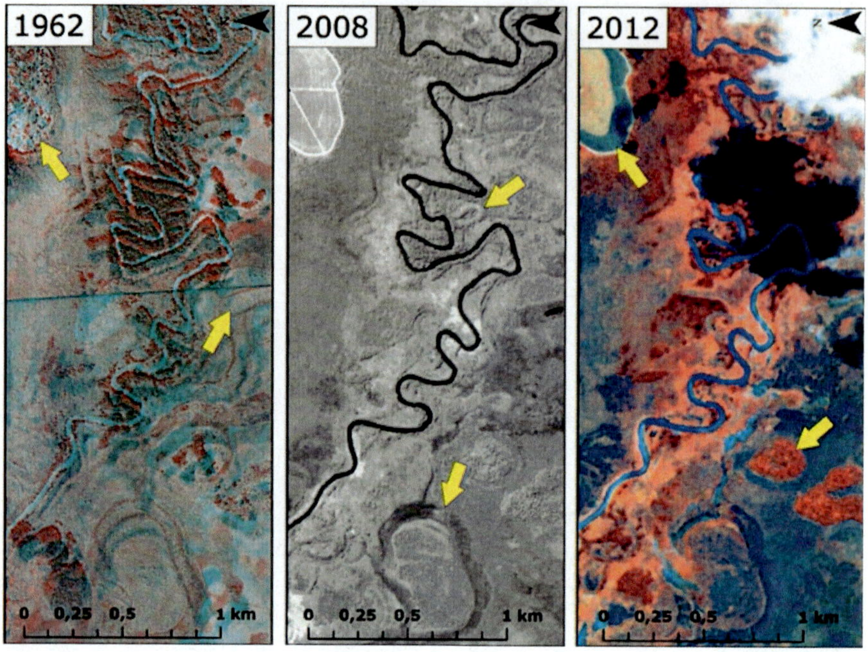

Fig. 4 Overlay of the products used for the lower Jacaré-Guaçu River—3D aerial photographs, Cbers-2B, and Rapideye—to vectorize the river at different times and identify alluvial valley forms and processes. (1962) Arrows indicating land use and former river bed (stereoscopic pairs); (2008) arrows indicating difference in size and morphology between meanders at low terrace level (T1). At bottom, arrow indicating larger size meander, and at top, recently abandoned meander (CBERS2B-HRC); (2012) false-color composition highlighting areas of riparian vegetation in shades of red, with arrows indicating difference in use between periods (upper arrow) and riparian vegetation in stronger shades of red at low terrace level (lower arrow) (Rapideye image, false-color composition 5-4-3 RGB)

The terraces occur discontinuously along the valley, and in the sections with strong lithological and structural control, terraces do not occur. In the crystalline basement, up to two levels of terraces occur: [i] the terraces of the pre-littoral section, associated directly with the general base level of the hydrographic basin, the Atlantic Ocean; [ii] the terraces of the inland crystalline sector are associated to the regional base levels as the anticlinals and synclinals of the Itapicuru Greenstone Belt. In the sedimentary sector, the terraces present up to four levels (Fig. 3) and have relations with the Inhambupe fault system, which limits the sedimentary basin with the precoastal crystalline sector.

The oldest terraces (T4 and T3) are positioned about 30 and 20 m, respectively, above the current river level. They are terraces whose surface cover material presented a muddy sand and sand texture, with absence of stratification (Fig. 5). These levels presented OSL ages of ~ 22,100 years BP (T4) and ~ 9,800 years BP (T3), and are currently in a dissection process. The T2 level occurs only in two sectors. The spatial

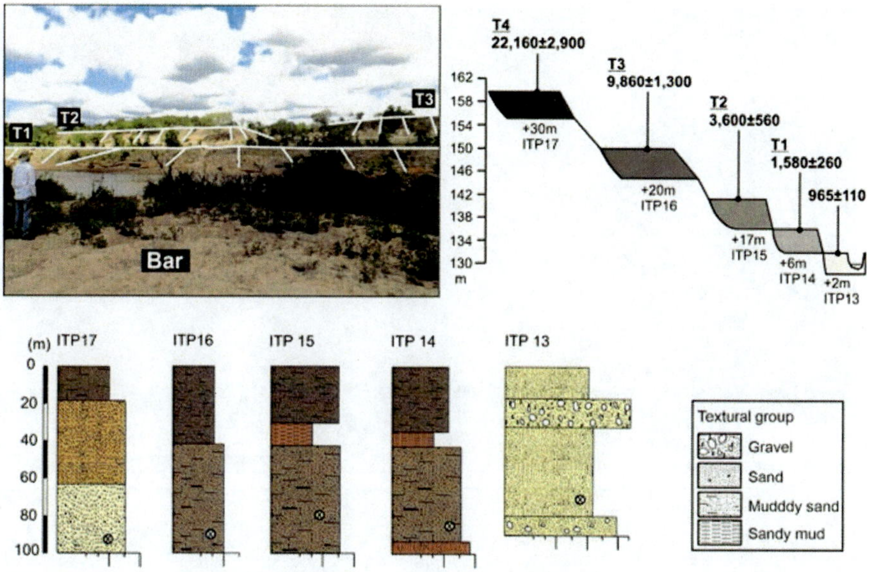

Fig. 5 Representative reach of the Itapicuru River in the sedimentary sector with four terrace levels and sidebar in the floodplain, with OSL ages chronologically consistent with topographic position

pattern identified in the mapping demonstrated that in the sedimentary sector, the T2 levels constitute the most preserved surfaces of the terraces. They present a muddy sand texture with sandy mud texture intercalations. The ages obtained are positioned between ~ 3,000 and ~ 2,000 years BP. In the pre-littoral crystalline sector, T2 level is remnant, occurring in two fragments with age ~ 8,000 years BP and muddy sand texture. The T1 level occurs in all sectors, being narrower and with ages around ~ 2,000 and ~ 800 years BP. They show predominantly muddy sand texture along with the profiles, interspersed by sandy mud or muddy mud texture units. Paleochannels of drainage occurs on the surface of the terraces (Fig. 6), indicating changes in the hydrological pattern in this time interval.

Alluvial plains occur discontinuously along the Itapicuru in the form of pockets. However, they are well developed in parts of the sedimentary sector, where they present features such as dykes, lateral bars, and abandoned channels (Fig. 4), indicative of the current dynamics of the Itapicuru. Dike deposits evidence of sandy-textured allostratigraphic units interbedded with muddy sandy-textured units whose OSL ages are ~ 570 years BP. The lateral bars are composed of sandy deposits interspersed with gravelly units and clayey layers and are representative of deposition in the bar itself or of bedrock that migrated by avulsion. They indicate the lateral migration of the Itapicuru River by oscillations in river discharge over the last 800 years, according to the OSL ages.

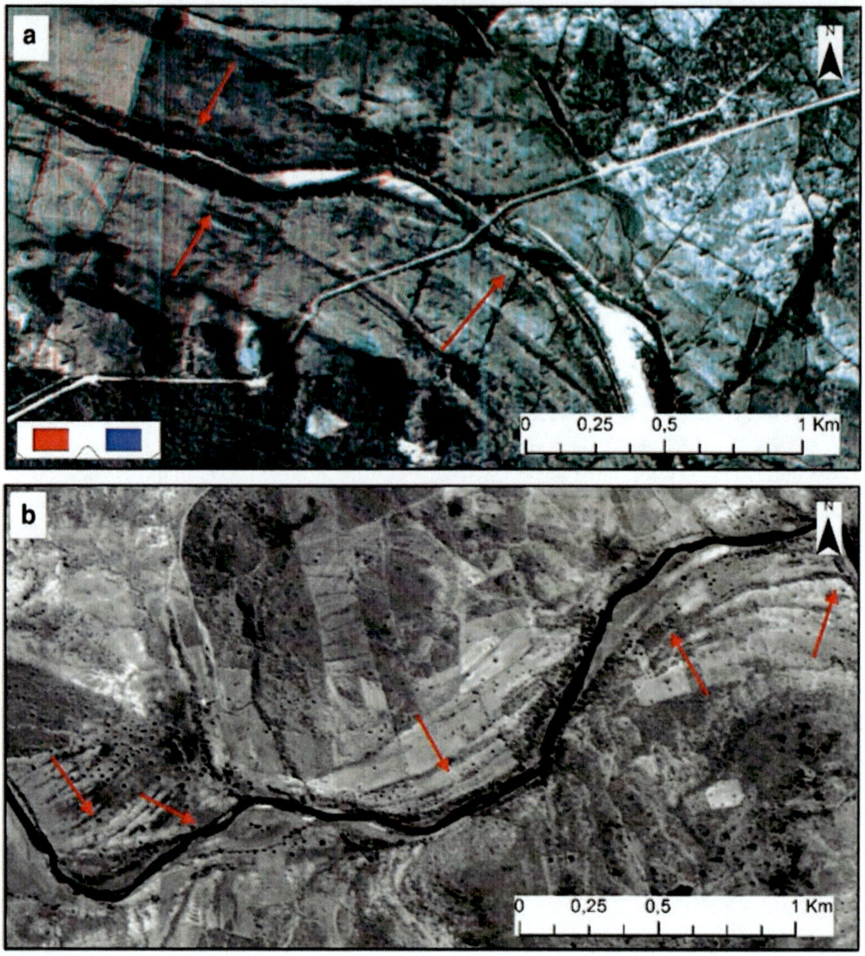

Fig. 6 Sections of the Itapicuru River with indications of lateral migration dynamics in the last 2000 years: paleochannels with preferential direction of river migration at T1 levels; and lateral bars in the floodplain, identified in stereoscopic pairs of aerial photographs (**a**); paleochannels at T1 levels and abandoned beds in the floodplain with preferential direction of migration, identified in Cbers-2B images (**b**)

5 Jacaré-Guaçu and Itapicuru: Holocene Evolution

The identification of the forms by the use of the different techniques allowed, in plant, the recognition of landscape patterns by the images and the connection of the forms to the processes, with previous understanding of the functioning of the alluvial valley in these different environments.

The constituent materials of the T1 level indicate a change in the energy pattern of the Jacaré-Guaçu River, associated with possible lateral migration, when the level becomes filled by overbank deposits, such as the clays and silt that dominate the surface horizons, and the chemical alteration of organic matter into peat in depressed areas of the former floodplain (Corrêa et al. 2016). Deposits correlated to periods of higher transport energy (blocks and pebbles), present in the former riverbeds at the low terrace level, would be linked to the river incision process and the abandonment of the former floodplain. The absence of sedimentary structures in the overlying sandy package may indicate the rapid transformation of deposits into soils by chemical and biological alteration in warm and humid environments (Celarino and Ladeira 2017), contrasting with the Itapicuru River profiles.

The forms arranged in the river plain are also uncharacterized by climatic conditions, as well as by flood flows and new conditions of use, which resignify the hydrological role of the river plain. Added to this is the reduction in sinuosity in the last 50 years, contrasting with the morphology of the abandoned channels at low terrace level, which are essentially sandy and less sinuous. Both reductions in sinuosity indicate a change in the processes and behavior of the river, with emphasis on the passage from an intermediate sinuosity to a smaller one in the lower reaches, and are also interpreted as a consequence of the self-organization of the river in relation to the new energy balance (Langbein and Leopold 1970; Timár 2003).

The absolute ages of the T1 low terrace level reinforce the transition of environments around ~ 8,000 years BP, as reinforced by another dating (OSL and ^{14}C) at the same plateau in the area (Valezio 2016; Cheliz and Gianinni 2020). The shape mapping and geochronology of these low terrace cover materials combined indicated the periods of river incision and lateral migration throughout the Holocene. As present in Fig. 3D, in the dating of the distal floodplain (not yet properly refined) and verified in the field, there is the possibility, in the recent period, of a new incision of the Jacaré-Guaçu riverbed and transformation of the present extensive floodplain into a new low-level fluvial terrace.

The paleochannels still preserved were identified by the orbital and non-orbital images, and those of the Jacaré-Guaçu River are morphologically and morphometrically different from the current ones, pointing to different hydraulic conditions. In addition, the asymmetry identified in the Itapicuru River and Jacaré-Guaçu River would be linked to processes of tectonic order, capable of influencing the erosion or pleasanting of the alluvial valley (Leeder and Alexander 1987; Latrubesse and Kalicki 2002; Kane et al. 2010), such as the presence of low terraces preponderantly on one of the banks.

In the Itapicuru River, the variation in the textural groups of the terraces indicated changes in the fluvial energy pattern over time achieved by absolute ages (Lima 2017; Lima et al. 2021). However, more significant variations were identified only in the recent deposits, which correspond to the marginal dykes and lateral bars. According to Tricart (1958) and Tricart and Silva (1968), significant changes occurred in the regional climatic pattern during the Holocene, which contributed to the intense deposition of coarse material at the bottom of the Itapicuru valley in the dry phases. Abandonment of the floodplains through channel incision would have occurred during the wet phases. The OSL ages obtained in this research demonstrated that the T2 and T1 levels were elaborated in a more recent time period than the one previously proposed, that is, during the Upper Holocene. Regional paleoclimatic models pointed out the current condition of semiaridity established in the last 4000 years (De Oliveira et al. 1999; Auler et al. 2004; Novello et al. 2012), but with the occurrence of small humid intervals observed mainly in the higher sectors of the Chapada Diamantina.

Thus, it is believed that the formation of the older terrace levels, T4 and T3, may have been triggered by changes in regional climatic conditions, as highlighted above. The levels of T2 and T1, in turn, may have been elaborated under conditions similar to the present ones; however, oscillations in river discharge as a result of increased precipitation upstream of the Itapicuru at decadal intervals would have been responsible for the intense lateral migrations of the river, as visualized in the orbital products. During regional wet events of the last two thousand years (Novello et al. 2012), it is possible that river dynamics were characterized by vertical incision simultaneously occurring with lateral migrations (Tofelde et al. 2019) at short time intervals (Limaye and Lamb 2016).

Over the last 800 years, lateral migration of the Itapicuru would have predominated until the current period, where evidence of this dynamic has been observed in the deposits of the sidebars and in the mapping carried out. However, this dynamic occurs in a spatially restricted manner as the current floodplain is narrow and discontinuous along the channel (Lima 2017). In several sectors, the T1 levels are in an advanced stage of lateral erosion with undermining at the base of the terraces and flooding of the bed, as a result of current anthropic interventions that favor greater vulnerability of the banks of the beds and the scarps of the T1 levels.

The techniques used to obtain the results, from aerial photographs to remote sensing, added to the dating, showed that the records of changes in the fluvial landscape of both rivers are Holocene. These characteristics identified for both rivers, given the application of the chosen methodologies, can be replicated for the study of fluvial environments in different climatic contexts, linking the identification, description, chronology, and interpretation of different levels of evolution of fluvial geomorphological systems.

6 Final Considerations

The work illustrates how the combination of different methodologies allows the interpretation of the geomorphological evolution of different river systems. The similarities of the techniques employed also opened space for adaptations of use for each type of environment, given the relationship of the forms and processes to be influenced by different dynamics, although on a long-term scale, the processes that triggered the formation of the terrace levels were based on variations in humidity and temperature. Both the morphology of the old meanders in the humid tropical sector and the river deposits in the hot and dry tropical sector demonstrate that hydrodynamic variations are still recorded in the landscape.

With different evolutionary pictures, attested by the absolute dating, the rivers responded to the extrinsic alteration factors according to the past and present regional characteristics. While the river is present in the state of São Paulo, the action of the humid tropical climate is preponderant in the configuration of the fluvial plain in the eight thousand years BP, having its last process of abandonment in the Upper Holocene, the river present in the state of Bahia has still preserved along its course four levels of low fluvial terraces, with sectors with greater tectonic imposition and with marked characteristic of the climate variations in the Holocene, having established itself more recently in hot and dry tropical climate.

Even with two fluvial units, their distinct environmental configurations, and complexity of geomorphological responses to different regional stressors over time, we can analyze and discuss them from the perspective of forms and processes. Semi-arid environments arranged in the same territory of humid tropical climates, as is the case of Brazil, reveal the potentiality of studies of integrated river systems still to be explored.

The fusion of techniques for better apprehension of the landscape should still be expanded with the arrival of new methodologies. The popularization and cheapening of more advanced procedures, without leaving aside established techniques, are and will be crucial for research to be further refined and deepened for the interpretation of these increasingly complex natural systems.

Acknowledgements We thank the São Paulo State Research Support Foundation (FAPESP), processes 2016/24390-0 and 2012/00145-6, and the National Council for Scientific Development (CNPq), process 408333/2013-8, for funding and enabling the research.

References

Auler AS, Wang X, Edwards RL, Cheng H, Cristalli PS, Smart PL, Richards DA (2004) Quaternary ecological and geomorphic changes associated with rainfall events in presently semi-arid northeastern Brazil. Journal of Quaternary Science 19(7):693–701.

Bisson M, Piccinini S, Zanchetta GA (2011) Multidisciplinary GIS-based approach for mapping paleoriver migration: A case study of the Serchio River (Lucca Alluvial Plain, Tuscany). GIScience & Remote Sensing 48(4):566–582.

Celarino ALS, Ladeira FSB (2017). How fast are soil-forming processes in Quaternary sediments of a tropical floodplain? A case study in southeast Brazil. Catena 15:263–280.

Celarino ALS, de Souza MM, Ladeira FSB, Branco FSRT (2013). Paleoenvironmental reconstruction of the lower Mogi Guaçu River basin (São Paulo state-Brazil), morphopedosedimentary records and fluvial processes. Catena 111:80–97.

Cheliz PM, Giannini PC (2020). Geomorphological, pedological and environmental transformations in the Lower Terraces of the Jacaré-Guaçu River (archaeological site Boa Esperança II, São Paulo) in the Late Pleistocene and Holocene. Revista Brasileira de Geografia Física 13(5).

Corrêa ICS, Weschenfelder J, Toldo Junior EE, Baitelli R, dos Santos-Fische CB (2016). Genesis of a Peat Deposit on the Continental Shelf of Rio Grande do Sul-Brazil. Quaternary and Environmental Geosciences 6(2).

Costa ALC. (2005) Estudo de vulnerabilidade à erosão com a aplicação da Equação Universal de Perda de Solo na Alta Bacia Hidrográfica do rio Jacaré-Pepira, utilizando GIS/SPRING. (Master's Dissertation), Institute of Geosciences and Exact Sciences, Unesp, Rio Claro, SP.

De Oliveira PE, Barreto AMF, Suguio K (1999) Late Pleistocene/Holocene climatic and vegetational history of the Brazilian caatinga: the fossil dunes of the middle São Francisco River. Paleogeography, Paleoclimatology, Paleoecology 152:319–337.

Hamilton SK, Kellndorfer J, Lehner B, Tobler M (2007) Remote sensing of floodplain geomorphology as a surrogate for biodiversity in a tropical river system (Madre de Dios, Peru). Geomorphology 89(1–2):23–38.

Hughes K, Croke J, Bartley R, Thompson C, Sharma A (2015) Alluvial terrace preservation in the Wet Tropics, northeast Queensland, Australia. Geomorphology 248: 311–326.

IPT—Instituto de Pesquisas Tecnológicas (2003). Diagnosis of the current situation of hydric resources and establishment of technical guidelines for the elaboration of the Tietê/Jacaré Hydrographic Basin Plan. Report 40.674.

Kane IA, Catterall V, Mccaffrey WD, Martinsen OJ (2010) Submarine channel response to intrabasinal tectonics: The influence of lateral tilt. AAPG Bulletin 94(2):189–219.

Kosin, M., Angelim, L.A.A., Souza, J.D., Guimaraes, J.T., Teixeira, R.L., Martins, A.A.M., Bento, R.V., Santos, R.A., Vasconcelos, A.M., 2004. Carta Do Brasil Ao Milionésimo, Folha Aracaju SC, vol. 24. CPRM, Brasília, 2004.

Langbein WB, Leopold LB (1970) River meanders and the theory of minimum variance. Rivers and river terraces. Palgrave Macmillan, London.

Larson PH, Dorn RI, Skotnicki SJ, Seong YB, Jeong A, Deponty J (2020) Impact of drainage integration on basin geomorphology and landform evolution: A case study along the Salt and Verde rivers, Sonoran Desert, USA. Geomorphology 371:107439.

Latrubesse EM, Kalicki T (2002) Late Quaternary palaeohydrological changes in the Upper Purus basin, southwestern amazonia, Brazil. Zeitschrift fur Geomorphologie, 129:41–59.

Leeder MR, Alexander J (1987) The origin and tectonic. Sedimentology 34:217–226.

Lima KC (2017) The Upper Holocene in the Itapicuru river basin (Bahia/Brazil): proposition of a scenario for the lower course and coastal plain. (Doctoral Thesis). Instituto de Geociências, Universidade Estadual de Campinas, Campinas, SP, Brazil.

Lima KC, Lupinacci, CM (2019) Cartography of anthropic changes in semi-arid drainage networks: possibilities for assessment using Google EarthTM imagery. Ar@cne 23:1–18.

Lima KC, Perez Filho A, Lupinacci CM, Valezio EV, Góes LM (2021) Fluvial responses to external and internal forcing: Upper Holocene dynamics in a low latitude semi-arid region in South America. Journal of South American Earth Sciences 112:103545.

Limaye AB, Lamb MP (2016) Numerical model predictions of autogenic fluvial terraces and comparison to climate change expectations. Journal of Geophysical Research: Earth Surface 121(3):512–544.

Molliex S, Jouet G, Blard PH, Moreau J, Demartini J, Storms JE, Vella C, Team A (2021) Quaternary evolution of the Golo river alluvial plain (NE Corsica, France). Quaternary Geochronology 61:101115.

Morais ES, Cremon EH, Santos ML, Rocha PC (2020) Late Pleistocene-Holocene landscape evolution in the lower Peixe River, Brazil: A meandering river valley. Journal of South American Earth Sciences 102:102664.

Murray A, Wintle AG (2000) Luminescence dating of quartz using an improved single-aliquot regenerative-dose protocol. Radiation Measurements 32:57–73.

Norton KP, Schlunegger F, Litty C (2016) On the potential for regolith control of fluvial terrace formation in semi-arid escarpments. Earth Surface Dynamics 4:147–157.

Novello VF, Cruz FW, Karmann I, Burns SJ, Stríkis NM, Vuille M, Cheng H, Lawrence Edwards R, Santos RV, Frigo E, Barreto EA (2012) Multidecadal climate variability in Brazil's Northeast during the last 3000 years based on speleothem isotope records. Geophysical Research Letters 39(23).

Piégay H, Arnaud F, Belletti B, Bertrand M, Bizzi S, Carbonneau P, Dufour S, Liébault F, Ruiz-Villanueva V, Slater L (2020) Remotely sensed rivers in the Anthropocene: State of the art and prospects. Earth Surface Processes and Landforms 45(1):157–188.
Riccomini, C (1997) Arcabouço estrutural e aspectos do tectonismo generador e deformador da Bacia Bauru no Estado de São Paulo. Revista Brasileira de Geociências 27(2):153–162.
SEI—Superintendência de Estudos Econômicos e Sociais da Bahia (1999). Balanço hídrico do estado da Bahia. Salvador.
Sridhar A (2007). Mid-late Holocene hydrological changes in the Mahi River, arid western India. Geomorphology 88(3–4):285–297.
Straffin E, Blum M, Colls A, Stokes S (1999) Alluvial stratigraphy of the Loire and Arroux rivers (Burgundy, France). Quaternaire 10(4):271–282.
Timár G. (2003). Controls on channel sinuosity changes: a case study of the Tisza River, the Great Hungarian Plain. Quaternary Science Reviews 22(20):2199–2207.
Tofelde S, Savi S, Wickert AD, Bufe A, Schildgen TF (2019) Alluvial channel response to environmental perturbations: fill-terrace formation and sediment-signal disruption. Earth Surface Dynamics, 7(2):609–631.
Tricart J (1958). Some geomorphological problems of Bahia. Notícia geomorfológica, Campinas, 1(2):27-32.
Tricart J, Silva TC. (1968). Estudos de geomorfologia da Bahia e Sergipe. Salvador: Foundation for the Development of Science of Bahia.
Valezio EV (2016). Balance in geomorphology: geosystems, floodplains and morphodynamics of the Jacaré-Pepira and Jacaré-Guaçu rivers (SP). (Master's Dissertation). Institute of Geosciences, State University of Campinas, Campinas, SP, Brazil.
Wintle AG, Murray AS (2006) A review of quartz optically stimulated luminescence characteristics and their relevance in single-aliquot regeneration dating protocols. Radiation Measurements 41:369–391.

Hydrogeomorphology of Brazilian Springs: Between Diversity and Lack of Knowledge

Mirella Nazareth de Moura and Miguel Fernandes Felippe

Abstract The springs are understood as complex environmental systems, in which hydrogeomorphological processes are engendered in an inter-scalar manner, under the influence of regional factors (such as climate and geology) and local factors (such as the depth of the alteration mantle and the position of rocky outcrops). The uniqueness of the springs from an ecological, hydrological, and social point of view has drawn the attention of researchers from various fields of knowledge, who are unanimous about the urgency of establishing protection actions for the springs, both globally and associated with the environmental policies of each country. However, a major obstacle on this path is the lack of methodologies for recognition and evaluation of springs that do not obliterate their physiographic and physiological diversity. Based on the continental dimensions and landscape variability of Brazil, this chapter brings to light the problem of classifying the springs, considered one of the first steps toward establishing management and conservation actions for these systems.

1 The Springs as Complex Environmental Systems

The academic literature is consistent in defending the importance of springs. These complex systems are noteworthy, since they are configured as unique and heterogeneous environments, endowed with functions and processes not only geomorphological and hydrological but also ecological and social (Valente and Gomes 2005; Springer and Stevens 2009; Felippe and Magalhães 2014; Moura 2020). Felippe and Magalhães (2009) highlight the importance of springs for society since rainwater is ephemeral, consequently, it falls on the perennial springs (fed by baseflow) the task of maintaining the flows of rivers and streams, even in dry seasons.

M. N. de Moura (✉)
Federal University of Minas Gerais, Belo Horizonte, Brazil
e-mail: mirellanm@ufmg.br

M. F. Felippe
Federal University of Juiz de Fora, Juiz de Fora, Brazil
e-mail: miguel.felippe@ich.ufjf.br

G. Barbosa dos Santos et al. (eds.), *Geomorphology of Brazil: Complexity, Interscale and Landscape*, Springer Proceedings in Earth and Environmental Sciences,
https://doi.org/10.1007/978-3-031-05178-4_6

However, such relevance has not been manifested in significant advances in the management of springs, so that there is an insufficient regulatory framework for the protection of springs in Brazil (Carmo et al. 2014). Added to this is the fact that, worldwide, research that takes them as objects of study is still scarce (Springer and Stevens 2009), despite being urgent (Stevens et al. 2021; Cantonati et al. 2020). Therefore, several topics for understanding and protecting springs are unclear and unanswered, not only by the complexity of this problem but also by the lack of theoretical and methodological bases. Considering the global threat to spring ecosystems illustrated by Stevens et al. (2021), the reality of Brazilian springs is no exception to the rule.

Brazil is worldwide renowned for its apparent water abundance. The climatic characteristics of most of its territory result in an imposing discharge of freshwater to river systems and, consequently, a robust water production; except in the semi-arid region of the Northeast, which occupies approximately 12% of the country (Sant'anna Neto et al. 2015). These climatic characteristics are consistent with a wide variety of landscapes, from the subtropical hills of the South to the dense equatorial forests of the Amazon, to the forest-covered mountain systems of the Atlantic façade and the semi-humid savannas of Central Brazil. The country is home to some of the largest rivers in the world and aquifers of great continental importance (such as the Guarani and Alter do Chão systems), with a relevant discharge of underground water.

In this context, it is estimated that the density of springs in the territory varies considerably, with works showing from 1.9 springs per km^2, in the semi-humid karstic depression, to more than 28 springs per km^2 on the Atlantic plateau.

However, little is known about the springs in Brazil. With continental dimensions, much of the territory lacks systematized information, especially in the interior of the country. Furthermore, governmental efforts at cataloging, mapping, and monitoring are incipient and disparate. Of the few studies already published that take the springs as objects of investigation, most deprive themselves of a more robust and sophisticated discussion, dodging the integration of them with the landscape that shelters them. The most recurrent theme is the quality of the springs, usually linked to the conformity of the permanent preservation areas and the contamination of their waters. Almost always, the springs are understood in a limited way, without recognizing the physiographic and physiological diversity of these systems. In the field of geomorphology, the aridity of investigations is even more notorious.

It is at this conjuncture that an international mobilization for the protection of springs has begun. Researchers from different countries have raised an urgent plea to incorporate springs as objects of cross-cutting studies between ecology, geology, and geography in order to build the necessary foundations for the efficient management of these systems. In summary, it can be said that on a global scale, there are major gaps in water exploitation policies, the mapping of springs is inadequate, the assessment protocols are insufficient and inconsistent, leading to fragile inventories on the springs, even in countries with tradition in research on the subject (Gerecke et al. 2011; Cantonati et al. 2020; Stevens et al. 2021). In the Brazilian context, to all these weaknesses can be added the mismatch between public policies on the environment and water resources, the lack of investment in basic research (especially in the socio-environmental area), the great socio-cultural and physical-geographic diversity of the territory, the regional inequality of knowledge about the springs and a shallow and mistaken idea of water abundance (Magalhães and Felippe 2012).

However, in spite of the unfavorable situation, Brazilian researchers are making commendable efforts that are gradually contributing to the advancement of the study of springs. It is already known that the Brazilian springs (Fig. 1) are mostly small discharges and configured hydroecologically as helocrene and reocrene, fed by local

Fig. 1 Diversity of Brazilian springs: **a** healthy wetland spring—Pantanal; **b** karstic spring used for agriculture and livestock—Semi-humid Central Brazil; **c** limnocrene spring under pressure of livestock—Semi-humid Central Brazil; **d** small helocrene spring—Atlantic Tropical Brazil; **e** pristine piping spring in protected area—Atlantic Tropical Brazil; **f** helocrene spring—Subtropical Brazilian Coastline. Photos: Miguel F. Felippe

underground flows, and vary in their perenniality according to multiple climatic contexts (Felippe and Magalhães 2014; Carvalho et al. 2015). A very strong indication of the importance of the rainfall regime for the hydrological dynamics of the springs is that most Brazilian springs have water between 12 and 60 years of return period, configuring themselves as modern waters (Felippe 2013).

Even though the Brazilian Forest Code mandates a 50 m radius of permanent preservation area for springs throughout the country, this is recurrently ignored. Studies in rural areas show that over 70% of springs are of low environmental quality (Gonzalez and Schiavinato 2019; Rezende and Luca 2017). An example of source degradation is that of Belo Horizonte (Brazil's sixth-largest city), where even urban parks report 35% of springs with moderate or worse integrity, and 24% with the presence of Salmonella sp. in the waters (Magalhães and Felippe 2012). The main pressures on Brazilian springs result from urban expansion, livestock and agricultural use, mining, and deforestation. However, the biggest challenge regarding the protection of springs in Brazil is the awareness of their diversity, functioning, and importance.

Part of these challenges is based on the difficulty of understanding the springs as a systemic whole that goes beyond the water that drains into them. In both technical-scientific and popular circles, the centrality of the "production" of water from the springs means that other elements that are extremely important for their functioning are obliterated. However, the spring is more than the water that flows from it, it is a whole "environmental system in which the upwelling of groundwater naturally occurs temporarily or perennially, and whose hydrological flows in the surface phase are integrated into the drainage network" (Felippe and Magalhães 2013). They thus engender a morphological subsystem and a hydrological subsystem that support the life (ecosystem) of the springs (Fig. 2).

Moreover, the permanent input and output of matter and energy of the spring, promotes new dynamics between the elements of the system, which brings up the debate of complexity in the understanding of the springs (Moura 2020). In the understanding of Morin (2015), any situation within a system can occur, as it may simply not occur, since complex systems are endowed with full autonomy, where the frequent instability creates opportunities for movement in the system, generating new forms of behavior. They are able to create opportunities for actions, interrelationships, and recursions, producing new ways for subjects to relate, subjectivating them, and producing new modes of existence (Alves and Seminotti 2006).

It is assumed here that the unstable and autonomous behavior of the spring system is configured as disorderly, unpredictable, unbalanced, and even chaotic, like any complex system (Moura 2020). For this reason, linear visions based on the balance of matter (water) are fragile and insufficient to understand the springs.

The Brazilian context, of great environmental and social diversity in its territory, only reinforces this assertion. Obviously, the springs in the Pantanal, on the thick Quaternary sedimentary packages in a sub-humid climate, will be different from those on the karstic terrains under the hot semi-humid climate of the São Francisco Depression. Therefore, we advocate the indispensable need to know the physiography and physiology of the springs, contextualize them in their regional environment, and

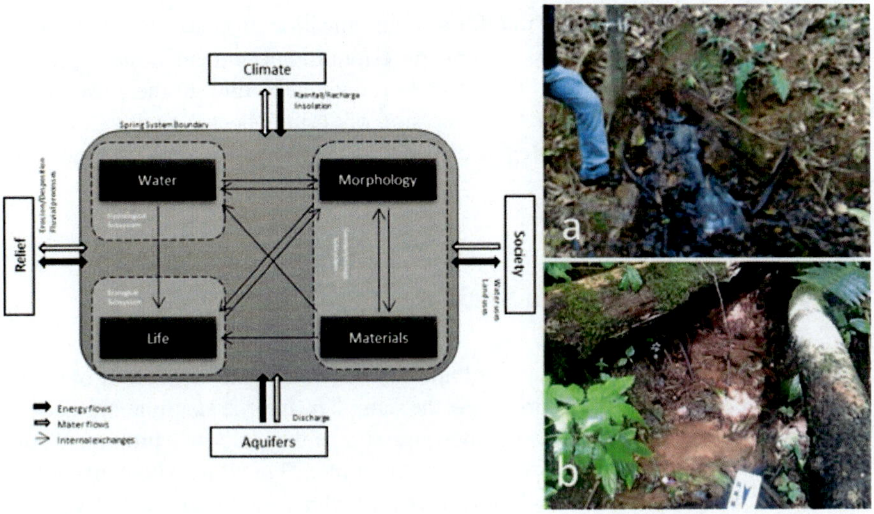

Fig. 2 Representation of the spring system based on Felippe (2013). **a** Point spring in piping, located in the Quadrilátero Ferrífero Mountains, in Belo Horizonte, MG. **b** Point spring in erosion channel, located in Serra dos Órgãos, Petrópolis, RJ. Photos: Miguel Felipe

understand the dynamics of the inter-scalar processes that constitute them, in order to identify the human pressures, they suffer and thus think about effective protection of these systems. One of the most important scientific artifices used for this is the systematization of heterogeneous elements into groups or classes of similarities.

The international literature traces back to Bryan's efforts (1919) the first broader attempt to systematize springs. For almost a century, this hydrogeological view persisted as the main key to classifying springs in the world, until Springer and Stevens (2009) compiled efforts made for years in springs in the southwestern United States. From this work, a classification based on the hydro-ecology of the springs was disseminated, becoming widely used in the international literature. For the Brazilian context, based on empirical data combined with fuzzy statistics, Felippe (2009) presents a systematization of the hydrogeomorphology of the springs in a new classification.

Thus, we are faced with two major problems: The urgency of systematization of knowledge about springs and the many typologies of springs that already exist. This leads us to a major question: which typology of spring, or rather, which classification key is the most appropriate for studying Brazil's springs?

Therefore, in this work, it was seen as very appropriate to make a comparison between the three types of springs, since, in the technical field, knowing well the springs, their characteristics, and classifications, are important tools for better management and environmental planning. From an academic perspective, this comparison allows the understanding of the different ways of systematizing knowledge about the springs, giving the researcher the opportunity to learn about the different theoretical and methodological spheres covered by the spring system.

However, it should be noted that for any classification proposals, knowledge of a spring comes first from its physiographic description. Subsequently, the application of a typology of springs allows them to be grouped according to their similarities. Such an artifice is essential for the protection of springs because with an efficient classification, management strategies can be drawn up to ensure their environmental protection.

2 Recognizing Brazilian Springs

The main contribution that geomorphological science can make to the study of springs in Brazil is to show that not all springs are the same. The inter-scalar relations between the elements of the vertical and horizontal structure of the landscape promote distinct hydrogeomorphological processes in space and time. Thus, it is to be expected that the set of springs in the South Amazonian crystalline depression, for example, is different (in hydrogeological and hydrogeomorphological terms, but also in ecological, economic, and socio-cultural terms) from the springs in the sedimentary plateaus of Central Brazil (as shown in Fig. 1).

However, within the same landscape unit, springs can occur under different physiographies due to local factors. For this reason, it is relatively common in a field reconnaissance to identify springs with completely different hydrogeomorphological and ecological aspects, even if they are spatially close. Added to this is the lack of understanding of the natural evolution of the springs (from the development of the drainage network, for example), which can cause two "neighboring" springs to be in different temporal stages of evolution (under the logic of homeostasis of the environmental system).

As seen in Fig. 2, the influences of the external elements on the spring system provide matter and energy to be worked on in the internal subsystems. Furthermore, regional factors are determinants in understanding the springs, although the local scale cannot be ignored. Thus, free and granular aquifers have a different influence from fissured and deep aquifers. More or less friable rocks tend to produce more or less thick alteration mantles, which is another extremely relevant factor. In addition, climate, together with baseflow, will be determinant for the flow of springs, not only due to the occurrence of specific precipitation events but especially due to the seasonality of the rainy season and its relevance in the recharge of aquifers. Finally, relief and society come together to direct water flows and control the distribution of water within the hydrological cycle (Felippe and Magalhães 2014).

In Brazil, the climatic factor is preponderant not only in feeding the springs but also in acting together with the geology to shape the alteration layers. Most of the springs are partially or totally supplied by the water contained in the soil. In less wavy reliefs this becomes evident since the surface coverings are over 100 m deep. On the other hand, environments with shallow soils tend to shelter temporary springs, the perennial ones being conditioned to the outcropping of fractured rocks. Surface excavation associated with overland flow is another relevant element, causing springs to occur in erosive gullies and ravines, especially on more undulating slopes (Felippe and Magalhães 2014). All this alternates regionally in dialogue with the vertical structure of the Brazilian landscapes.

This complexity associated with the physiographic diversity of the springs only reinforces the need for systematization studies to advance knowledge about these systems in Brazil. However, before characterizing and classifying the springs according to the three typologies chosen for this study, it is quite appropriate to present a brief description of the classification keys to be worked on, as well as where they were applied.

2.1 Classification Keys

As previously stated, three proposals for classification of springs are put in dialogue: Bryan (1919), Springer and Stevens (2009), and Felippe (2009).

In general terms, it should be taken into account that the construction of Bryan's (1919) typology of springs is based on two factors: the origin of the water and the structure of the rock that brings it to the surface, and one can notice strongly present and defined hydrogeological highlights throughout its characterization.

For Springer and Stevens (2009) the criteria used include geomorphic considerations, forces that emerge the water to the surface, flow properties, habitats, spring biota, management, and use of springs. That is, the authors focus on the ecology and manifestation of water, fostering, with a high degree of subjectivity, its typology.

Felippe (2009), on the other hand, based on qualitative empirical data, developed a typology of springs that would enable the use of the groupings created in different environmental dynamics, clearly emphasizing in its classification, a hydrogeomorphological nature. For him, the sample universe will define the groupings of springs (inductive reasoning), and the same spring can fit into more than one class (fuzzy logic).

2.2 Bryan's Classification (1919)

Bryan's (1919) typology of springs are divided into two types: *deepwater springs* and *shallowwater springs*. Deepwater springs are subdivided into *volcanic springs* and *fracture springs*.

The volcanic springs are associated with current or past volcanism and originate from water expelled from the action of the underlying magma or surface water in contact with very heated rocks. The fracture springs, on the other hand, generally have a strong and constant flow, without annual oscillations, high temperature, and are often well mineralized.

As for shallow water springs, Bryan (1919) subdivides them into more categories. The first is that of the *springs in depression* (or porous rocks), formed where the saturation zone reaches the soil surface. They have a smooth flow, being slowly and continuously replenished, normally in the form of swamps. The depressed springs are divided into four categories, according to their topographic position: *dimple springs,; valley; channels; on the edge* (slope breaks).

The *contact springs*, on the other hand, are those whose porous rocks overlap the impermeable material, directing the water to the surface through the stratigraphic contact. The shape and altitude of the surfaces adjacent to this impermeable material determine the subtypes when the contact will be more *regular and horizontal, inclined regular* or *irregular*.

Artesian water has water contained in the pore spaces of a permeable bed, situated between impermeable strata. Within this category are listed *Plunging springs*, in bedded rocks, inclined and eroded in such a way that the porous bed receives water from the rain or from channels in its upper end (the lower end remains exposed on the surface); *siphoned springs*, originating from folds, where a porous stratum constitutes itself as an inverted siphon for the transport of water; *springs without an impermeable layer*, which occur in unconsolidated deposits, where the porous material is exposed, so as to receive water at a high level and discharge this water at a lower level; *fracture springs,* which do not depend on the outcrop of the saturated porous bed in its lower portion for the exfiltration of water, but rather on fractures, openings that carry the water to the surface.

2.3 Springer and Stevens' Classification (2009)

Springer and Stevens (2009) list 12 types of springs based on hydroecological aspects. *Cave springs* are characterized by the exfiltration of water in underground cavities of a well-developed karstic system. Also associated with karstic systems, *exposed springs* correspond to where the water can seep out of caves, rock shelter fractures, or drains, where unconfined aquifers are exposed close to the ground surface.

The *fountain springs*, on the other hand, are configured as artesian springs with pressurized CO_2 in a confined aquifer. *Geyser springs* show an explosive flow of hot water of aquifer confinement. The *jet springs* are also punctual but show a discrete flow, which gushes from a cliff wall, originating from an unconfined and overlying aquifer.

Hanging garden springs, have drip flow generally horizontally along with a geologic contact in a cliff wall of an unconfined, perched aquifer. *Helocrene springs* emerge from low gradient wetlands, often indistinct or multiple springs, exfiltrating from shallow unconfined aquifers. The *hillside springs come* from confined or unconfined aquifers on a slope (30–60°), often showing indistinct or multiple exfiltrations. The *hipocrene ones* are found buried where the flow does not reach the surface, usually due to very low discharge and high evaporation or transpiration. *Limnocrene* springs, on the other hand, are marked by the exfiltration of water from confined or unconfined aquifers into lakes or wells. The penultimate type of spring, the hill springs, are those where the water emerges from a *mineralized hill,* often in magmatic or fault systems. Finally, the *reocrens*, encompass fluid springs, originating from one or more transmission channels.

2.4 Felippe's Classification (2009)

According to Felippe's (2009) proposal, six fundamental types of springs are contemplated, from the relationship of exfiltration with the local morphology and the hydrological feeding system.

The *phreatic springs* have diffuse exfiltration, with considerable variation of the water table and granular aquifer overlaid on a fissure aquifer, resulting in a low average annual flow. The *dynamic* ones have high energy in all their processes involved, being perennial and usually with high flow. They exfiltrate in a punctual or multiple ways, in outcrops or ducts. The *floating springs*, on the other hand, are defined by the fluctuation of the phreatic level throughout the year, promoting the longitudinal movement of the spring on the slope, normally with diffuse exfiltration resulting from contacts of the rock with the mantle. Like the phreatic springs, they are morphologically characterized by concavities and low flow.

Two classes of seasonal springs (with considerable hydrological variation between dry and wet periods) are placed. The *seasonal erosive* type is characterized by seasonal interception of the water table by erosive features, (gully slopes or drainage channel slopes). They are intermittent, punctual, and of low flow. The *seasonal hillside* springs occur on slopes with deep layers of weathering and are intermittent. Its morphology is usually in vertical or horizontal ducts, the flow is low, and the type of exfiltration is punctual.

The last type of spring is related to the *anthropogenic* ones, i.e., originated by human intervention. Identifying a spring as anthropogenic can be complex, due to the absence of information prior to the intervention carried out in that space. Thus, anthropogenic springs may possess any of the characteristics of the other types, but present anomalies in their dynamics, which would not be verified in a "natural" spring.

The main difference between Felippe's (2009) proposal is that, despite the possibility of fitting into these six standard types, it is assumed that a spring can possess characteristics of several types. With this, hybrid profiles can be created contemplated (e.g., dynamic-floating; phreatic-erosive, etc.).

2.5 The Application of the Classification Keys

For the comparison of classification keys, 13 springs were chosen as illustrative cases. All had already been duly characterized by Oliveira et al. (2013), Dias et al. (2014), Moura et al. (2016) and Moura (2020), with the survey of the following basic parameters: depth of the soil and surface coverings; types of land use and occupation in the PPA of the spring and in the contribution basin; lithology of the contribution basin of the springs; lithology of the aquifer of the spring; flow rate of the spring; slope of the first-order channel; water uses of the spring; morphology of the spring; type of exfiltration; mobility and seasonality of the springs.

The classification was based on primary data obtained in the aforementioned studies, organized and corrected according to the same system. Another essential tool for the characterization and classification of the springs was the use of geoprocessing and remote sensing, which allowed the observation of characteristics of the springs that were not always clear in the field. After this compilation, from the physiographic and physiological elements described for each spring, we observed the fundamental descriptive elements for the classification of the springs in the classification keys of Bryan (1919), Springer and Stevens (2009), and Felippe (2009).

The springs are located on the campus of the Federal University of Juiz de Fora, under Cwb climate with an average annual rainfall index of 1,572.8 mm (Machado 2010). The local relief is configured as a set of high slopes predominantly convex, configuring successive headwaters that drain to the same base level, formed by Manacás Lake. The mamelonization of the surface over thick and evolved alteration mantles is regionally common in the context of the Ribeira Belt, in the orogenic system called Mantiqueira Province (western terrain, in the limits of the Minas Gerais central-south plateau system), developed during the Brazilian-Pan African Neoproterozoic Orogeny (Heilbron et al. 2004).

The rocks that underlie the study area date back to the Neoproterozoic, in the Andrelândia Megassequence, in which banded biotite gneiss with intercalations of quartzite and sillimanite-granada-biotite gneiss stand out; sillimanite-granate-biotite gneiss with intercalations of orthopyroxene-granate-biotite gneiss; banded biotite gneiss; quartzite and calcisilic rocks (Duarte et al. 2003).

The springs on campus are fed by granular and fissure aquifer systems, with dynamic communication between the shallow granular cover, associated with interaction mantles and colluvial deposits, and the underlying fissural aquifer.

Therefore, having in question a free aquifer system (due to the water pressure on the bordering surface) and a granular-fissural system (due to the water transmission capacity), it is common to observe a water table formed by the permeability rupture between the higher hydraulic conductivity of the granular aquifer and the lower hydraulic conductivity of the fissural aquifer (Costa 2008). Thus, it is possible to observe throughflow, oscillating seasonally (a condition often observed in the UFJF springs), and baseflow associated with deeper waters of the fissural aquifer.

The occupation of the site by the university accompanied by landfills, cuts, and earthworks has promoted a typically technogenic relief, with consequent changes in the dynamics of hydrological flows, which directly influence the physiographic and physiological characteristics of the springs (Fig. 3).

In order to present the studied springs and the respective typologies in which they were classified, it is necessary to first specialize all 28 springs on campus within the perimeter, 13 of which were studied, as shown in Fig. 3, and then briefly describe these 13 studied springs.

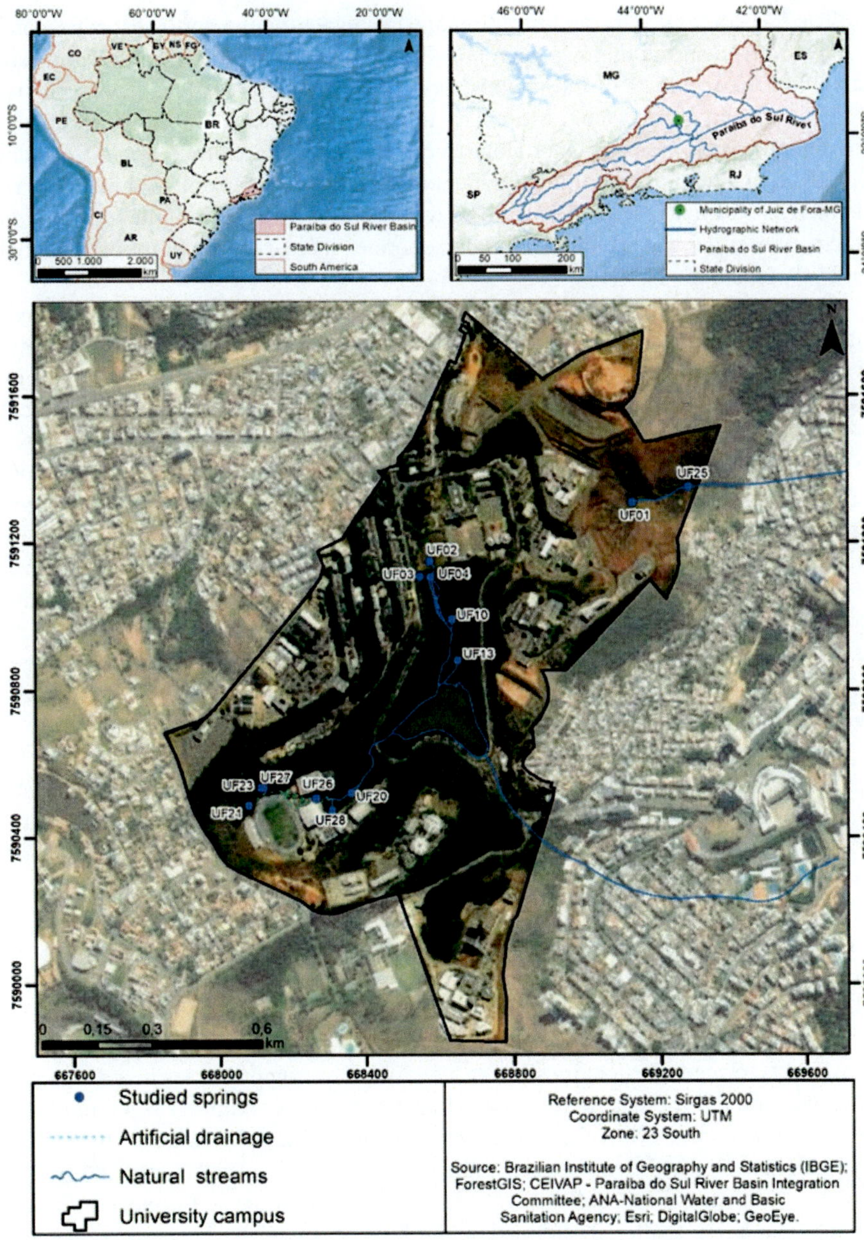

Fig. 3 Location of the springs studied on the UFJF campus. *Source* Elaborated by the authors

The UF01 spring is located in a drainage headwater on the northwest edge of the campus. It is a spring that has been heavily altered by the population for domestic supply of the residents of the neighborhood. Thus, part of its flow was channeled and dammed in a water tank. Moreover, this spring comes from the abrupt change of slope between the headwater slope and its hollow. Its exfiltration is diffuse and comes from a humid area of low gradient (a markedly swampy environment poorly defined and with a very thick mantle of weathering). It is greatly influenced by the fluctuation of the water table, which corroborates its marked mobility on the slope and low annual flows.

The UF02 spring emerges from a low gradient wetland area with a thick layer of weathering. Furthermore, it is possible to verify a seasonal interception of the water table by ravines. It is also characterized as a spring of cavity morphology, diffuse exfiltration, low flow, mobile and perennial. It is noteworthy the strong human pressure on this spring, which is located in the main recreational area on campus.

The UF03 spring is manifested on a slope by a PVC pipe after being drained upstream. Although its morphology was severely modified, thus being considered a channeled spring, its exfiltration is punctual and perennial.

The UF04 spring is located in a break in the slope, very close to the base level. Its exfiltration is diffuse and in concavity, it has low flow and is located in a humid area of low gradient and deep mantle of weathering.

The UF10 spring, on the other hand, is a point, occurring in an erosive furrow in one of the lowest points of the slope. It is intermittent and has very low flow, with notable oscillation of the water table throughout the year.

The UF13 has a channel morphology, punctual and fixed exfiltration, and as for its seasonality, it is perennial. It is a very fluid spring, originating from two or three transmission channels (depending on precipitation events). It is noteworthy that the morphology of this spring is in a channel today because it is the result of anthropic intervention.

The UF20 and UF21 springs are used for public supply. They are formed in a channel, excavated by anthropic intervention. Both are formed by water erosion, especially due to cavitation in breaks in the channel slope. The exfiltration of the UF20 is diffuse while the UF21 is punctual. Both are fixed, perennial, and of low flow.

UF23 is formed by water erosion, as well as excavation processes in the area, and channelization of the spring itself. Its exfiltration area is configured as a humid area with thick mantle and low flow. The spring is channeled; however, its channel has hydraulic contact with groundwater indistinctly.

The UF25 spring has suffered interference to facilitate water collection by the population, being artificially drained. It has high energy of hydrological processes, where it is clearly noticeable the erosion of the water downstream. Thus, its morphology can be considered a channel, with punctual, fixed, and perennial exfiltration, with high flow for the standards of the sample universe.

Finally, the UF26, UF27, and UF28 springs are located in breaks in the slope of their respective slopes. They were originated from previous excavations, linked to the installation of urban infrastructure. However, the UF27 and UF28 springs are multiple, fluid, and have a concavity morphology, while the UF26 spring is punctual, with a channel morphology, perennial, and fixed.

Table 1 shows the framework of the 13 studied springs in the respective classification keys; the relative frequency of each type is shown in Fig. 4.

Firstly, it should be noted that in none of the classification keys all types were verified. Based on Bryan's (1919) classification, the springs on campus are divided into valley and dimple springs. It is emphasized that both types are part of the shallow water springs group and the subgroup of springs in depressed areas. This homogeneity is explained by the fact that the study area is not very large, where the springs are fed by the same type of aquifer system and also over the same geological-geomorphological domain. Since the UFJF context is a large concavity, surrounded by an interfluve, the water coming from the recharge zones of the campus is already very close to the discharge area, where the base level controls not only the geomorphological base level but the phreatic level itself, which is already shallow. In this way, all the water has the same path, engendering a very specific geomorphological configuration.

However, considering the initial motivation to carry out a systematization of the springs, it does not seem very fortuitous to have as a result a grouping in so few classes. In practice, this would result in little or no difference in terms of conservation and environmental recovery strategies. Notably, for large areas and of vast

Table 1 Framework of the studied springs, according to the typologies of Bryan (1919), Springer and Stevens (2009), and Felippe (2009)

Types of Springs			
Springs	Bryan (1919)	Stevens and Springer (2009)	Felippe (2009)
UF01	Valley Spring	Helocrene	Floating
UF02	[a]	Helocrene	Erosive seasonal
UF03	Dimple spring	Hillslope	Dynamic
UF04	Valley Spring	Helocrene	Hillslope seasonal
UF10	Valley Spring	Rheocrene	Erosive seasonal
UF13	Dimple spring	Rheocrene	Phreatic
UF20	Dimple spring	Rheocrene	Phreatic
UF21	Dimple spring	Hypocrene	Anthropogenic
UF23	Dimple spring	Hypocrene	Phreatic
UF25	Dimple spring	Gushet	Dynamic
UF26	Valley Spring	Hillslope	Anthropogenic
UF27	Valley Spring	Rheocrene	Anthropogenic
UF28	Valley Spring	Rheocrene	Anthropogenic

[a] Due to the drainage of the spring to a location different from the original one, it is not feasible to perform its classification by Bryan's key (1919), since there are no safe records of the previous conditions

Source Prepared by the authors

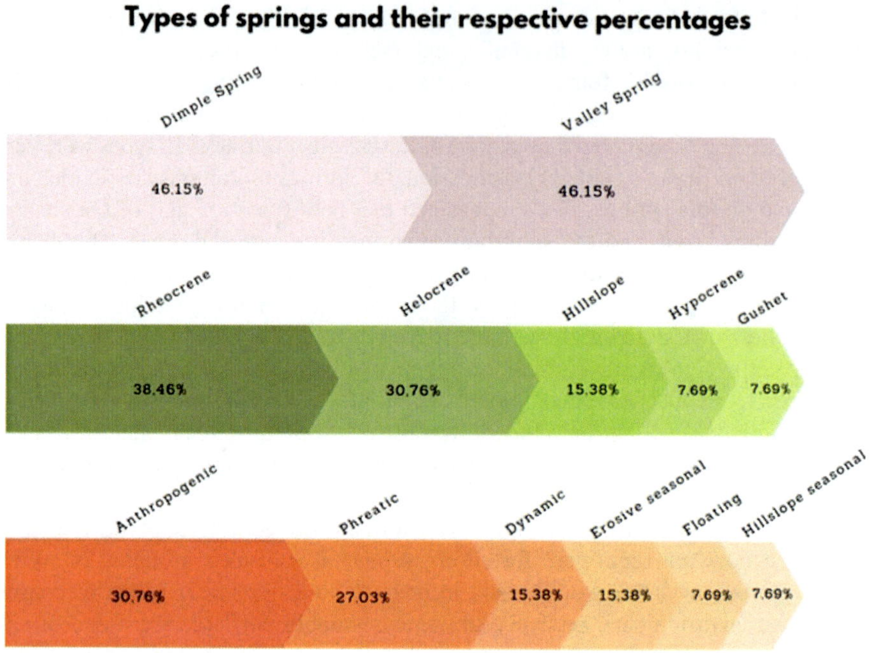

Fig. 4 Graph referring to the three classifications studied and the percentages referring to the typologies identified. *Source* Prepared by the authors

hydrogeological diversity, Bryan's (1919) classification seems useful, which does not apply to localized contexts.

Observing the classification keys of Springer and Stevens (2009) and Felipe (2009), it was possible to see a greater heterogeneity of types (Fig. 5). However, even in the face of a certain heterogeneity, when taking into account the typology of Springer and Stevens (2009), it can be observed that the Reocrena type springs (38.46%) are more recurrent while taking into account the typology of Felippe (2009), the anthropogenic springs appear more frequently (30.76%).

The recurrence of anthropogenic springs is associated with the action of topographic modification by clippings and earthworks in a context of shallow water tables. The concavity in which the campus is inserted facilitates the triggering of fast and short flows, which converge toward the Manacás Lake, causing the campus to have a water table very close to the surface. Not coincidentally, the phreatic springs were also very frequent in Felippe's classification key (2009). Precisely because they are associated with the shallow aquifer, their occurrence will be conditioned by the surface morphology. The concave segments of the slopes are coherent shelters for this type of spring. However, if there is interception by erosive channels, the tendency is for a seasonal erosive or floating spring to form, depending on the severity of the oscillation of the water table.

Fig. 5 Types of springs found for the classification keys of Bryan (1919), Springer and Steves (2009) and Felippe (2009), respectively: **a** UF01—Valley Spring/Helocrene/Floating; **b** UF03—Dimple spring/Hillslope/Dynamic; **c** UF27 Valley Spring/Rheocrene /Anthropogenic; **d** UF13 Dimple spring/Rheocrene/Phreatic. *Source* Mirella Moura

For Springer and Stevens (2009), Reocrena springs were the most common on campus. Due to the local morphology (sequence of headwaters) and the cuttings in the terrain for civil construction, the linear erosive processes are intense, engendering unrestricted furrows. As already mentioned, with the water table close to the surface, only a few centimeters of vertical incision are sufficient for water exfiltration to occur. The helocrens were also very present, having been described in situations where the topographic gradient is not conducive to the formation of channels, thus promoting diffuse exfiltration in small marshes.

It should be noted that these classifications have different parameters, so they emphasize different traits, which means that there is no correspondence and pairing between them. They are just different ways, with different attributes, to characterize a spring. Thus, much before opposing each other, the classification keys presented complement each other. The exercise carried out in a limited sample universe shows that Bryan's classification (1919) is the most limited of the three. On the other hand, both Springer and Stevens (2009) and Felippe's (2009) classification proved effective in highlighting the diversity of springs. The former, however, highlights aspects related to groundwater manifestation and may be useful for ecological studies.

The latter seems more suitable for understanding human pressures since it has the hydrogeomorphological processes as its main focus.

3 Reflections and Recommendations for Studies on the Brazilian Headwaters

The studied springs illustrate a reality far beyond that of their own spatial cutout (Fig. 6). This is because the springs are located in the context of Tropical Atlantic Brazil, with an elevated relief, a thick mantle of alteration, and a humid climate, regional conditions that are repeated over much of the territory. Another element that reinforces this aspect is the direct and indirect human influences on hydrological processes that generate disturbances in the dynamics of the springs, something recurrent in the Brazilian urban and periurban context.

The intention here is in no way to construct arguments from an irresponsible and generalist induction. However, the recurrence of the aforementioned aspects is undeniable, which can be easily recognized in the literature produced on springs in different regions of Brazil, as in Felippe and Magalhães (2014), Carmo et al. (2014), Marques and Felippe (2017), Silva et al. (2017), Gonzales and Schiavinato (2019), Pieroni et al. (2019), Schiavinato and Gonzales (2020). Thus, the types found in this study reflect the reality of springs in Tropical Atlantic Brazil. Considering the preponderance of factors related to local morphology, climatic seasonality, and the characteristics of the surface coverings, and adding the literature reports in case studies scattered throughout the country, the correspondence of the sample universe of this research with cases recognized in the Atlantic facade of southern Brazil, in the crystalline domain of Central Brazil, in the semi-humid context of the Sertanejo Residual Plateaus, among others, is notorious.

An important reflection is the demystification of the point source, perennial and of great flow, mistakenly reported by the media. Nor are the artesian and volcanic springs recurrently reported in foreign literature. Brazilian research is unanimous in agreeing that the country's springs are mostly of small magnitude, often diffuse, fed by shallow flows, and depending on the climatic context, intermittent.

In view of this, it does not seem useful for the Brazilian case to focus exacerbatedly on the hydrogeological aspects of the springs, as recommended by Bryan (1919), under the penalty of making a systematization that will reflect few classes. On the other hand, the typology of Springer and Stevens (2009) and that of Felippe (2009) present a good resolution for the commonplace of springs in Brazil. Therefore, depending on the objective of the studies one or the other (or even both) can be used.

The approach of Springer and Stevens (2009) seems to be more effective with regard to environmental restoration practices. With more precise description of the aquatic and terrestrial ecosystems that border the springs, this classification key subsidizes the choice of appropriate specimens for plant recomposition, for example.

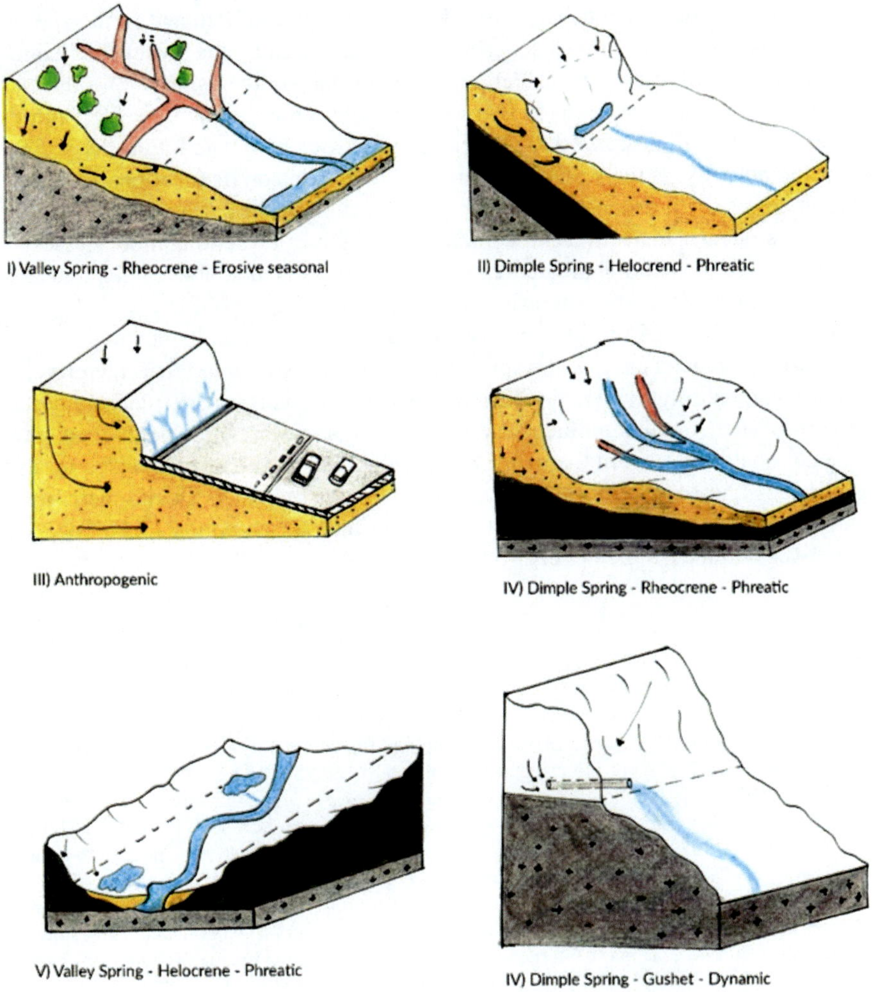

Fig. 6 Types of springs most recurrent in the Brazilian reality. *Valley spring—Rheocrene—Erosive seasonal* (springs formed by vertical incision promoted by rainfall runoff in climates of double seasonality). *Dimple spring—Helocrene—Phreatic* (springs located on predominantly concave terrain, with the formation of swamps and humid areas, from which the springs originate). *Anthropogenic* (springs whose exfiltration is promoted by anthropic action, normally as a consequence of cuts or severe changes in the terrain. *Dimple spring—Rheocrene—Phreatic* (preferably springs in concave morphologies with channels in the form of erosive furrows, normally dry, until reaching the phreatic level, where exfiltration occurs). *Valley spring—Helocrene—Phreatic* (springs located in the valley bottoms, originating from marshes and humid areas that feed the main channel). *Dimple spring—Gushet—Dynamic* (*springs* in concave relief, normally with steep slopes, promoting punctual exfiltration of concentrated subsurface flows)

On the other hand, it contributes little to the prevention of damage to the springs, since the processes are in the background, overshadowed by the biocenoses habitat.

Meanwhile, the typology of Felippe (2009) stands out. Besides being the classification key that showed the best resolution within the sample universe, with the largest number of classes, it advocates the processes associated with the springs, focusing not only on the structure of the system but also on its functionality. Therefore, it seems the most appropriate for the protection of the springs, in the sense of defining strategies to avoid their degradation. Moreover, presenting more classes has the advantage of enabling greater detailing of management actions, which require a close approach to the idiosyncrasies of each spring.

Finally, we defend the idea that beyond the choice of a classification key, it is necessary to understand that the springs are diverse systems and that, therefore, no single technical solution will be efficient in all of them. This is particularly important in view of the Brazilian political-environmental context, in which legal regulations on a national level govern the management of springs in the most diverse physical-geographic and socio-cultural contexts. The classification keys should therefore be understood as tools to better understand the springs. Based on this knowledge, any initiatives will have a greater chance of success.

References

Alves MC, Seminotti N. (2006). The small group and the complexity paradigm in Edgar Morin. *Psicologia USP*, 2(17), pp. 113–133. https://doi.org/10.1590/S0103-65642006000200006

Bryan K. (1919). Classification of Springs *Journal of Geology*, 27, pp. 522–561.

Cantonati M, Stevens LE, Segadellis S, Springer AE, Goldscheider N, Celico F, Filippini M, Ogata K, Gargini A. (2020). Ecohydrogeology: The interdisciplinary convergence needed to improve the study and stewardship of springs and other groundwater-dependent habitats, biota, and ecosystems. *Ecological indicators,* 105803 (110). https://doi.org/10.1016/j.ecolind.2019.105803

Carmo LG, Felippe MF, Magalhães Jr AP. (2014). Áreas de preservação permanente no entorno de nascentes: conflitos, lacunas e alternativas da legislação ambiental brasileira. *Boletim Goiano de Geografia,* 2(34), pp 275–293. https://doi.org/10.5216/bgg.v34i2.31733

Carvalho CM, Da Silva LA, Da Rocha NS, Cruz RC, Da Silva Borba MF. (2015). Avaliação de impacto ambiental em nascentes do Alto Camaquã a partir de uma análise multivariada. Revista Ibero-Americana de Ciências Ambientais, 6(1), pp. 202–212.

Costa W. (2008). Hidrogeologia dos meios fissurados. In: Feitosa FAC, Manoel Filho J, Feitosa EC, Demétrio JGA (eds). Hidrogeologia: Conceitos e aplicações, 3rd edn. Rio de Janeiro, Rio de Janeiro, pp. 121–150.

Dias JS, Moura MN, Barbosa MJ, Felippe MF. (2014). Proposed classification of the Hydrosystems of the UFJF campus based on their Hydrogeomorphological characteristics. *Revista Geonorte,* 20 (5), pp 437–441.

Duarte BP, Heilbron M, Valladares C, Nogueira J.R, Tupinambá M, Eirado LG, Almeida, JC, Almeida, GC. (2003). Geology of the Juiz de Fora and Chiador sheets. In: Soares ACP, Noce CM, Trouw R, Heilbron M (eds). Projeto Sul de Minas, 1 edn. Belo Horizonte, Minas Gerais, pp 153–258.

Felippe MF. (2009). Caracterização e tipologia de nascentes em unidades de conservação de Belo Horizonte com base em variáveis geomorfológicas, hidrológicas e ambientais. Dissertation. Master's in Geography, Universidade Federal de Minas Gerais.

Felippe MF. (2013). Gênese e dinâmica de nascentes: contribuições a partir da investigação hidrogeomorfológica em região tropical. Ph.D. thesis, Federal University of Minas Gerais.

Felippe MF, Magalhães Júnior AP. (2014). Development of a hydrogeomorphological typology of springs based on multivariate cloud statistics. *Brazilian Journal of Geomorphology,* 3(15), pp. 93–409. https://doi.org/10.20502/rbg.v15i3.463

Felippe MF, Magalhães Júnior AP. (2009). Consequências da ocupação urbana na dinâmica das nascentes em Belo Horizonte-MG. *Encontro nacional sobre migrações*, 6, pp. 1–19.

Gerecke R, Cantonati M, Spitale D, Stur E, Wiedenbrug S. (2011). The challenges of long-term ecological research in springs in the northern and southern Alps: indicator groups, habitat diversity, and medium-term change. *Journal of Limnology,* (70), pp. 168–187. https://doi.org/10.3274/JL11-70-S1-13

Gonzalez AZD, Schiavinato VMS. (2019). Environmental assessment of springs in the sub-basin of the córrego das Pitas-MT, Brazil. *Revista equador,* 3 (8), pp. 260–278.

Heilbron, M, Pedrosa-Soares AC, Campos Neto MC, Silva LC, Trouw R, Janasi VA. (2004). Mantiqueira Province. In: Mantesso-Neto V, Bartorelli A, Carneiro CDR, Brito-Neves BB (eds). Geology of the South American Continent: evolution of the work of Fernando Flávio Marques de Almeida. São Paulo, São Paulo, pp. 203–236.

Machado PJO (ed). (2010). Diagnóstico Físico—Ambiental da Bacia Hidrográfica do Córrego São Pedro: Um exercício acadêmico de Gestão dos Recursos Hídricos, Ubá, Minas Gera.

Magalhaes Júnior AP, Felippe MF. (2012). The Importance Of River Springs In Sustainable Water Management: The City Of Belo Horizonte, Brazil. In: Bilibio C, Hensel O, Selbach JF (eds). Sustainable water management in the tropics and subtropics - and case studies in Brazil. Jaguarão, Rio Grande do Sul, pp. 299–346.

Marques LO, Felippe MF. (2017). Hydrogeomorphological diversity of springs in the domains of hillsides: cases of the botanical garden of UFJF (Juiz de Fora-MG). *The Challenges of Physical Geography at the Knowledge Frontier,* (1), pp. 3293–3298. https://doi.org/10.20396/sbgfa.v1i2017.2305

Morin E. (2015). Introduction to complex thinking. Porto Alegre, Brazil.

Moura MN, Alves MJB, Dias JS, Almeida Neto JO, Felippe MF. (2016). Degree of environmental alteration of the hydrosystems of the UFJF campus. *Journal of Geography* 2(6), pp.173–183.

Moura MN. (2020). Environmental quality of springs: methodological proposal. Dissertation. Master's in Geography, Federal University of Juiz de Fora.

Oliveira MCP, Ávila BT, Dias JS, Moura MN, Silva BM, Silva SVB, Felippe MF. (2013). Macroscopic evaluation of the quality of springs on the campus of the federal university of juiz de fora. *Journal of Geography,* 1 (3), pp. 1–7.

Pieroni JP, Rodrigues Branco KG, Valle Dias GR. Ferreira GC. (2019). Assessment of the conservation status of springs in microwatersheds. *Geosciences*, 1(38), pp. 185–193.

Rezende JH, Luca, MV. (2017). Avaliação rápida de rios e nascentes como instrumento de análise ambiental urbana e rural. Revista Ibero- Americana de Ciências Ambientais, 8(4), pp. 85–100.

Sant'anna Neto JL, Galvani E, Vieira BC (2015). Climates of Brazil: past and present. In: Landscapes and landforms of Brazil. Springer, Dordrecht, pp. 33–41.

Schiavinato VMS, Gonzalez AZD. (2020). Degradação de áreas de nascentes na sub-bacia hidrográfica do córrego das Pitas- MT: o que pensam os pequenos produtores rurais? *Pathways of Geography,* 74 (21), pp. 295–312. https://doi.org/10.14393/RCG217448882

Silva GA, Silva RE, Rodrigues SC. (2017). Descriptive analysis of São Francisco River springs in Chapadão Diamante, Serra da Canastra-MG. *The Challenges of Physical Geography at the Knowledge Frontier* (1), pp. 562–573. https://doi.org/10.20396/sbgfa.v1i2017.1826

Springer AE, Stevens LE. (2009). Spheres of discharge of springs. *Hydrogeology Journal,* (17), pp. 83–93. https://doi.org/10.1007/s10040-008-0341-y

Stevens LA, Aly AA, Arpin SM, Apostolova I, Ashley GM, Barba PQ, Barquim J, Beauger A, Benaabidate L, Bhat SU, Bouchaou L, Cantonati M, Carrol TM, Death R, Dwire KA, Felippe MF, Fensham RJ, Fryar AE, Garsaball RP, Gjoni V, Glazier DS, Goldscheider N, Gurrieri JT, Guomundsdóttir R, Guzman AR, Hájek M, Hassel K, Heartstill-Scalley T, Herce JS, Hinterlang

D, Holway JH, Ilmonen J, Jenness J, Kapfer J, Karaouzas I, Knight RL, Kreiling AK, Lameli CH, Ledbetter JD, Levine N, Lyons MD, Mace RE, Mentzafou A, Marle P, Moosdarf N, Norton MK, Pentecost A, Pérez GG, Perla B, Saber AA, Sada D, Segadelli S, Skaalsveen K, Springer AE, Swanson SK, Schwartz BF, Sprouse P, Tekere M, Tobin BW, Tshibalo EA, Voldoire O. (2021). The Ecological Integrity of Spring Ecosystems: A Global Review. *Reference Module in Earth Systems and Environmental Sciences*. https://doi.org/10.1016/B978-0-12-821139-7.00111-2

Valente OF, Gomes MA. (2005). Conservação de nascentes: hidrologia e manejo de bacias hidrográficas de cabeceiras. Viçosa, Minas Gerais.

Fluvial Morphometry Applied to Studies of Drainage Rearrangement Processes in the Iron Quadrangle—Brazilian Atlantic Plateau, Southeastern Brazil

Felipe Gabriel Silva Alves, Antônio Pereira Magalhães Junior, and Jhonathan Felip Magalhães Reis

Abstract The work seeks to illustrate the applicability of morphometry in the investigation of processes of drainage rearrangement and relief configuration in the Iron Quadrangle, one of the compartments of the Brazilian Atlantic Plateau. To this end, the studies were concentrated in the Conceição River basin, a tributary of the Upper Doce River, southeastern Brazil. Indices that point out drainage migration processes and that allow inferences about the lithostructural and morphotectonic control of the hydrographic network were applied, namely: RDE (Declivity-Extension Ratio), FABD (Drainage Basin Asymmetry Factor), and FSTT (Transverse Topographic Symmetry Factor). In a complementary way, aspects of the morphology and geometry of the drainage were surveyed using satellite imagery. The results show evidence of model evolution from first-order *knickpoints* and channel migration as a result of structural rearrangements. An old tributary of a neighboring watershed (Barão de Cocais River) would have been pirated by a direct tributary of the Conceição River, indicating greater denudational aggressiveness of this basin.

1 Introduction

The dynamics of the drainage network is one of the main vectors of configuration and transformation of the relief. Its mechanisms involve different processes that include rearrangements of the position and routes of springs and watercourses and, consequently, the reorganization of hydrography at the watershed level. These modifications can be conditioned by the climatic framework, in terms of the dynamics of the hydrological cycle and balance at different scales, and by the geological framework, in terms of different levels of rock resistance to denudational processes, the degree of lithological deformation, the density and characteristics of lines of weakness (structures) and also the tectonic dynamics, as well as by anthropic factors that may interfere with the stability of river systems (Magalhães Júnior and Barros

F. G. S. Alves (✉) · A. P. Magalhães Junior · J. F. Magalhães Reis
Federal University of Minas Gerais—UFMG, Belo Horizonte, Brazil
e-mail: felipegabriel.silvaalves@yahoo.com.br

G. Barbosa dos Santos et al. (eds.), *Geomorphology of Brazil: Complexity, Interscale and Landscape*, Springer Proceedings in Earth and Environmental Sciences,
https://doi.org/10.1007/978-3-031-05178-4_7

2020). In any case, changes in the relationship between denudational and depositional processes of a river system, internally and comparatively about neighboring systems, can generate pictures of greater or lesser erosive "aggressiveness", with acceleration or decrease of processes of remontant erosion, retreat of headwaters and connection-disconnection of hydrographic arteries. Diversions and relatively quick changes of direction of fluvial segments can also result from physical-environmental dynamics and lead to drainage rearrangements. Commanding these transformations from the regulation of fluvial processes, the base levels of different scales of action are essential elements in the approach of the theme.

Drainage network rearrangement processes comprise a traditional focus of geomorphological studies internationally (Vieux 2016; Roberts 2019; Schoenbohm 2019; Struth et al. 2020; Hooshyar et al. 2020). However, there is a usual difficulty in obtaining robust field evidence to underpin investigative processes, such as fluvial deposits that indicate changes in transport capacity/competence or even changes in position/location of watercourses. Not coincidentally, most of the evidence addressed comes from the parallel analysis of the geometry and organization of hydrography, as well as morphology, from the application of geoprocessing and remote sensing techniques (Cherem et al. 2009; Wilson et al. 2012; Wu et al. 2019; Oyedotun 2020). The techniques of treatment and analysis of aerial images have been evolving rapidly in recent decades, opening the range of options for application in geomorphological studies, including those related to morphometry, as in the case of Digital Elevation Models (DEM) and Terrain Models (TTM) whose development was driven by the research efforts of SRTM—*Shuttle Radar Topography Mission* (Brubacher et al. 2012; Ibanez 2012).

Although Brazil has, in relative terms, few works on the subject, there has been significant development in scientific production in recent decades, particularly those referring to evidence of drainage rearrangement in large watershed dividers (Cherem et al. 2013; Salgado et al. 2018; Silva et al. 2019; Nascimento et al. 2019; Paixão et al. 2020). The application of morphometric indices and comparative techniques for investigating denudation rates in neighboring basins, such as 10Be isotope analyses, has been contributing to this scenario (Salgado et al. 2004; Varajão 2009; Cherem et al. 2012; Salgado et al. 2012).

The Iron Quadrangle (QF) is one of the most studied geological domains of the Brazilian Atlantic Plateau (eastern portion of the country), a fact aided by its mineral wealth such as iron and gold. However, the region still lacks studies and information on the evolution of its drainage network and the role of fluvial processes in the configuration of the relief. On the other hand, some research already points to evidence of spatial derangement processes of the regional hydrography in the QF, particularly in the Maracujá and Ribeirão Preto river basins (Salgado et al. 2007; Magalhães Júnior et al. 2012; Barros et al. 2019; Costa et al. 2018; Magalhães Júnior 2020; Lopes et al. 2020). The case of the Ribeirão Preto basin, located in the Upper Rio Doce basin, is particularly rich as it involves hydrographic, morphological, and sedimentary evidence. Medina et al. (2005), Fabri et al. (2008), and Barros (2012) raised hypotheses of fluvial capture in the area, but there is a need for further research and survey of evidence for its confirmation. In this context, the present work seeks to illustrate the applicability of morphometry in the study of evidence of drainage rearrangement in the Ribeirão Preto basin, signaling the possibilities of studies on the evolution of the fluvial network and the relief configuration in the Iron Quadrangle and the Brazilian Atlantic Plateau.

2 Study Area

The study area is located in the Brazilian Atlantic Plateau, Southeastern Brazil, particularly in the Iron Quadrangle domain (Minas Gerais state)—Fig. 1. The Preto stream is a left margin tributary of the Conceição River, a tributary of the Doce River, having its headwaters located in Serra do Gandarela—the eastern topographic divisor of the QF, specifically in the Conceição do Rio Acima district, municipality of Santa Bárbara/MG. The Conceição River also presents headwaters in the same region, known as the Conceição River Anticline Valley and where a large part of its valley is situated. The springs outcrop at the foot of the crest of the Ouro Fino

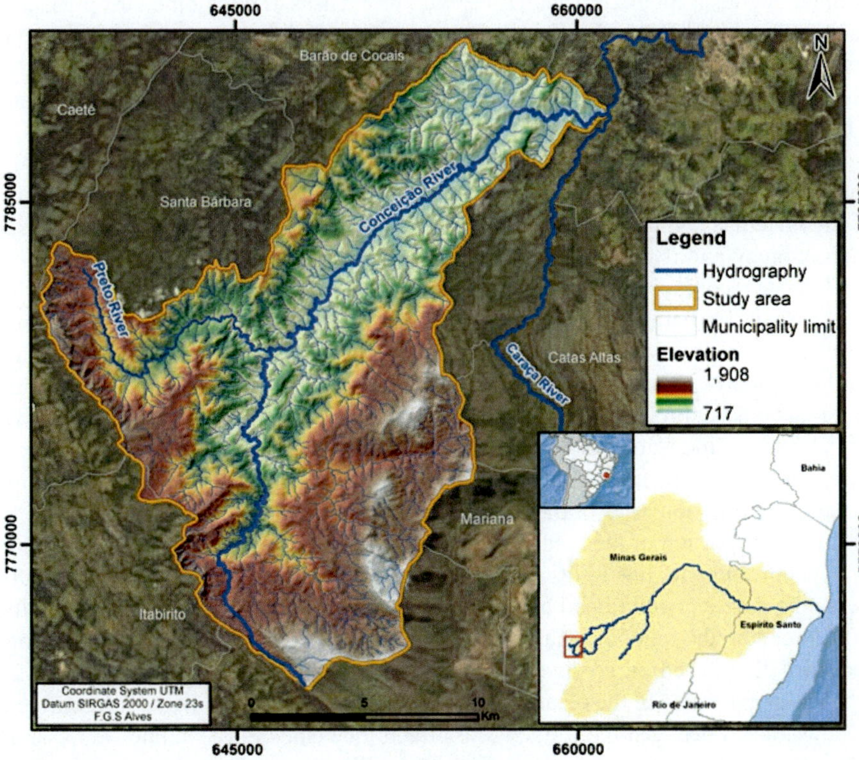

Fig. 1 Contextualization map of the study area. *Source* Prepared by the authors

Mountains syncline, following the NE-SW axis in a south-south direction, passing through the hanging depression of the Gandarela syncline, where, after about 90 km from the source, it flows into the Una River downstream of the city of São Gonçalo do Rio Abaixo/MG (Saadi et al. 2005; Barros 2010, 2012).

According to the geological characterization of Alkmim and Marshak (1998), the lithological substrate is mostly part of the Rio das Velhas Supergroup, specifically the Nova Lima Group, which consists mainly of green shales of metasedimentary and metavolcanic origin, phyllites with intercalations of quartzites, grauwackes, dolomites, talc schists, and iron formations. However, there are important areas of headwaters in the rocks of the Minas Supergroup, represented by the Caraça and Itabira groups, besides rocks of the crystalline basement in the lower course. In the Caraça Group, the Moeda Formation is composed of conglomerates, quartzites, and phyllites, while the Batatal Formation has sericite phyllites and carbonatic sediments. In the Itabira Group, the Gandarela Formation is composed of carbonate rocks (mainly dolomites), itabirites, and phyllites, while the Cauê Formation, at the base of the group, is composed predominantly of dolomitic and amphibolitic itabirites, with small lenses of phyllites. In the northern portion of the basin, there is a small area in the Espinhaço Supergroup, represented by the Cambotas Formation, which is formed by Mesoproterozoic quartzites, with thin lenses of conglomerates present in the Cambotas and Tamanduá mountain ranges. It is also worth mentioning the occurrence of Quaternary superficial formations such as alluvial deposits and cangas, as well as diabase dikes that fill faults in the upper Ribeirão Caraça basin (Alkmim and Marshak 1998).

As described by Endo and Fonseca (1992) and Ferreira Filho and Fonseca (2001), the Conceição basin comprises two main structural systems. The Fundão-Cambotas Fault System is composed of two major westward pushing faults dated to the Brasiliano orogeny and is confined by the anachronistic Caeté Dome. The Água Quente Fault System composes a set of reverse and thrust structures responsible for the uplift of the Santa Bárbara Metamorphic Complex over supracrustal rocks of the Minas and Rio das Velhas supergroups. The area is situated in the Suspended Depression of the Gandarela Syncline, where the headwaters of Preto creek occur, presenting external flanks delimited by escarpments with expressive levels (300–400 m), carved in itabirites of the Cauê Formation (Fig. 2). The carved relief of the interior of the syncline has itabirite ridges lowered and dismantled, in part, by the evolution of the drainage of the tributaries of the Conceição River.

In the SE portion of the area is the Caraça Massif, characterized by a succession of syncline and anticline faults with imbricated blocks and imposing scarps of subvertical quartzite, conditioned by SE oriented structures (Medina et al. 2005). In the NW direction, the relief gradually loses altitude, with evidence of a geomorphological evolution based on erosive processes and tectonic movements (Barros 2012). The anticlinal valley of the Conceição River represents an extensive semi-rectilinear valley situated between the Caraça Massif and the Gandarela Syncline Depression. This excavated anticlinal goes back to the post-Pliocene erosive phase that inverted the folded structures of the Iron Quadrangle resulting in levels that vary between 180 and 300 m, with declivities between 30° and 40°. Nevertheless, the Marginal

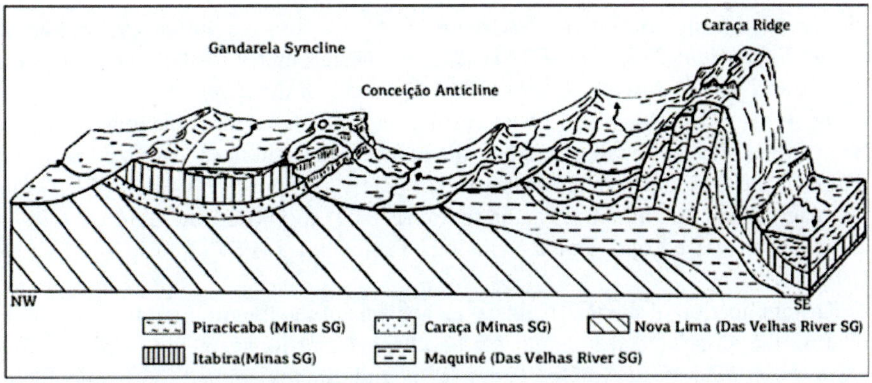

Fig. 2 Schematic geological profile of the study area. *Source* Adapted from Barros (2012)

Depression of Piracicaba River, situated between the middle and the lower course of Conceição River, has a contrasting relief with the one existing upstream the basin. An undulated model predominates, with hills of convex and convex-concave geometry, with little dissection, with levels between 40 and 100 m and slopes between 10° and 20° (Medina et al. 2005).

The regional climate is characterized as semi-humid tropical according to the Brazilian Institute of Geography and Statistics (IBGE), with hot and humid summers, and dry winters with mild temperatures conditioned by altitude (Barros 2010). According to the Köppen classification, the area features a high altitude tropical climate—CWb—in the higher zones, with rainy summers, while the lower areas have the high altitude tropical climate—Cwa, with higher temperatures (annual average of 20.6 °C).

According to Mourão and Stehmann (2007). The regional vegetation features native remnants of the Semideciduous Seasonal Forest, transition areas between the Atlantic Tropical Forest and the Cerrado biome (savanna), as well as herbaceous mountain vegetation. In this last case, there are rare species associated with and adapted to iron crusts ("cangas") formed by the hematite concentration.

3 Materials and Methods

The initial stage consisted of a bibliographic survey about the theme, involving the methodological procedures employed in studies on hydrographic rearrangements. Then it was performed the analysis of the geometry of the drainage network through images. Thus, the Geographic Information System (GIS) software ArcGis (ESRI) in version 10.5, the software QuantumGis in version 3.14, and the software Google Earth Pro in version 7.3.3 (Google) were used. Satellite images provided by Google Inc. were used, as well as satellite images provided by the *National Aeronautics and Space Administration* (NASA) with 30 m^2 spectral resolution by the *Shuttle Radar Topography Mission* (SRTM) satellite interferometry method, as well as the *Advanced Land Observing Satellite* (ALOS Palsar) of the *Japan Aerospace Exploration Agency* (JAXA), with 12.5 m^2 resolution, which uses the same image acquisition method. The application stage of the morphometric indices involved the calculation of the Declivity-Extent Ratio Index (RDE), proposed by Hack (1983), the Drainage Basin Asymmetry Factor Index (FABD), proposed by Gardner (1983), and the Transverse Topographic Symmetry Factor Index (FSTT), proposed by Cox (1994).

The application of the RDE aims at an analysis of the longitudinal fluvial profile, relating the slope and the length of the channel and its segments using the tool "*Knickpoint Finder*" of ArcGis. The index is traditionally applied for the identification of "anomalies" in the longitudinal profile using the formula $RSL = (RSLs/RSLt)$, where $RSLs$ is the slope-to-segment length ratio, $RSLt$ the slope-to-total length ratio, and RSL the slope-to-channel length ratio (Etchebehere 2006; Barros et al. 2012; Araújo and Rodrigues 2016; Bueno 2016; Sordi 2018). For this, an SRTM satellite image with spatial resolution of 30 m^2 limited by the interfluve of the Conceição River basin was used.

The FABD signals the degree of lateral displacement of the main watercourse, perpendicular to its axis, and may indicate tectonic influences and internal fluvial processes expressed by the formula: $FA = 100 * (ar/at)$, where FA is the Asymmetry Factor, ar the basin area on the right margin and at the total basin area (Salamuni 2004). When the result is a value close to or equal to 50, the index indicates little or no tectonic activity; for values higher than 50, the scenario indicates probable right margin tilting, while values lower than 50 denote probable left margin tilting.

The FSTT represents the distance of displacement of the main river course of the basin by the formula $T = Da/Dd$, where Da—the distance from the midline of the drainage basin axis to the midline of the current channel and Dd—the distance of the midline of the basin in relation to the hydrographic divider. If there are no changes in the topographic profile, T is close to zero, whereas if the asymmetry advances, the value approaches one (Firmino 2016; Moreira and Perez Filho 2018).

4 Results

The drainage network of Ribeirao Preto shows two abrupt changes of direction with angulations close to 90°. These drainage elbows are commonly associated with structural conditioning (Barros 2012), and may also reflect the lithological and tectonic context. In this sense, elbows are indicators often used to signal possibilities of drainage rearrangement processes by fluvial captures (Bishop 1995), which in this work would involve drainage segments of the Barão de Cocais river basin.

The FABD was calculated for the Ribeirão Preto basin, and the data were distributed by the contributing basins of the main river (Fig. 3). Near the downstream drainage elbow, there is a tendency for the main channel to shift moderately

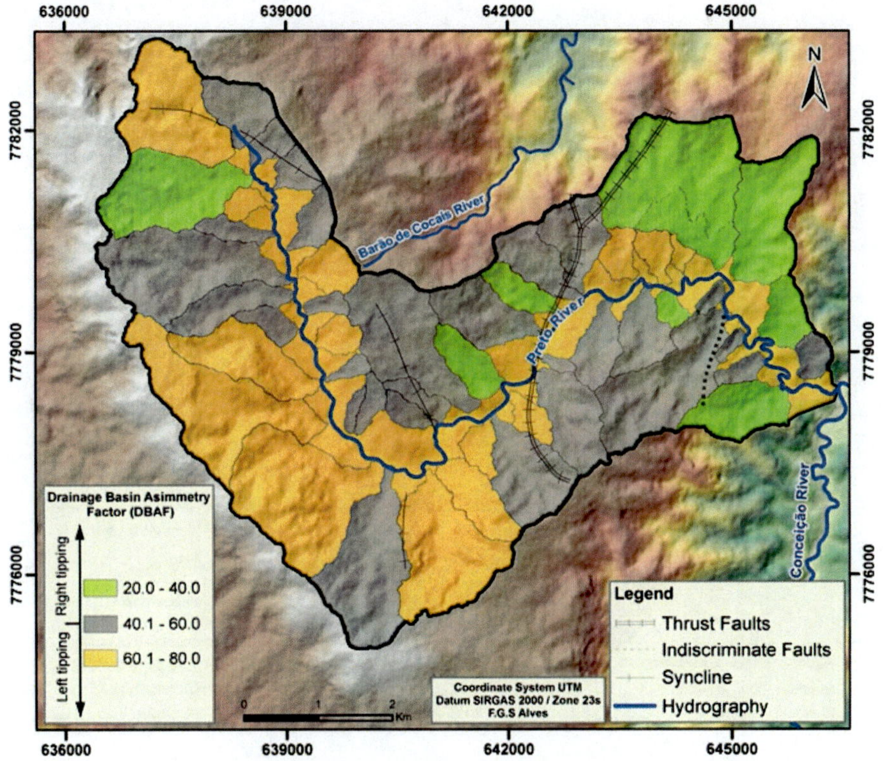

Fig. 3 Drainage Basin asymmetry factor (DBAF). *Source* Prepared by the authors

toward the right bank of the valley in an SW direction. Just downstream of the elbow occurs an indistinct fault (Codemig 2018) on the right bank of the Ribeirão Preto valley. In turn, upstream of the elbow occurs a thrust fault that cuts the valley in the N–S direction. The induction of stream migration processes by structural control can lead to drainage network rearrangement processes (Searle 1987).

In the most upstream sector of the basin, where another drainage elbow occurs, the migration trend of the channel is opposite, that is, moderately toward the left bank of the valley, in the NE direction. The elbow coincides with synclines that cut through the valley on both banks.

The Transverse Topographic Symmetry Factor was elaborated from transversal profiles along the Conceição River basin and, in more detail, the Ribeirão Preto one (Fig. 4). The calculations for the Conceição basin aimed to provide subsidies for the interpretation of the FSTT data of the Ribeirão Preto basin since the behavior of the base level conditions that of its tributaries. The profiles were based on the basin morphology and geometric symmetry as delimitation criteria; the profiles must be perpendicular, and their tracing must have opposite ends in the basin of contribution. The results show average anomaly degrees (0.41–0.8) in the middle and lower Conceição, with higher values near the mouth (0.63 and 0.76). The valley presents

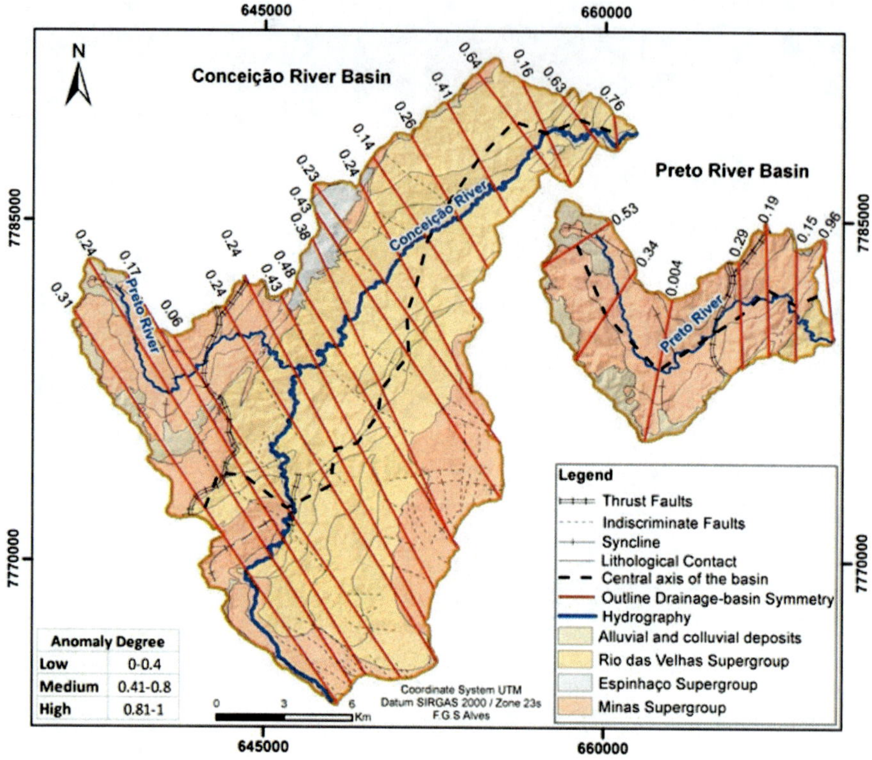

Fig. 4 Transversal topographic symmetry factor (TSTF) for the Conceição River and Preto creek basins. *Source* Adapted from Codemig (2018)

several structural systems and lithological units, with the area of the mouth marked by the occurrence of shales and banded ferritic formations. This fact may be linked to the preferential direction of the watercourse along with banded iron formations.

In the middle part of the basin, near the confluence of the Preto stream, the Conceição River demonstrates a dissonant behavior concerning the central axis of the basin, with a displacement toward the left margin in the SE–NW direction. In this sector, there is the presence of a complex of faults of the same direction, transversal to the axis of the channel, which, together with the lithological framework, maybe direct the watercourse. In the upper basin, the faults present NE–SW and NW–SE directions, and the FSTT values are low.

In the case of the Ribeirão Preto basin, the lithological factor is of paramount importance in the behavior of the watercourse, since the entire valley bottom is carved in dolomites, except near the mouth where itabirites outcrop. However, the presence of a contractional fault that cuts the mid-valley should be highlighted and that may have conditioned its change in direction, since, after contact with the fault, the stream starts to migrate in the contact zone between itabirites and dolomites. In the Ribeirão Preto basin, there is also a predominance of medium anomaly degrees, but with the highest asymmetry values in the upstream (0.53) and downstream (0.96) extremes. In the first case, moderate asymmetry is perceived from the stream to the left bank, coinciding with a contact zone between shales and itabirites on the right bank and the presence of dolomites on the left bank (CODEMIG 2018). In turn, the area of the mouth (confluence with the Conceição) coincides with an indistinct fault and with lithological contacts between dolomites, itabirites, and phyllites, rocks with different levels of resistance to denudation. However, when one compares these values with those obtained for the whole Conceição basin, one notices that, in this case, the degrees of asymmetry are much lower and do not indicate important anomalies (0.32 and 0.39). This illustrates the important influence of the scale of the analysis and the morphological and geological factors on the calculation of the index.

The RDE pointed out several 2nd order *knickpoints* in the Preto river valley and a 1st order anomaly in the contact between the ferritic and quartzite units that sustain the highland relief of the upper-middle stream, and the shales present in the lower course (including the confluence zone with the Conceição river)—Fig. 5. In fact, this anomaly coincides with a craggy zone that configures an important morphological change in the valley and that is cut by indistinct faults. The anomalies shown by

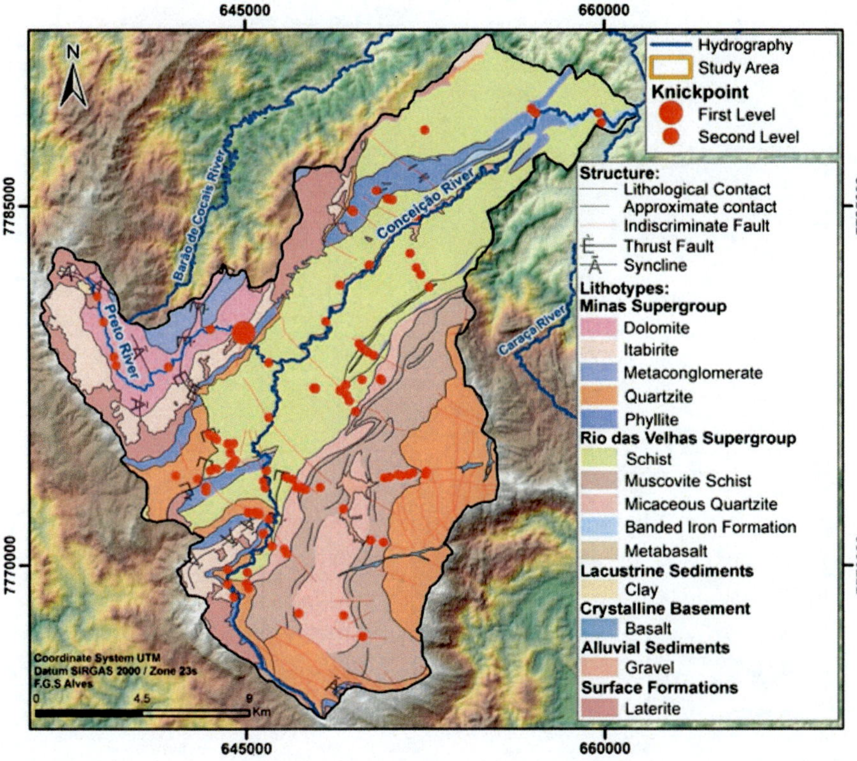

Fig. 5 Anomalies of 1st and 2nd orders evidenced by the application of RDE. *Source* Cartographic base adapted from Codemig (2018)

the RDE are generally associated with lithological contacts, structural lineaments, confluences, or tectonic activities, the latter mainly in the case of 1st order anomalies (Etchebehere 2004; Stevaux et al. 2009; Firmino 2015).

The hypotheses regarding the occurrence of rearrangement processes of the drainage network in the Ribeirão Preto basin, throughout the Quaternary, involve an ancient direct tributary of the Conceição River that would have pirated transversely, from the remontant retreat of the drainage headwaters, an ancient tributary of the Barão de Cocais River that ran in the SW–NE direction (Fig. 6). Barros (2012) pointed out that the geological context of the Conceição basin involves thrust faults and ridgeline *nappes* that condition the formation of ramps with variable dips and high slope angles, caused by the compressive *stress* between the structures. This complex framework would favor an important dynamism to the drainage network throughout the Quaternary.

In this sense, Medina et al. (2005) propose the opening of two epigenetic gorges that truncated the itabirite ridges of the eastern flank of the Gandarela Syncline and provided the fluvial capture by remontaneous erosion processes of the former tributary of Barão de Cocais, by a tributary of Conceição. The authors also state

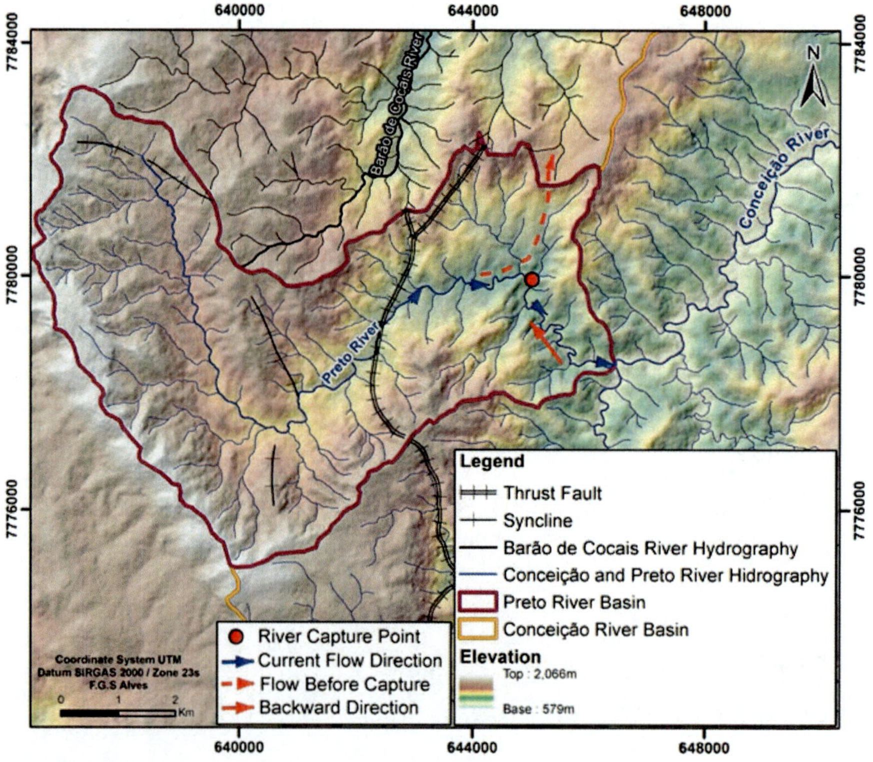

Fig. 6 Illustrative scheme of the fluvial capture hypothesis according to the literature. *Source* Prepared by the authors

that in the southwestern portion of the Gandarela Syncline Suspended Depression (DSSG) there is an aggressive process of adjustment of the drainage network of the Ribeirão Preto stream to the base level (Conceição River), reflected in a typically hilly relief quite eroded by gully processes, with gradients from 40 to 80 m and slopes around 5° to 10°. This context would be responsible for the emptying process of this area and could indicate the evolution of the drainage elbow further upstream. In this sense, Fabri et al. (2008) indicate that the preponderant factor for the opening of the epigenetic gorges was the fitting of the Black Creek along an ancient thrust fault, enhanced by the fragile dolomites of the Gandarela Formation (Minas Supergroup).

The two elbows in the Preto river and the anomalous FSTT values at its mouth reinforce these ideas. The confluence presents lithological contacts (dolomites and itabirites) that may have conditioned the preferential direction of flow toward the Barão de Cocais river basin, anachronistic to the appearance of the drainage elbow. The complex local structural framework must also have contributed to the drainage rearrangement process, given the presence of an important contractional fault in the mid-valley. However, the available mapping bases do not detail the presence of faults near the confluence. The RDE indicated a 1st order *knickpoint* located in the perimeter bordering the contact between the lithologies, which corroborates the capture proposal, besides, through the FSTT, the section presents the greatest anomaly in the channel migration dynamics. In the headwaters of the Ribeirão Preto, the mean FSTT values indicate an anomaly in the migration dynamics of the main channel, and the FABD values show its tendency to migrate to the left bank.

5 Conclusion

Geomorphological theories are often complex and face methodological limitations for their proof. The research on processes of rearrangement of the drainage network faces, in this sense, the challenge of obtaining evidence of hydrographic reorganization over time. To this end, the investigation of case studies is an essential strategy, and the Iron Quadrangle emerges as a prominent stage due to the complex geological and geomorphological evolution throughout the Quaternary period.

The results of the applied morphometry and the analysis of the drainage network geometry contribute to reinforcing the proposal of fluvial capture already indicated in the literature, that is, a former tributary of the Barão de Cocais River would have been pirated by a direct tributary of the Conceição River. In this scenario, the Conceição basin presents greater denudational aggressiveness than the neighboring basin, illustrating how the Quaternary fluvial dynamics contribute to the configuration of the regional relief.

Acknowledgements To the National Council for Scientific and Technological Development (CNPq) for the research scholarship and the research group RIVUS, Geomorphology and Water Resources, of the UFMG.

References

ARAÚJO MS, RODRIGUES SC (2016) Application of the RDE index (slope x extent ratio) and longitudinal profile of the Mata stream, Serra da Canastra—MG. National Symposium of Geomorphology 11: 15–21

Alkmim FF, Marshak S (1998) Transamazonian Orogeny in the Southern Sao Francisco Craton Region, Minas Gerais, Brazil: evidence for Paleoproterozoic collision and collapse in the iron quadrangle. Precambrian Research vol 90: 29–58

BARROS ACM, MEIRA DA, MONTEIRO KA, CORRÊA ACB (2012) Application of the RDE index in the Riacho Grande basin—PB. National Symposium of Geomorphology 9: 21–24

BARROS LFP (2012) Sedimentary Events and Morphodynamic Evolution of the Conceição River Valley—Iron Quadrangle/MG: 112. https://doi.org/10.20502/rbg.v13i3.268

BARROS LFP, MAGALHÃES JR APO (2019) Role of the Rio Doce Basin in the Geomorphological Configuration of the Iron Quadrangle/MG. Bulletin of Geography vol 37: 145–167. https://doi.org/10.4025/bolgeogr.v37i1.36117.

BARROS LFP, MAGALHÃES JR AP, BARROS PHCA (2010) Litological and Tectonic Conditioning in the Morphometry of the Conceição River Basin-Quadrilátero Ferrífero/Mg. Revista de Geografia. Recife: UFPE—DCG/NAPA

BISHOP P (1995) Drainage rearrangement by river capture, beheading and diversion. Progress in Physical Geography 19: 449–473. https://doi.org/10.1177/030913339501900402

BRUBACHER GGO, GUASSELLI LA, LUERCE TD (2012) Evaluation of SRTM bases for extraction of morphometric and drainage variables. Geosciences 31: 381–393

BUENO KEM (2016) Application of the RDE index to identify debris flow deposition regions in the Jacareí river basin—Serra do Mar Paraná State

CHEREM LFS, MAGALHÃES JR AP, FARIA SD (2009) Morphological analysis of drainage network extracted from MDE-SRTM. National Geomorphology Symposium 11: 15–21

CHEREM LFS, VARAJÃO CAC, SALGADO AAR, VARAJÃO AFDC, BRAUCHER R, BOURLÉS D, MAGALHÃES JR AP, NALINI JR HÁ (2012) Chemical denudation and relief lowering in interplanaltic edges with granitic substrate: two examples in the SE of Minas Gerais. Brazilian Symposium on Remote Sensing 14(1): 25–30. https://doi.org/10.20502/rbg.v13i1.344

CHEREM LFS, VARAJÃO CAC, MAGALHÃES JR AP, VARAJÃO AFDC, SALGADO AAR, OLIVEIRA LAF, BERTOLINI WZ (2013) The Role of Fluvial Catchments in the Morphodynamics of Interplaintic Edges of Southeastern Brazil. Revista Brasileira de Geomorfologia 13: 73–84. https://doi.org/10.20502/rbg.v14i4.325

CODEMIG (2018) MAPEAMENTO GEOLOGICO DE OURO PRETO

COSTA RVF, MATSCHULLAT J, LEITE MGP, NALINI JR HA, LEÃO LP (2018) Geochemical mapping of potentially hazardous elements in surface waters and stream sediments of the Iron Quadrangle, Brazil. Geochimica Brasiliensis 32(2): 243–267. https://doi.org/10.21715/GB2358-2812.2018322243

COX RT (1994) Analysis of drainage-basin symmetry as a rapid technique to identify areas of possible Quaternary tilt-block tectonics: an example from the Mississippi embayment. Geological Society of America Bulletin 106(5): 571–581. https://doi.org/10.1130/0016-7606(1994)106<0571:AODBSA>2.3.CO;2

ENDO I, FONSECA MA (1992) Fundão-Cambotas shear system in the Iron Quadrangle, MG: geometry and kinematics. Revista Escola de Minas 45: 15–17

ETCHEBEHERE ML (2004) Application of the "Relação Declividade-Extensão—RDE" Index in the Rio do Peixe Basin (SP) for Detection of Neotectonic Deformations. Revista do Instituto de Geociências—USP 4(2): 43–56. https://doi.org/10.5327/S1519-874X2004000200004

ETCHEBEHERE MLC, SAAD AR, SANTONI GC, CASADO FC, FULFARO VJ (2006) Detection of probable neotectonic deformations in the Rio do Peixe valley, western region of São Paulo state by applying RDE (Relation Declivity-Extension) indices in drainage segments. Geosciences 25: 271–289

FABRI FP, MARENT BR, MAGALHÃES JR AP, SALGADO AAR (2008) Classification of River Streams of Conceição River Basin—Iron Quadrangle/MG: preliminary results. National Symposium of Geomorphology 7

FERREIRA FILHO FA, FONSECA MA (2001) Deformation partitioning in contractional regimes with obstacles in the foreland: example of the Água Quente Fault System, eastern edge of the Iron Quadrangle, MG. Revista Brasileira de Geociências 31: 267–278

FIRMINO IG (2015) Evaluation of Cartographic Models of Morphometric Indices aimed at Neotectonic analysis. Brazilian Geographical Journal 6:168–182

FIRMINO IG (2016) Identification of drainage anomalies in the Soap River Basin (PR) by means of the RDE index and the topographic symmetry factor (FTT). Geography Bulletin 33: 61–73. https://doi.org/10.4025/bolgeogr.v33i0.31927

GARDNER TW (1983) Experimental study of knickpoint and longitudinal profile evolution in cohesive, homogeneous material. Geological Society of America Bulletin 94: 664–672. https://doi.org/10.1130/0016-7606(1983)94<664:ESOKAL>2.0.CO;2

HACK JT (1983) Stream-profile analysis and stream-gradient index. Journal Research of the U. S. Geol. Survey 1: 421–429

HOOSHYAR M, ANAND S, PORPORATO A (2020) Variational analysis of landscape elevation and drainage networks. Proceedings of the Royal Society A: Mathematical, Physical and Engineering Sciences 476. https://doi.org/10.1098/rspa.2019.0775

IBAÑEZ DM (2012) Analysis of the Relief of the Central Amazon using Digital Elevation Model and Fractal Geometry. Course in Geochemistry and Geotectonics: 172. https://doi.org/10.11606/T.44.2012.tde-06112013-093004

LOPES FA, LANA CE, CASTRO PTA, LANA CC (2020) Paleomorphology of the northwestern of the Iron Quadrangle (central Brazil): Stratigraphic and geochronological evidence of a Pleistocene alluvial fan system. Stratigraphic and geochronological evidence of a Pleistocene alluvial fan system. Quaternary International 542: 30–40. https://doi.org/10.1016/j.quaint.2020.02.025

MAGALHÃES JR AP, BARROS LFP (org.) (2020) Hidrogeomorfologia—formas, processos e registros sedimentares fluviais. Bertrand Brazil 1: 417

MAGALHÃES JR AP, BARROS LFP, RAPOSO AA, CHEREM LFS (2012) Quaternary Fluvial Dynamics of the Maracujá River, Iron Quadrilateral (MG). Revista Brasileira de Geomorfologia 13: 3–14. https://doi.org/10.20502/rbg.v13i1.337

MEDINA AI, DANTAS ME, SAADI A (2005) Projeto APA Sul RMBH—Estudos do Meio Físico 6

MOREIRA VB, PEREZ FILHO A (2018) Application and correlation of the morphometric indices Transverse Topographic Symmetry Factor and Drainage Basin Asymmetry Factor in rivers draining the Uberlândia-Uberaba/MG plateau. XII National Symposium of Geomorphology

MOURÃO A, STEHMANN JR (2007) Levantamento da flora do campo rupestre sobre canga hematítica couraçada remanente na mina do Brucutu, Barão de Cocais, Minas Gerais, Brazil. Rodriguesia 58: 775–786. https://doi.org/10.1590/2175-7860200758404

NASCIMENTO FA, SALGADO AAR, GOMES AAT (2019) Evidence for Fluvial Rearrangements in the Amazon-Essequibo Interfluve—Northern Amazonia. Brazilian Journal of Geomorphology 20: 663–671. https://doi.org/10.20502/rbg.v20i3.1520

OYEDOTUN TDT (2020) Quantitative assessment of the drainage morphometric characteristics of Chaohu Lake Basin from SRTM DEM Data: a GIS-based approach. A GIS-based approach. Geology, Ecology, and Landscapes. 01: 01–14. https://doi.org/10.1080/24749508.2020.1812147

PAIXÃO RW, SALGADO AAR, FREITAS MM, ALMEIDA JCH (2020) Possibility of Endorheism and Fluvial Capture in the Morphogenesis of the Paraíba do Sul River Basin. Revista Brasileira de Geomorfologia 21: 821–834. https://doi.org/10.20502/rbg.v21i4.1779

ROBERTS GG (2019) Scales of Similarity and Disparity Between Drainage Networks. Geophysical Research Letters 46: 3781–3790. https://doi.org/10.1029/2019GL082446

SAADI A MEDINA A I M DANTAS M E (2005) Projeto Apa Sul RMBH: Estudos Do Meio Físico: escala 1:50 000. Belo Horizonte: Ministry of Mines and Energy, p 60

SALAMUNI E (2004) Morphotectonics of the Curitiba Sedimentary Basin. Revista Brasileira de Geociências 4(34): 469–478

SALGADO AAR, VARAJÃO CAC, COLIN F, BRAUCHER R, VARAJÃO AFDC, NALINI JR HA, CHEREM LFS, MARENT BR, BRINDUSA CB (2007) Estimation of the Erosion Rates of the Highlands of the Upper Rio das Velhas Basin in the Iron Quadrangle: implications for relief evolution. Revista Brasileira de Geomorfologia 8(2): 3–10. https://doi.org/10.20502/rbg.v8i2.89

SALGADO AAR, SOBRINHO LCG, CHEREM LFS, VARAJÃO CAC, BOURLÉS D, BRAUCHER R, MARENT BR (2012) Study of the Evolution of the Escarpment between the Doce/Paraná Basins in Minas Gerais through the Quantification of denudation Rates. Revista Brasileira de Geomorfologia 13(2): 213–222. https://doi.org/10.20502/rbg.v13i2.280

SALGADO AAR, CHEREM LFS, SORDI MV (2018) Large Fluvial catches in Brazil: synthesis of new findings. Quaternary Studies 19: 23–31. https://doi.org/10.30893/eq.v0i19.176

SALGADO AAR, COLIN F, NALINI JR HA, BRAUCHER R, VARAJÃO AFDC, VARAJÃO CACO (2004) Role of Geochemical Denudation in the Process of Differential Erosion in the Iron Quadrangle/MG. Revista Brasileira de Geomorfologia 5(1): 55–69. https://doi.org/10.20502/rbg.v5i1.32

SEAGREN EG, SCHOENBOHM LM (2019) Base Level and Lithologic Control of Drainage Reorganization in the Sierra de las Planchadas NW Argentina. Journal of Geophysical Research: Earth Surface 124(6): 1516–1539. https://doi.org/10.1029/2018JF004885

SEARLE MP, WINDLEY BF, COWARD MP, COOPER DJW, REX AJ, REX D, TINGDONG Li XUCHANG Xiao, JAN MQ, THAKUR VC (1987) The closing of Tethys and the tectonics of the Himalaya. Geological Society Of America Bulletin 98(6): 678–701. https://doi.org/10.1130/0016-7606(1987)98<678:TCOTAT>2.0.CO;2

SILVA TA, SALGADO AAR, SOUZA FCR (2019) MORPHOGENESIS OF THE TOCANTINS/SÃO FRANCISCO INTERFLUVE IN THE PARANÁ RIVER REGION. Revista Brasileira de Geomorfologia 20 (2): 273–285. https://doi.org/10.20502/rbg.v20i2.1514

SORDI MV (2018) Fluvial Rearrangement as a Mechanism of Relief Evolution in Passive Margin Escarpments: Serra Geral Catarinense South Brazil. Programa de Pós-Graduação em Análise Ambiental, Universidade Federal de Minas Gerais—UFMG, p 260

STEVAUX JC, PAES RJ, FRANCO AA, ETCHEBEHERE ML, FUJITA RH (2009) Morphodynamics in the confluence of large regulated rivers: the case of Paraná and Paranapanema rivers. Latin American Journal of Sedimentology and Basin Analysis 16 (2): 101–109

STRUTH L, GIACHETTA E, WILLETT SD, OWEN LA, TESÓN E (2020) Quaternary drainage network reorganization in the Colombian Eastern Cordillera Plateau. Earth Surface Processes and Landforms 45: 1789–1804. https://doi.org/10.1002/esp.4846

VARAJÃO CAC, SALGADO AAR, VARAJÃO AFDC, BRAUCHER R, COLIN F, NALINI JR HA (2009) Study of landscape evolution in the Iron Quadrangle (Minas Gerais Brazil) by measuring erosion rates (10Be) and pedogenesis. Revista Brasileira de Ciência do Solo 33(5): 1409–1425. https://doi.org/10.1590/S0100-06832009000500032

VIEUX BE Watersheds and Drainage Networks In: VIEUX BE (2016) Distributed Hydrologic Modeling Using GIS. Springer, Dordrecht

WILSON JSJ, CHANDRASEKAR N, MAGESH NS (2012) Morphometric Analysis of Major Sub-Watersheds in Aiyar & Karai Pottanar Basin Central Tamil Nadu India Using Remote Sensing & GIS Techniques. Bonfring International Journal of Industrial Engineering and Management Science 2(1): 8–15

WU T, LI J, LI T, SIVAKUMAR B, ZHANG G, WANG G (2019) High-efficient extraction of drainage networks from digital elevation models constrained by enhanced flow enforcement from known river maps. Geomorphology 340: 184–201.https://doi.org/10.1016/j.geomorph.2019.04.022

Geomorphological Map of the Rio de Janeiro city (Scale 1:25,000): The Challenge of Mapping the Technogen

Marcelo Eduardo Dantas and Loury Bastos Mello

Abstract The geomorphology of the Rio de Janeiro city reveals a mosaic of landscapes conditioned by the Cenozoic tectonics and by the variation of the relative sea level, where coastal massifs surrounded by fluviomarine lowlands and marine plains near the coast stand out. The methodological assumptions for the elaboration of the geomorphological map are described in Dantas (Geodiversity of the state of Maranhão. CPRM, Teresina, pp. 133–140), based on the use of the Relief Patterns Library (Dantas in Library of relief patterns: susceptibility chart to gravitational mass movements and flooding. CPRM, Rio de Janeiro, 2016. http://rigeo.cprm.gov.br/jspui/handle/doc/16589). For the definition of relief patterns, the 3rd and 4th taxons of the methodology of Ross (Rev Dep Geogr 6:17–29, 1992) were adopted, based on interpretation of digital orthophoto mosaic coupled with digital terrain model at scale 1:25,000. The morphostructural and morphosculptural units were also individualized. Finally, it is necessary to map the changes printed on the physical landscape, "creating" favorable space for urban expansion. In synthesis, the mapping of the Technogen is of utmost importance for the geomorphological mapping of the Rio de Janeiro city.

1 Introduction

The city of Rio de Janeiro occupies a peculiar urban site "squeezed" between sea and mountain (Bernardes and Segadas-Soares 1987), presenting a remarkable geological-geomorphological diversity (Guerra 1965; Pinto 1965). However, the flat and well-drained spaces are relatively scarce. Since the mid-nineteenth century and throughout

M. E. Dantas (✉) · L. B. Mello
SGB/CPRM Serviço Geológico do Brasil, Pasteur Ave., 404, Urca, Rio de Janeiro, RJ 22290-040, Brasil
e-mail: marcelo.dantas@cprm.gov.br

L. B. Mello
e-mail: loury.mello@cprm.gov.br

G. Barbosa dos Santos et al. (eds.), *Geomorphology of Brazil: Complexity, Interscale and Landscape*, Springer Proceedings in Earth and Environmental Sciences,
https://doi.org/10.1007/978-3-031-05178-4_8

practically the entire twentieth century, when the city experienced a vertiginous population growth, excluded sectors of society were forced to occupy the steep slopes of the foothills of the coastal massifs or the swampy terrains of the alluvial and fluviomarine lowlands (Segadas-Soares 1965). The occurrence of socio-environmental disasters (not natural) resulting from intense rainfall has been recorded in the Rio de Janeiro city since the beginning of the twentieth century (Brandão 1992).

Historically, the process of growth of the urban network involved works of conquest of the urban space via rectification and channelization of rivers, filling of mangroves and lagoons; cutting of slopes and opening of tunnels; and even more radical changes such as removal and razing of hills and large embankments over the Guanabara Bay. However, the geomorphology of the city of Rio de Janeiro has such a relevant importance in the configuration of the Carioca society that, in a certain way, influenced the socio-spatial segregation of the city since the end of the XIX century, where the wealthier layers of the population followed the tram lines toward the South Zone and the more humble population and the excluded sectors of society followed the train tracks, toward the North Zone (Abreu 1987). The Tijuca Massif imposed itself as a great natural boundary between the two distinct urban expansion fronts and as an important obstacle for the expansion of the metropolis toward the Jacarepaguá lowlands until the 1970s.

Thus, in a scenario of a territory drastically changed for the implementation of the metropolis of Rio de Janeiro, there is the need to highlight and map the deep anthropogenic changes printed on the physical landscape that produced the urban site conducive to the expansion of the city. In synthesis, the mapping of the Technogen is of utmost importance for the geomorphological mapping of the Rio de Janeiro city, with emphasis on the mapping of the embankments on the Guanabara Bay; of the sanitary landfills; of the mining fronts (mainly quarries and gravel pits for the supply of aggregates for civil construction) and terraces of the hillside dismounting (highlighting the Castelo, Santo Antônio and Cruz Vermelha terraces, located in the central area of the city and Inhangá, the latter, in Copacabana).

The geomorphological map of the Rio de Janeiro city (Mello and Dantas 2019), which covers an area of 1255 km^2, was mapped on a scale of 1:10,000 and presented on a scale of 1:25,000 to meet the demands of mapping the mass movement and flooding susceptibility with a cartographic level precision of great detail that was compatible with the great accumulation of knowledge generated by reputable municipal institutions as Geo-Rio and Rio Águas and by several public universities, UFRJ, UERJ, UFF, and the Rural University. Therefore, this mapping was developed to be the most detailed geomorphological cartography product of the municipality of Rio de Janeiro, in order to be applicable for the most diverse environmental management and territorial planning studies.

2 Study Area

The geomorphology of the municipality of Rio de Janeiro, as well as that of its hinterland, presents a notable inheritance coming from a Cenozoic tectonics, a fact already observed since the studies of Ruellan (1944). This famous french professor previously pointed out, with the rudimentary cartographic resources of his time, the significant topographic unevenness existing between the coastal massifs, the Baixada Fluminense and the Serra do Mar, besides the indentations represented by the Guanabara and Sepetiba bays. Ruellan affirmed then that the Serra do Mar consisted a dissected front of fault block.

Later, reports of structural geology and geotectonics executed in the scope of the Remac Project (Asmus and Ferrari 1978), as well as in-depth studies on the geological and tectonic evolution during the Cenozoic carried out by Almeida (1976), Riccomini et al. (1989) and Riccomini, Sant'Anna and Ferrari (2004), contextualized the relief of the Atlantic seaboard of the Rio de Janeiro metropolitan region as a system of *horst* and *grabens* fault blocks of a passive margin of the South American Plate. Ferrari (1990, 2001) refines this tectonic contextualization and enframe a large part of the metropolitan region of Rio de Janeiro in the so-called Guanabara Graben, embedded between the coastal massifs and the Serra do Mar front (Silva et al. 2015). In the Rio de Janeiro city, the coastal massifs of Tijuca and Pedra Branca, supported by igneo-metamorphic rocks of Neoproterozoic age of the Ribeira Belt; the Mendanha-Gericinó massif, constituted by a cretaceous pluton of alkaline rocks, stand out; and, from these mountainous terrains, spread out the fluviomarine lowlands (*baixadas*) of Guanabara, Jacarepaguá, and Sepetiba, all dotted with residual reliefs, such as hills (*morros*) and small isolated ridges (*serras*); and finally, the marine plains and beaches that border all the coastline, from Leme to the Marambaia sandy barrier. The filling processes of these fluviomarine lowlands, as well as the formation of the marine terraces are closely associated with the hydroeustatic fluctuations of the relative sea level during the Upper Quaternary (Amador 1997; Muehe and Valentini 1998; Fernandez and Rocha 2015; Muehe and Lins de Barros 2016).

3 Methodology

The geomorphological map of the Rio de Janeiro city was produced on semi-detail scale (1:25,000) (Fig. 1) based on the identification and compartmentalization of the Rio de Janeiro municipality in morphostructural units, morphosculptural units, and geomorphological units, following the methodology recommended by Ross (1992). The geomorphological cartography is, in turn, based on a morphological and morphometric compartmentalization of homologous units, with Relief Pattern Library employment, emphasizing a systematic relief amplitudes measurement, slope gradients, and drainage density (Dantas 2013, 2016). The morphological analysis was obtained from the empirical evaluation of the various sets of shapes and patterns of relief positioned at different topographic levels, through field observations and analysis of remote sensors (Orthophotos and MDE with spatial resolution of 5 m—IBGE 2010).

In order to improve the visualization of the terrain, was sought to highlight the areas of plains, as well as the areas on top of the elevations. Through the GlobalMapper 7.0 *software* tools, it was applied an overlay of the Digital Elevation Model

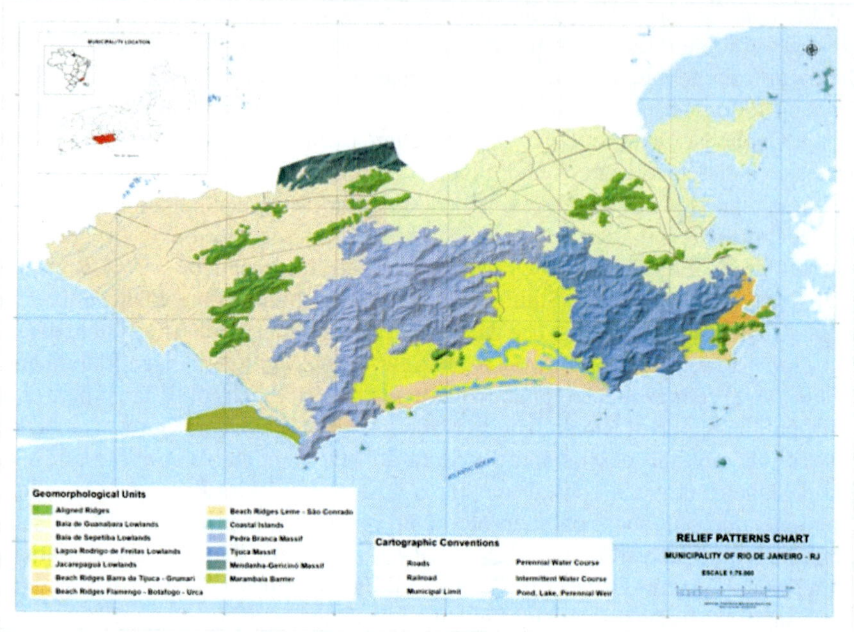

Fig. 1 Map of geomorphological units of the Rio de Janeiro city

(DEM), pseudo illuminated (*slope shader*), with the orthophotos generating greater contrasts between these two areas of the relief, resulting in an easier separation of these units, especially in the lowlands and valley bottoms domains (Shinzato et al. 2012). The geomorphological units were digitized on canvas, using the ArcGis 10.2 *software*. The products generated from the digital treatment of images and DEM were: slope; hydrography; contour lines with equidistance of 5 m; shaded relief with two directions of view—NE and NW; and hypsometry, as already proposed by Dantas et al. 2014.

Finally, highlights a peculiar methodological approach to proceed with the historical reconstitution of the Rio de Janeiro physical space transformation resulting from successive and cumulative human interventions carried out due to the metropolis expansion. Thus, the mapping of a densely urbanized area extreme transformed by human action over five centuries requires new approaches, which were introduced for the preparation of this geomorphological map, highlighting:

(a) the intensive use of Google Earth, Google Maps, and Street View.
(b) the use of a thorough bibliographic and cartographic research about the municipality of Rio de Janeiro, with emphasis on the following authors: Maia et al. (1984), Amador (1997), Abreu (1987), Roncarati and Neves (1976), Roncarati and Barrocas (1978), IPP (2008), Pereira et al. (2012) and the Embrapa pedogeotechnical map (Mendonça-Santos et al. 2009).

Fig. 2 Photograph of the Copacabana neighborhood, in the year 1920, highlighting the former low elevations location of the Morro do Inhangá, in the beach arc central portion. These elevations were razed throughout the twentieth century. *Source* Morro do Inhangá (2010)

Fig. 3 Photograph of Praça da Bandeira, in 1940, showing the occurrence of floods that periodically reach a consisting mangrove embankments area in a confluence of rivers that flowed into the Saco de São Diogo in the past. *Source* Malta, 29 Jan. (1940)

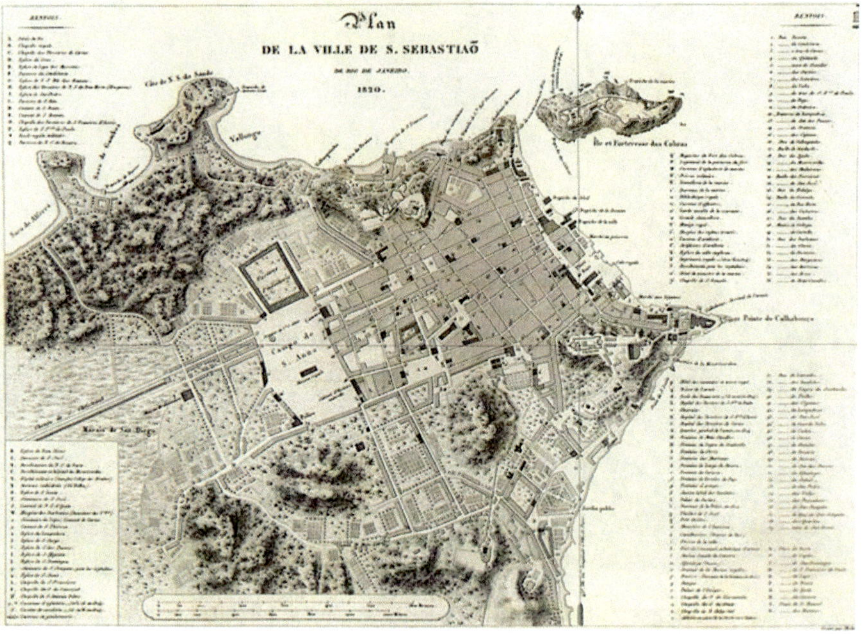

Fig. 4 Topographic plan of São Sebastião village of Rio de Janeiro, drawn up in the year 1820 and showing, with rich detail, the original coastline contour and the location of all hills, at a time before the embankments and razing of hills, throughout the twentieth century. *Source* Freycinet [ca. 1824]

(c) the use of photographs (Figs. 2 and 3), images, and historical maps (Fig. 4) to reconstitute the natural environments, before the radical transformations undertaken by successive urbanization events.

(d) the use of historical map platforms and graphic animations (ImagineRio 2021) that allow reconstituting the urban site's historical evolution and the environmental transformations resulting from the urbanization process of the Rio de Janeiro city.

4 Results and Discussions

The Rio de Janeiro city is compartmentalized into seven geomorphological units that will be systematically described below, emphasizing dominant relief patterns, their genesis and morphogenetic evolution, and technogenic transformations:

4.1 Tijuca Massif

The Tijuca Massif occupies the central-eastern portion of the city and consists of an imposing mountainous massif, surrounded by the Guanabara, Jacarepaguá, and South Zone lowlands. It is supported by a diverse geological substrate composed of paragneisses, orthogneisses, and granitoid rocks of Neoproterozoic age (Heilbron et al. 2016, 2020). The highest peaks, (Pico da Tijuca, 1021 m; Pico do Papagaio, 989 m; Pedra da Gávea, 842 m) are constituted by granite dome-shaped elevations or laccoliths that stand out amidst the rugged and forested relief of steep slopes, often covered by deposits of colluvium and talus (Fig. 5). A network of channels with high drainage density and marked structural control stands out, as observed along the valleys of the Cachoeira and Maracanã rivers, both conditioned by a lineament of NE-SW direction.

This massif divides the southern, northern, and western zones, acting as a water dispersion zone between the Guanabara and Jacarepaguá lowlands and is characterized by alignments of WSW-ENE direction ridges, including the Serra da Carioca where the peak of Corcovado (704 m) is located. Small associated mountain alignments are also remarkable, such as the alignment Morro dos Cabritos (379 m)—Pão de Açúcar (395 m), disposed of parallel to Serra da Carioca (Motta 2017) (Fig. 6). Extensions in NE direction, near Santa Teresa and Glória neighborhoods, emphasize the very dissected aspect of this massif, already undone in a relief of aligned hills (Dantas 2000).

Fig. 5 Detail view of the Gávea summit (842 m high), which presents a relatively flat top due to the granitic laccolith intrusion more resistant to weathering than the underlying gneissic rock. *Photograph* Marcelo Ambrosio Ferrassoli

Fig. 6 Panoramic view of the Sugar Loaf (395 m high), the iconic rocky elevation, of peculiar and slender rounded shape that guards the entrance to the Guanabara Bay, as a historic sentinel of the Rio de Janeiro city. *Photograph* Loury Bastos Mello

Virtually the entire area of the Tijuca Massif was devastated in the early nineteenth century, for planting coffee plantations, and reforested around 1860, due to serious environmental problems arising from deforestation. Noteworthy in this context: the erosion of slopes, silting up of rivers, and the subsequent scarcity of water for urban supply. This secondary forest of 150 years of existence is the scene of many geomorphological and geoecological studies, which demonstrate the importance of the structure and functionality of the forest cover in regularizing the slopes hydrology and controlling mass movements (Coelho Netto 1992, 2005, 2007; Rocha Leão 1997; Fernandes et al. 2006; Negreiros et al. 2006, among others).

Like the other coastal massifs, the Tijuca Massif presents a high potential of vulnerability to erosion events and mass movements. During the extreme events of 1966/1967 and 1988, were recorded in Carioca and Maracanã river basins (Mousinho and Silva 1968); in 1996, were recorded in Quitite and Papagaio river basins, in Jacarepaguá (Cruz et al. 1998; Fernandes et al. 2001, 2004; Vieira and Fernandes 2004); and in 2010, the slopes of Santa Teresa were the most affected (Coelho Netto et al. 2012). Thus, neighborhoods located at the foothills of steep slopes, including Jardim Botânico, Humaitá, Cosme Velho, Santa Teresa, Tijuca, Grajaú, and Freguesia, as well as communities like Rocinha and Vidigal are subject to suffer the impact of torrential rains, landslides and mass runs of great magnitude from the slopes of the mountain massif.

Due to the fact that the massif is close to the historic core of the metropolis of Rio de Janeiro, it suffers an intense urban pressure from all sectors, resulting in the vegetation cover degradation. This process is developed through deforestation, fires, or slums, with greater intensity on the northern slope (urban network expansion around the Tijuca and Méier neighborhoods) and, more recently, in the western slope (urban network expansion of Jacarepaguá) (Fernandes et al. 1998; Zaú et al. 2007).

Despite the significant human transformation of the natural landscape of the Tijuca Massif, such changes were not sufficient to generate technogenic formations.

4.2 *Pedra Branca Massif*

The Pedra Branca Massif is located in the western part of the city and, like the Tijuca Massif, consists of an imposing mountainous massif, surrounded by the Sepetiba and Jacarepaguá lowlands. It is supported predominantly by tarditectonic granitoid rocks of Cambrian age (Heilbron et al. 2016, 2020). In this massif, the highest point of the municipality of Rio de Janeiro (Pico da Pedra Branca, 1024 m) stands out like a granite dome-shaped elevation amidst steep and forested slopes. Its southwestern extension consists of an extensive alignment of SW-NE direction that directly reaches the ocean, close to the Grumari and Barra de Guaratiba beaches. The isolated mountain ranges of Serra do Cantagalo (254 m) and Serra da Paciência (202 m), in WSW-ENE direction, are also remarkable, being detached elevations of the Pedra Branca Massif within the Sepetiba lowlands (Dantas 2000).

Extensive Pedra Branca Massif areas were deforested in favor of the coal cycle during the nineteenth century, intended to provide the city with charcoal (Oliveira and Fraga 2011). From 1960, the Pedra Branca Massif began to suffer the increase of urban pressure in its lower hills, with high susceptibility to mass movements and floods, especially in its eastern (Jacarepaguá) and northern (Realengo, Bangu and Campo Grande) slopes. On the northern slope, the forest has been completely replaced by grass vegetation.

Quarries for gravel supply are registered at the foothills of the Pedra Branca Massif steep slopes, which denotes, locally, a radical alteration of the natural landscape, representing technogenic formations mappable on a scale of 1:25,000. These quarries, active or abandoned, also occur to a lesser extent in the Tijuca Massif and other small hills and mountains. The virtual urban expansion toward abandoned quarry terraces may result in the generation of new risk areas. However, old quarries deactivated many decades ago, are now fully incorporated into the urban fabric of Rio de Janeiro. Many imposing historic buildings erected using ashlar elements between the nineteenth century and the first half of the twentieth century were built from material extracted from these old quarries, mainly using the iconic Facoidal Gneiss, "the most Carioca of rocks" (Castro et al. 2021).

4.3 Mendanha-Gericinó Massif

The Mendanha-Gericinó massif occupies a northern portion of the city, bordering Mesquita and Nova Iguaçu municipalities (Motta, 2017). It consists of an intrusive and isolated mountainous massif, with dome shape, elongated in the WSW-ENE direction, reaching 940 m high, and is abruptly delimited by the surrounding fluviomarine lowlands, dotted by isolated hills (Dantas 2000). It is supported by an igneous rock pluton of alkaline composition from Neocretaceous age (Heilbron et al. 2016). The Marapicu hill (620 m) integrates this unit and consists of an intrusion located slightly southwest of the main intrusive body, presenting a morphology of a subvolcanic neck, preserving ring structures. The unit presents a high potential of vulnerability to erosion and mass movement events, which are often triggered by the degradation of the forest cover, initially as a result of agricultural exploitation and later due to the urban increasing pressure to which the massif is submitted. This process is verified in its southern slope (urban networks expansion of Bangu and Campo Grande), where the lower slopes deforestation has been increasing.

4.4 Sepetiba Lowlands and Marambaia Sandy Barrier

The Sepetiba lowlands are situated in the western part of the city, west of the Pedra Branca Massif, being generated by cumulative depositional events of Pleistocene and Holocene ages, which filled the inland Sepetiba lowlands. Such formation process of this vast lowland generated a complex mosaic of depositional environments that include tidal plains (mangroves); fluviomarine plains (marshes) and floodplains, besides the alluvial-colluvial lowlands of the Guandu, Prata, Guandu do Sena, Piraquê and Portinho river basins. The Sepetiba lowlands consist of an extensive flat surface with shallow to sub-flooding water tables, dotted with hills, aligned hills, and small mountain ridges between the neighborhoods of Pedra de Guaratiba and Sepetiba to Santa Cruz and Campo Grande.

The Sepetiba lowlands experiences, from the mid-twentieth century, an accelerated process of urban network expansion, resulting in the installation of neighborhoods in flood risk areas, being Jardim Maravilha the most emblematic example. Despite this urbanization process, extensive mangroves and apicuns remain at the bottom of Sepetiba Bay, between Pedra and Barra de Guaratiba, and the inner portion of the Marambaia sandy barrier. This extensive beach-ridge, which is a unique geomorphological unit, is characterized by an extensive coastal sandy marine deposit that separates Sepetiba Bay from the ocean, with areas with dune fields resulting from the aeolian remobilization of the sand barrier.

4.5 Jacarepaguá Lowlands

The Jacarepaguá lowlands occupy the southern portion of the West Zone of the municipality, between the Tijuca and Pedra Branca massifs (Fig. 7). Just like the other fluviomarine plains, the events of regression and transgression of the relative sea level are determinants of its original morphological configuration. Thus, the Jacarepaguá lowlands display a diverse mosaic of depositional environments where it stands out, near the coastline, corresponding to Barra da Tijuca, a lagoon system with two sandy barriers anchored between the Joá and Prainha promontories, interspersed with lagoon plains (marshes) that represent floodable lowlands resulting from the filling of part of the current lagoons. Behind it, in Jacarepaguá, there is a floodplain composed of the Camorim, Grande, Anil, Arroio Pavuna, and Arroio Fundo river basins, dotted with hills and isolated rocky elevations.

The Jacarepaguá lowlands experienced, as of 1970, an accelerated process of the urban network expansion, being characterized as the main growth vector of the Rio de Janeiro city (Dantas 2000). The disorderly urban expansion generated serious socio-environmental problems in communities such as Cidade de Deus and Rio das Pedras, located in areas susceptible to flooding. Serious geotechnical problems have also arisen in the construction of buildings groups over the "soft soils" on lagoon plain, expressed by means of sinking. However, it is the lagoon bodies that suffer the most intense degradation process in the region, as a result of the large amounts of sewage and waste discharge, associated with the low renewal capacity of their waters.

Fig. 7 Overall view of the Jacarepaguá alluvial-colluvial plain. In the background, the beauty silhouette of the Tijuca Massif west flank. *Photograph* Marcelo Ambrosio Ferrassoli

4.6 Guanabara Lowlands

The Guanabara lowlands, located in the northern part of the city, cover its eastern portion north of the Tijuca Massif, being also generated by Quaternary depositional events that filled the Guanabara Bay recess. The depositional environments that model the Guanabara lowlands are represented by tidal plains (mangroves); fluviomarine plains (marshes) and floodplains of Sarapuí, Pavuna, Acari, Irajá, Faria-Timbó, and Maracanã river basins. However, it is emphasized as the dominant morphological feature, the extensive areas covered by alluvial-colluvial lowlands. These terrains consist of depositional surfaces of mixed origin that fill the lowered sectors of the Guanabara Graben (Ferrari 1990), in the Sepetiba and Guanabara lowlands and that extends throughout the Baixada Fluminense (Dantas 2000) and can be correlated to the Caceribu Formation, described by Amador (1980). The Guanabara lowlands are dotted with isolated low hills, besides small ridges like the Engenho Novo hills (180–210 m) that divide the Maracanã and Faria-Timbó river basins and the Misericórdia hills (170–250 m), where the Alemão and Penha slum complexes are located.

Vast embankments over the Guanabara Bay were implanted since the beginning of the XX century, highlighting the disappearance of the old Inhaúma inlet, Maria Angu beach, and the creation of Fundão Island and the Galeão airport, representing technogenic deposits mappable on a scale of 1:25,000.

In general, all these fluviomarine lowlands present an advanced stage of socio-environmental degradation, presenting serious problems as result from the significant population growth verified throughout the twentieth century. The installed degradation comes from the bad disposal of solid waste, the lack of basic sanitation, the slope deforestation, the silting up of the channels, and the inadequate soil occupation, among the main problems (Dantas 2000).

4.7 Downtown and South Zone Lowlands

The lowlands located in the Center and South Zone of the city, densely urbanized and object of greater real estate valuation, suffered the deepest changes in their geographic space. So, the geomorphological mapping of these terrains becomes very difficult in response to the extreme difficulty of reconstituting the past depositional environments. We highlight tidal plains (mangroves); fluviomarine plains (marshes); floodplains of Carioca and Macacos river basins; and marine plains of the Flamengo, Leme-Copacabana, and Ipanema-Leblon beach arcs.

Important embankments over the Guanabara Bay's water surface were implemented since the beginning of the twentieth century, highlighting the disappearance of Saco de São Diogo and the construction of Santos Dumont Airport, Aterro do Flamengo, Cais do Porto, Urca neighborhood (Fig. 8) and the surroundings of Rodrigo de Freitas lagoon (Fig. 9), in addition to the enlargement of the Copacabana beach. Such embankments were built from the dismantling of the past hills of

Fig. 8 Aerial view of the Urca neighborhood, built from embankment over the waters of Guanabara Bay. *Photograph* Marcelo Eduardo Dantas

Fig. 9 Panoramic view of the Rodrigo de Freitas lagoon, obtained from the Corcovado hill, where Christ the Redeemer statue was erected. Its water body has been successively reduced due to land reclamation works on its margins. *Photograph* Loury Bastos Mello

Fig. 10 Photograph of the esplanade of Castelo hill in 1930, a few years after its dismantlement, where the immense urban void stands out amidst the nineteenth- and early twentieth-century houses. This esplanade is positioned 5–8 m above the base level of the surrounding plains. It can be seen, on the left, an extensive area that has just been reclaimed along the Guanabara Bay. *Source* Holland (1930?)

the downtown consisting of thick regolith of weathered biotite-gneiss rock (Castelo, Santo Antônio, and Senado) (Fig. 10). These esplanades constituted excellent sites for urban expansion in the city center itself, being represented by the road axes of the Antônio Carlos and Chile avenues and by the Cruz Vermelha square. All these technogenic formations are also mappable at a scale of 1:25,000.

5 Conclusions

The Rio de Janeiro city presents a remarkable geomorphological diversity represented by imposing coastal massifs, with steep and predominantly forested slopes, supported by lithologies of distinct compositions, as well as by extensive fluviomarine and marine plains constituted by a complex mosaic of Quaternary depositional environments that may present gravitational, alluvial, marine or transitional genesis.

However, the execution of a geomorphological mapping in a densely urbanized metropolis that presents a secular accumulation of successive human interventions on the physical environment proves to be extremely difficult, requiring the use of new and some unusual analysis tools, such as: the use of programs and applications like

Google Earth and Street View; and the historical rescue of the natural space transformation process with the use of historical sources (cartographic or documentary), old photographs and, if possible, historical map platforms.

The reconstitution of the original physical environment of the Rio de Janeiro city and the determination and mapping of that technogenic formations are of great relevance to support studies of environmental management and territorial planning, including the Municipal Master Plan. From this reconstitution, it is possible to determine areas with higher susceptibility to flooding and mass movement events or areas of greater environmental fragility, as well as to provide subsidies for engineering geological studies aiming at the expansion or consolidation of the urban network.

References

Abreu MA (1987) Evolução urbana do Rio de Janeiro. Instituto Pereira Passos—IPP, Rio de Janeiro

Almeida FFM (1976) The system of continental rifts bordering the Santos Basin, Brazil. An Acad Bras Ciênc 48(Supl):15–26

Amador ES (1980) Cenozoic sedimentary units of the Recôncavo of Guanabara Bay. An Acad Bras Ciênc 52(4):742–762

Amador ES (1997) Baía da Guanabara e ecossistemas periféricos: homem e natureza. Ed. Reproarte Gráfica e Editora, Rio de Janeiro

Asmus HE, Ferrari AL (1978) Hypothesis on the cause of Cenozoic tectonism in southeastern Brazil. In: Aspectos estruturais da margem continental leste e sudeste do Brasil. Petrobras, Rio de Janeiro, p 75–88 (REMAC Project, 4)

Bernardes LMC, Soares MTS (1987) Rio de Janeiro: cidade e região. General Department of Documentation and Cultural Information, Rio de Janeiro

Brandão AMPM (1992) Climate change in the Rio de Janeiro metropolitan area: a probable influence of urban growth. In: Abreu M A (org) Natureza e sociedade do Rio de Janeiro. Secretaria Municipal de Cultura, Turismo e Esportes, Departamento Geral de Documentação e Informação Cultural, Divisão de Editoração, Rio de Janeiro, p 143–200

Castro NF, Mansur KL, Frascá MHBO et al (2021) Heritage stone of Rio de Janeiro (Brazil): the Facoidal gneiss. Episodes 44(1):59–74

Coelho Netto AL (1992) O geoecossistema da floresta da Tijuca. In: Abreu MA (org) Natureza e sociedade do Rio de Janeiro. Municipal Secretariat of Culture, Tourism and Sports, Rio de Janeiro, p 104–142

Coelho Netto AL (2005) The forest-urban interface and water-related natural disasters in the Tijuca massif: challenges to urban planning from a socio-environmental perspective. Rev Dep Geogr 16:46–60

Coelho Netto AL, Avelar AS, Fernandes MC, Lacerda, WA (2007) Landslide susceptibility in a montainous geoecosystem, Tijuca Massif, Rio de Janeiro: the role of morphometric subdivision of the terrain. Geomorphology. https://doi.org/10.1016/j.geomorph.2006.03.041

Coelho Netto AL, Avelar AS, Sato AM, Dias MA, Schlee MB, Negreiros AB (2012) Vulnerability in mountainous geo-ecosystems and landslide disasters at the Forest-urban interface: geological, geomorphological and geoecological controls. In: Lacerda WA, Palmeira EM, Coelho Netto AL et al (eds) Desastres naturais: suscetibilidade e riscos, mitigação e prevenção, gestão e ações emergenciais, vol 1. COPPE/UFRJ, Rio de Janeiro, p 63–80

Cruz ES, Vilela CL, Azeredo M e Coelho Netto AL (1998) Influence of geomorphology and vegetation on scars generated by mass movements in the Tijuca massif-RJ. In: 2º Simpósio Nacional de Geomorfologia, Florianópolis, 1998. UGB, Florianópolis, p 359–363

Dantas ME (2000) Geomorfologia do estado do Rio de Janeiro. In: CPRM—Serviço Geológico do Brasil Estudo geoambiental do estado do Rio de Janeiro. CPRM, Brasília

Dantas ME (2013) Analysis of relief patterns: an instrument applied to the mapping of Geodiversity. In: Bandeira ICN (ed) Geodiversity of the state of Maranhão. CPRM, Teresina, p 133–140

Dantas ME (2016) Library of relief patterns: susceptibility chart to gravitational mass movements and flooding. CPRM, Rio de Janeiro. http://rigeo.cprm.gov.br/jspui/handle/doc/16589

Dantas ME, Shinzato E, Renk JFC, Moraes JM, Machado MF, Nogueira, AC (2014) The use of geomorphology to assess susceptibility to mass movement and flooding—Mimoso do Sul/ES. Rev Bras Geol Eng Ambient 4(2):23–42

Fernandes MC, Rosas RO, Coelho Netto AL (1998) Análise dinâmica do uso e cobertura do solo do maciço da Tijuca-RJ: subsídios ao entendimento dos vetores de transformação. In: 2° Simpósio Nacional de Geomorfologia, Florianópolis, 1998. UGB, Florianópolis, p 304–307

Fernandes MC, Avelar AS, Coelho Netto AL (2006) Domínios geo-hidroecológicos do Maciço da Tijuca/RJ. Anu Inst Geociênc 29(2):122–148

Fernandes NF, Guimarães RF, Gomes RAT, Vieira BC, Montgomery, DR, Greenberg H (2001) Geomorphological constraints of landslides on slopes: evaluation of methodologies and application of a model to predict susceptible areas. Rev Bras Geomorfol 2(1):51–71

Fernandes NF, Guimarães RF, Gomes RA, Vieira BC, Montgomery, DR, Greenberg H (2004) Topographic controls of landslides in Rio de Janeiro: field evidence and modeling. Catena 55:163–181

Fernandez GB, Rocha TB (2015) Holocene coastal barriers: geomorphology and depositional architecture on the coast of Rio de Janeiro. Rev Bras Geomorfol 16(2):301–319

Ferrari AL (1990) A geologia do "rift" da Guanabara (RJ) da sua porção centro-ocidental e sua relação com o embasamento Pré-Cambriano. In: 36th Brazilian Congress of Geology, Natal, 1990. Proceedings, vol 6. SBG, Natal, p 2858–2871

Ferrari AL (2001) Evolução tectônica do gráben da Guanabara. PhD Thesis in Geology, University of São Paulo—USP, São Paulo

Freycinet LCD ([ca. 1824]) Plan de la ville de S. Sebastião de Rio de Janeiro. Metal engraving, 21.5 cm × 29.5 cm. http://objdigital.bn.br/objdigital2/acervo_digital/div_cartografia/cart175727/cart175727.jpg

Guerra AT (1965) Paisagens físicas do Rio de Janeiro. Rev Bras Geogr 27(4):539–568

Heilbron M, Eirado LG, Almeida J (org) (2016) Mapa geológico e de recursos minerais do estado do Rio de Janeiro. Scale 1:400,000. CPRM, Belo Horizonte. http://rigeo.cprm.gov.br/jspui/handle/doc/18458

Heilbron M, Eirado LG, Almeida JCH, Tupinambá M, Peixoto C, Valeriano CM, Girão R. (2020) Proterozoic to Ordovician geology and tectonic evolution of Rio de Janeiro State, SE Brazil: insights on the central Ribeira Orogen from the new 1:400,000 scale geologic map. Br Journ Geol 50(2):1–25

Holland SH ([1930?]) Esplanada do Castello, Mercado Municipal, Baia de Guanabara etc. [*s. n.*], Rio de Janeiro. http://brasilianafotografica.bn.br/brasiliana/handle/bras/2855

IBGE (2010) Digital elevation model of the state of Rio de Janeiro. IBGE, Rio de Janeiro

ImagineRio (c2021) Historical overlays (1880). https://imaginerio.org/?utm_medium= website&utm_source=archdaily.com.br#en/1939/16/-22.904028215313716/-43.1879210472107/all/

Instituto Pereira Passos—IPP (2008) Areas of the city susceptible to flooding from sea level rise in the metropolitan region of Rio de Janeiro. Pereira Passos Institute, Rio de Janeiro

Maia MCAC, Martin L, Flexor JM (1984) Holocene evolution of the Jacarepaguá coastal plain. In: 33rd Brazilian Congress of Geology, Rio de Janeiro, 1984. vol 1. SBG, Rio de Janeiro, p. 105–118

Malta U (1940) Praça da Bandeira. 1 photograph, between, 9 cm × 12 cm. Jan 29th 1940. https://brasilianafotografica.bn.gov.br/brasiliana/handle/20.500.12156.1/5713

Mello LB, Dantas ME (2019) Relief patterns map: municipality of Rio de Janeiro, RJ. Scale 1:25,000. CPRM, [Rio de Janeiro]. https://rigeo.cprm.gov.br/handle/doc/21211

Mendonça-Santos ML, Antunes FS, Calderano S, Santos HG, Polivanov H, Campos TMP, Zaroni MP, Donagemma GK, Amaral CP (2009) Correlação pedológico-geotécnica do município do Rio de Janeiro. Série Documentos EMBRAPA, vol 117. Embrapa Solos, Rio de Janeiro

Morrotes do Inhangá ([192-]). In: BLOG curiosidades cariocas. [Rio de Janeiro], 16 Sep. 2010. http://rio-curioso.blogspot.com/2010/09/morrotes-do-inhanga.html

MOTTA M (2017) Sobre Rochas: Rio de Janeiro. Rio de Janeiro: Andrea Jakobsson, p 251

Mousinho MR, Silva JX (1968) Considerações geomorfológicas a propósito dos movimentos de massa ocorridos no Rio de Janeiro. Rev Bras Geogr 30(1):55–72

Muehe D, Valentini E (1998) The coast of the state of Rio de Janeiro: a physical-environmental characterization. Fundação de Estudos do Mar—FEMAR, Rio de Janeiro

Muehe D, Lins de Barros FM (2016) The beaches of Rio de Janeiro. In: Short AD, Klein AHF Brazilian beach systems. Springer, Dordrecht, p 363–396

Negreiros AB, Barreto ML, Miguel FLCG, Coelho Netto AL (2006) Forest restructuring and resulting hydrology in a landslide clearing: upper Macacos River valley, Tijuca National Park, RJ. In: 6 Simpósio Nacional de Geomorfologia, Goiânia, 2006. UGB, Goiânia

Oliveira RR, Fraga JS (2011) Social metabolism of a forest and a city: landscape, charcoal workers and social invisibility in 19th and 20th century Rio de Janeiro. GeoPUC 4(7):1–18

Pereira E, Braga PMC, Mendes CT, Bergamaschi S (2012) Quaternary sedimentation in the coastal plain of Jacarepaguá and Guaratiba (Rio de Janeiro State). In: Rodrigues MAC, Pereira SD, Santos SB (eds) Baía de Sepetiba: estado da arte. Editora Corbã, Rio de Janeiro, p 63–82

Pinto MN (1965) The city of Rio de Janeiro: physical and human evolution. Rev Bras Geogr 27(2):191–232

Riccomini C, Peloggia AUG, Saloni JCL, Kohnke MW, Figueira, RM (1989) Neotectonic activity in the Serra do Mar rift system (southeastern Brazil). J South Am Earth Sci 2(2):191–197

Riccomini C, Sant'anna LG, Ferrari AL (2004) Geological evolution of the continental rift of southeastern Brazil. In: Mantesso-Neto V, Bartorelli A, Carneiro CDR, Neves BBB (eds) Geologia do continente Sul-Americano: evolução da obra de Fernando Flávio Marques de Almeida. Editora Beca, São Paulo, p 383–405

Rocha Leão OM (1997) Potencialidades e limitações da revegetação no controle da hidrologia e erosão superficial, maciço da Tijuca-RJ. Dissertation, Master in Geography, Federal University of Rio de Janeiro—UFRJ, Rio de Janeiro

Roncarati H, Neves LE (1976) Projeto Jacarepaguá: estudo geológico preliminar dos sedimentos recentes superficiais da baixada de Jacarepaguá, município do Rio de Janeiro—RJ. Petrobras, Rio de Janeiro

Roncarati H, Barrocas SLS (1978) Projeto Sepetiba: estudo geológico preliminar dos sedimentos recentes superficiais da baía de Sepetiba—municípios do Rio de Janeiro, Itaguaí e Mangaratiba—RJ. Petrobras, Rio de Janeiro

Ross JLS (1992) The cartographic record of geomorphic facts and the question of relief taxonomy. Rev Dep Geogr 6:17–29

Ruellan F (1944) Geomorphological evolution of Guanabara Bay and the surrounding regions. Rev Bras Geogr 4(4):455–508

Shinzato E, Renk JFC, Dantas ME, Teixeira WG, Vargas L (2012) 3D geotechnology in the geomorphological compartmentalization of the watersheds of the rivers Macacu and Caceribu—recôncavo da Baixada da Guanabara/RJ. In: 9 Simpósio Nacional de Geomorfologia, Rio de Janeiro, 2012. UGB, Nova Iguaçu

Silva TM, Ferrari AL, Tupinambá M, Fernandes NF (2015) The Guanabara Bay, a giant body of water surrounded by mountains in the Rio de Janeiro metropolitan area. In: Vieira BC, Salgado AAR, Santos LJC (eds) Landscapes and landforms of Brazil. Springer, Dordrecht, p 389–399

Soares MTS (1965) Fisionomia e estrutura do Rio de Janeiro. Rev Bras Geogr 27(3):329–387

Vieira BC, Fernandes NF (2004) Landslides in Rio de Janeiro: the role played by variations in soil hydraulic conductivity. Hydrol Process 18:791–805

Zaú AS, Coelho Netto AL, Oliveira RR (2007) Forest cover: degradation, ecological consequences and edge effects in Tijuca National Park, Rio de Janeiro/RJ. In: 1 Encontro IALE-Br, Rio de Janeiro, Vale do Paraíba do Sul, São Paulo, Apr. 2007

Internal Sedimentary Architecture and Geochronology of a Regressive Holocene Coastal Plain Under Fluvial Influence: An Example from Rio de Janeiro Coast, SE—Brazil

Maria Emília Radomski Brenny, Thais Baptista da Rocha, Israeli Rodrigo Mathias dos Santos, and Guilherme Borges Fernandez

Abstract The aim of the present work is to investigate the internal sedimentary architecture (ISA) and geochronology along a regressive coastal plain located between Rio das Ostras and cape Búzios, Rio de Janeiro, SE, Brazil. This study considers the role of mean sea-level fluctuations during the Holocene and the influence of fluvial-estuarine processes on its evolution, based on geomorphological characteristics, shallow geophysics (GPR), and geochronology (OSL). The OSL ages ranged from 6.880 ± 630 years to 1.940 ± 180 years. The geomorphology of the coastal plain shows a predominant set of beach ridges surrounded by fluvial deposits. The ISA presents four radarfacies indicated by beach/foreshore; upper shoreface; washover/overwash process and fluvial sediments. The depositional architecture showed the transition from a transgressive to regressive pattern, dating from 5.800 ± 750 years, occurred close to Holocene Transgressive Maximum (MTH). After MTH the coastal plain prograding by formation of beach ridges under sea-level fall, where the reflectors identified from marine-influenced by fluvial process, suggesting the relationship between modern fluvial sediment input and sea-level behavior.

1 Introduction

The coastal plains observed on wave-dominated coasts present, not rarely, sedimentary environments formed by sandy materials deposited by wave action, with or

M. E. R. Brenny (✉)
Department of Territorial Management, Geological Service of Brazil—CPRM, Rio de Janeiro, Brazil
e-mail: maria.brenny@cprm.gov.br

T. B. da Rocha · I. R. M. dos Santos · G. B. Fernandez
Physical Geography Laboratory, Universidade Federal Fluminense, Rio de Janeiro, Brazil
e-mail: thaisbaptista@id.uff.br

G. B. Fernandez
e-mail: guilhermefernandez@id.uff.br

G. Barbosa dos Santos et al. (eds.), *Geomorphology of Brazil: Complexity, Interscale and Landscape*, Springer Proceedings in Earth and Environmental Sciences,
https://doi.org/10.1007/978-3-031-05178-4_9

without fluvial contribution (Davies and Hayes 1984). For a better understanding in terms of paleogeographic evolution, these plains are normally described by sea-level oscillations during the Quaternary. Frequently, along these coastal plains, is possible to identify a complex of different environments as for example coastal lagoons, inlets, and foredunes, depending on predominant coastal process. When these plains are formed essentially by sequences of *sandy ridges*, *beach ridges*, or *beach/foredunes* ridges, these features can be identified as reliquary features, so passive to be associated as *Quaternary coastal geoarchives*, and in general characterize as regressive coastal plains (*strandplains*), where prograding process are predominant.

This progradation process promotes the relative projection of the coastline related to the ocean and is responsible for increasing the continental area over the sea. This mechanism can occur either eustatic sea-level falling (forced regression) or when the sediment input rate is greater than the accommodation space (normal regression), occurring even under conditions of gradual sea-level rising (Roy et al. 1994; Hesp et al. 2005; Dillenburg and Hesp 2009; Otvos 2012, 2019). This kind of process is relatively common along Brazilian coastline (Fernandez et al. 2019), mainly regarding during middle and late Holocene sea-level fall (Angulo et al. 2006).

Is a fact that regressive barriers investigated along Brazilian coast are dated by Holocene deposits (Dillenburg and Hesp 2009), and are associated with relative sea-level falling in the last 5500 years (Angulo et al. 2006). Some of these works focus specifically investigate this geomorphological process in specific sites along Northeast, Southeast, and South of Brazil, whose analyses are mainly developed from multiproxy analyses, usually involving the use of subsurface (shallow geophysics and boreholes), surface (geological/geomorphological mapping; digital and elevation models) and geochronological methods, normally using carbon isotopes and Optically Stimulated Luminescence (Martin et al. 1993; Lessa et al. 2000; Angulo et al. 2006; Barboza et al. 2009, 2011, 2014, 2021; Dillenburg et al. 2011; Souza et al. 2012; Silva et al. 2014a; Nascimento et al. 2018, Figueiredo et al. 2018, 2021; Rocha et al. 2019).

Along Rio de Janeiro coast, the evolution of Holocene coastal deposits does not present a predominant regressive characteristic, even the littoral submitted a falling sea level during the last 5500 years (Fernandez and Rocha 2015). For example, on the north-central coast of Rio de Janeiro, Silva et al. (2014b) and Figueiredo et al. (2018), investigated the evolution of Holocene barriers in Maricá and Massambaba, respectively, do not identify the regressive reflectors in the internal sedimentary structure. In both cases, Holocene coastal barriers are relatively narrow, usually around 300 m in length. The geophysical data present, in fact, identify some transgressive deposits (Silva et al. 2014b) and not rarely aeolian sediment deposits on the top of the barriers (Figueiredo et al. 2018), indicating a typical morphostratigraphic aggradational pattern (Fernandez and Rocha 2015). Along Paraíba do Sul wave-dominated delta, in the northern part of Rio de Janeiro, the Holocene evolution is marked by transgressive and regressive deposits, depending on the presence of river discharge or shoreface sediment supply. Rocha et al. (2013) show that in the area that receives sediments from the shoreface, the barrier experiments a retrogradational pattern, derived from overwash process and washover deposits, a transgressive barrier. On the other hand, Rocha et al. (2019) and Figueiredo et al. (2021) identify regressive features influenced by river sediment input, and/or changes in the wave climate, becoming a decisive mechanism for beach ridges and spits observed. Specifically, with respect to the internal sedimentary architecture in these examples from Rio de Janeiro, most of the morphostratigraphic barriers include GPR reflectors, and sometimes boreholes. Less attention was made about the influence of the fluvial-estuarine sediments, probably because of difficulties to obtain clear GPR data, and/or urbanization on most of these coastal plains.

That is not the case along Brazilian coastline. Some studies identify the role of fluvial materials on the barrier evolution. On the Southern coast of Brazil, Silva et al. (2014a, b); Barboza et al. (2014) and Bogo et al. (2015) show that interdigitation between fluvial and coastal deposits presents a fundamental role to understand the geochronological evolution of regressive barriers. Furthermore, Hein et al. (2016) demonstrated the impact of estuarine sediments and fluvial process in the Holocene evolution of regressive barriers, in Santa Catarina.

In this sense, this work aims to investigate the coastal plain evolution, based on internal sedimentary architecture and geochronology of the regressive coastal plain, located between the rivers Una and São João, center part of Rio de Janeiro coast, considering the role of mean sea-level behavior during the Holocene and the influence of estuarine/fluvial processes on its evolution.

2 Study Area

The Rio da Janeiro coast presents different coastal environments, distributed by coastal plains, where is possible to observe for example coastal lagoons, coastal dunes, mangroves, estuaries, and coastal sandy barriers (Fernandez et al. 2019). The Precambrian rocks define the length of most of these plains (Fernandez et al. 2019). The fluvial influence is limited in part of the coastline. Some of the fluvial input comes from Serra do Mar and reach the ocean, without accommodation space to form coastal plain, as observed in Southern part of the state (Fernandez et al. 2019). The influence of fluvial material is clearly along Paraiba do Sul wave-dominated delta, and in a embayment between cape Buzios and Rio das Ostras.

The study area, or the coastal plain located in this embayment, is markedly by influence of fluvial deposits, represented by São João, Una, and Das Ostras river, and marine processes (beach ridges), surrounded by Precambrian and Alkaline rocks (Fig. 1). In more detail, the coastal plain presents two depositional units, well-marked by the occurrence of fluvial and marine deposits, associated directly with São João river that controls fluvial deposits, and sequences of beach ridges regarding wave process. These deposits are accommodated by geomorphological units, associated with *Precambrian hills* and *Precambrian flatted surfaces* (Fig. 1). Two other relief units were also identified, such as coastal *massifs* lithologic associated with Cretaceous alkaline intrusive rocks (*Alkaline Intrusive Massifs*). The southernmost of the

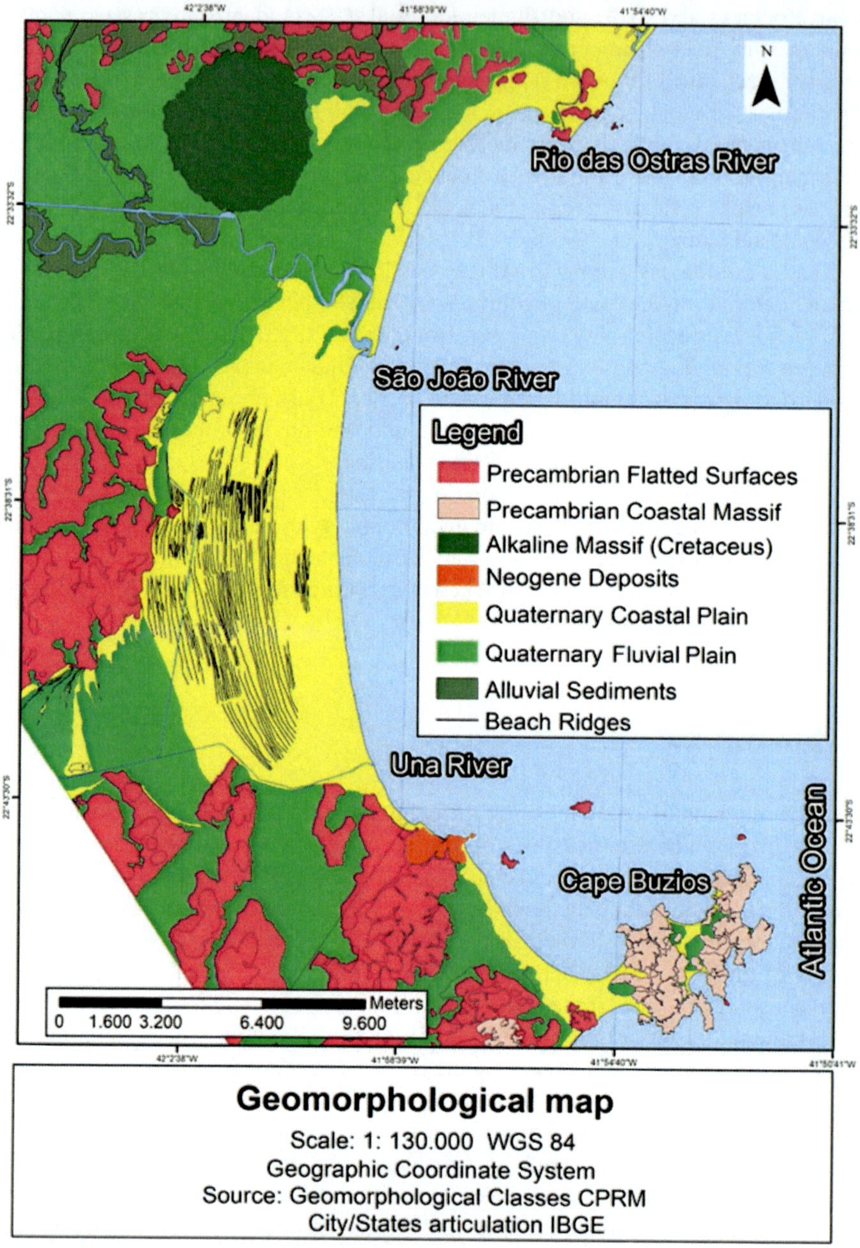

Fig. 1 Geomorphological map of the compartment between Cabo Búzios and Rio das Ostras with emphasis on the asymmetry of the extension between the coastal plains located to the South and North of the São João river. Adapted from classes mapping in scale 1:50,000: Brazilian Geological Survey—CPRM

area is markedly by tablelands developed over sediments associated with the Neogene (Fig. 1).

The coastal plain presents an intrigant asymmetry of fluvial-marine area, described by variable width, between north and south of the São João River. The marine sediments that compose the northern part extend for no more than one kilometer, while to the south, the wave-dominated plain (beach ridges) develops for approximately six kilometers (Fig. 1). This asymmetry probably is explained by the fluvial sediment contribution from São João river, toward the north by predominant northeast waves.

Regarding the wave and tide coastal processes, the area is characterized by a microtide regime, with maximum amplitude of 1.3 m. The main incidence waves come from north to northeast directions associated with fair-weather conditions. Waves from south and southwest indicate storm surges, formed by cold fronts.

Fernandez and Muehe (1995) describe the sedimentation of the inner continental shelf and found that is formed by reliquary sediments proximal to the coast, with occurrence of fine sediments from the 25 m bathymetric. The fine materials were associated with modern fluvial sedimentation, while the reliquary sands indicate probably exposed substrate of coarse sands related to the last Holocene marine transgression.

In a detail, Fernandez and Muehe survey the shoreface between Rio das Ostras and Cape Búzios, and show that fine and very fine sands close to beach, deposited in the southward part of the embayment, and associated from São João river source. On the other hand, coarse sands dominate the northern part of the area. This distribution defines the beach morphodynamics, which was classified as dissipative, due to the influence of fluvial fine sediments in the south, and reflective beaches domain the north, by coarse materials.

3 Methodology

This work is based on geomorphological, geochronological e geophysical methods. The geomorphology was determined by series of topographic profiles, to obtain the altitude of beach ridges for further adjustment of geophysical data. To obtain these profiles, we use Geodesic GPS and total station to extract the altimetric data directly from the terrain. The regional topographic was extracted in *Google Earth Pro*, and adjusted with the topography obtained in field.

During the geomorphological survey, series of ground-penetrating profiles was performed to obtain raw data for internal sedimentary structure analyzes (Fig. 2). The GPR lines were made in common-offset mode, using 270 and 400 MHz antennas. We survey the GPR and topographic profiles, in a cross-shore direction, i.e., from continental to the ocean.

Along the GPR profiles, geochronological samples were collected to be analyzed by the Optically Stimulated Luminescence (OSL) method (Fig. 2). The sample collection was carried out from manually opened trenches, at depths ranging from 1.2 m to 2.3 m and lateral dimension approximately of one meter. Seven samples were collected and distributed along the coastal plain (Fig. 2). In two points (sample OSL

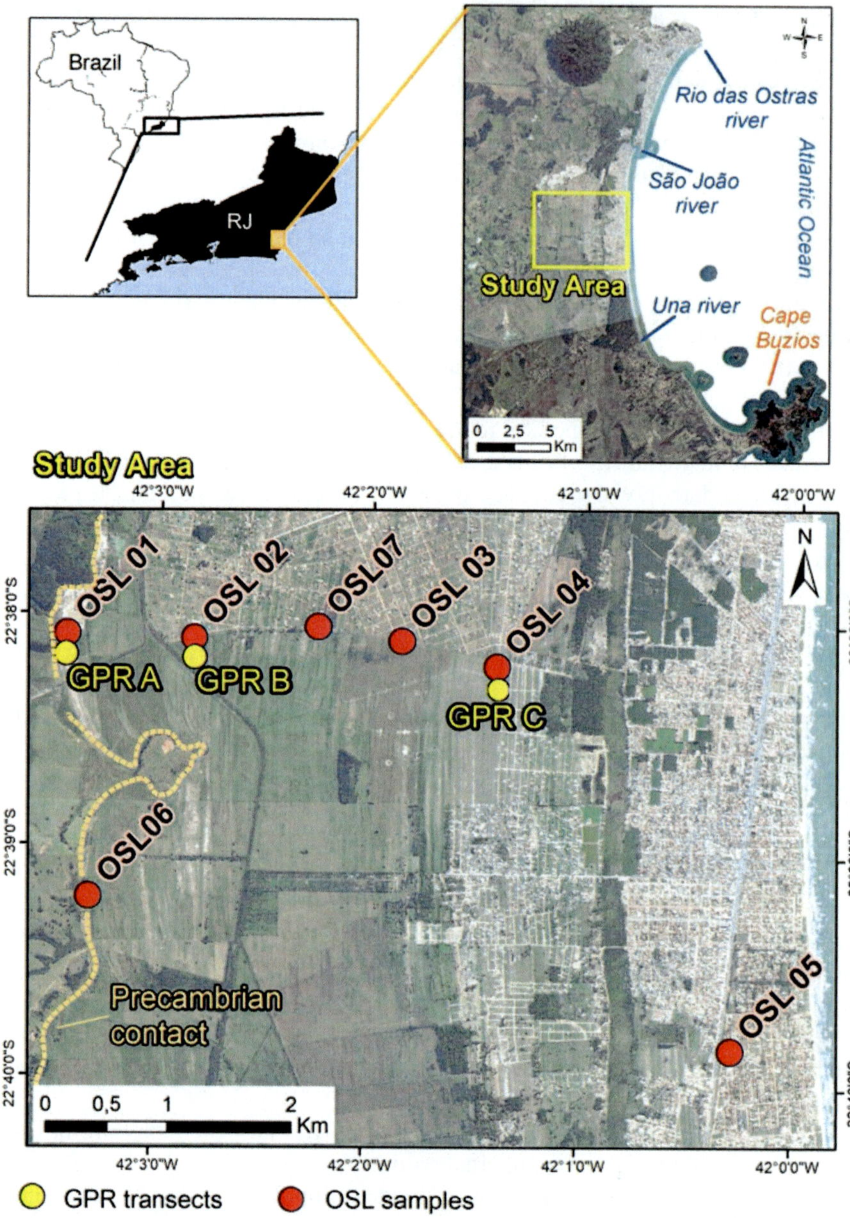

Fig. 2 Location of the GPR and topographic profiles and OSL samples, distributed along study area

1 and OSL 7), we collect two samples at different depth levels, to compare the data at two different depths. The geochronological analysis was defined according to the Single-Aliquot Regenerative-dose (SARs) protocol (Murray and Wintle 2003), using 24 aliquots.

In the laboratory, the processing of the GPR lines in RADAN™ 6.6 software for typical GPR procedures. To convert the depth obtained in GPR data from time (ns) into depth (m), we generate a velocity profile by Common-Mid-Point (CMP) mode with 80 MHz antenna, using the same procedures from Rocha et al. (2013). The velocity ranged from 0.15 to 0.06 m/ns, considered coherent for sandy sediments according to Nilsen and Clemmensen (2009). For the interpretation of the radarfacies, we consider some principles of stratigraphy interpretation, where we identify the morphology, terminations pattern, unconformities, dip direction, and continuity of the reflectors, as suggested by Neal (2004).

4 Results

4.1 Description and Interpretation of Radarfacies

Four radarfacies (Rf) were identified on radargrams showing the marine and fluvial process described in 6 m thickness. We select three GPR profiles, where the first one cover 230 m, second reach 185 m, and the last one 155 m. The radarfacies obtained are described above, and plotted in Figs. 3 and 4:

(a) Rf 1—Foreshore (berm and beachface): These radar-facies are characterized by sloping, sub-parallel geometry, slightly sigmoidal, moderately continuous reflectors, dipping toward the sea, and *downlap* termination. This Rf is interpreted as foreshore (berm and beachface). This Rf is identified in all three GPR line transects (*A*, *B*, and *C*), marking the upper limit of Rf 2, and may also be marked by erosional truncations related to the deposition of Rf 3 and Rf 4.

(b) Rf 2—Upper Shoreface: This Rf presents discontinuous reflectors, with irregular geometry, and may present a concave-convex pattern, with concordant upper limit in relation to Rf 1. In GPR A and GPR B lines, its occurrence tends to present an attenuated reflection, probably indicating the presence of shoreface fine sediments. It was not possible to establish the lower limit of these radarfacies due to the attenuation of the GPR signal, and limitations of the signal depth surveyed by high-frequency antenna used in this work.

(c) Rf 3—washover/overwash: This Rf presents sub-parallel to inclined geometry, predominantly dipping direction toward the continent. It presents upper and lowers limits identified by erosive surfaces, sometimes ending in *onlap*. Rf 3 was observed in GPR line A, whose transect is located near the contact between Precambrian and Quaternary deposits.

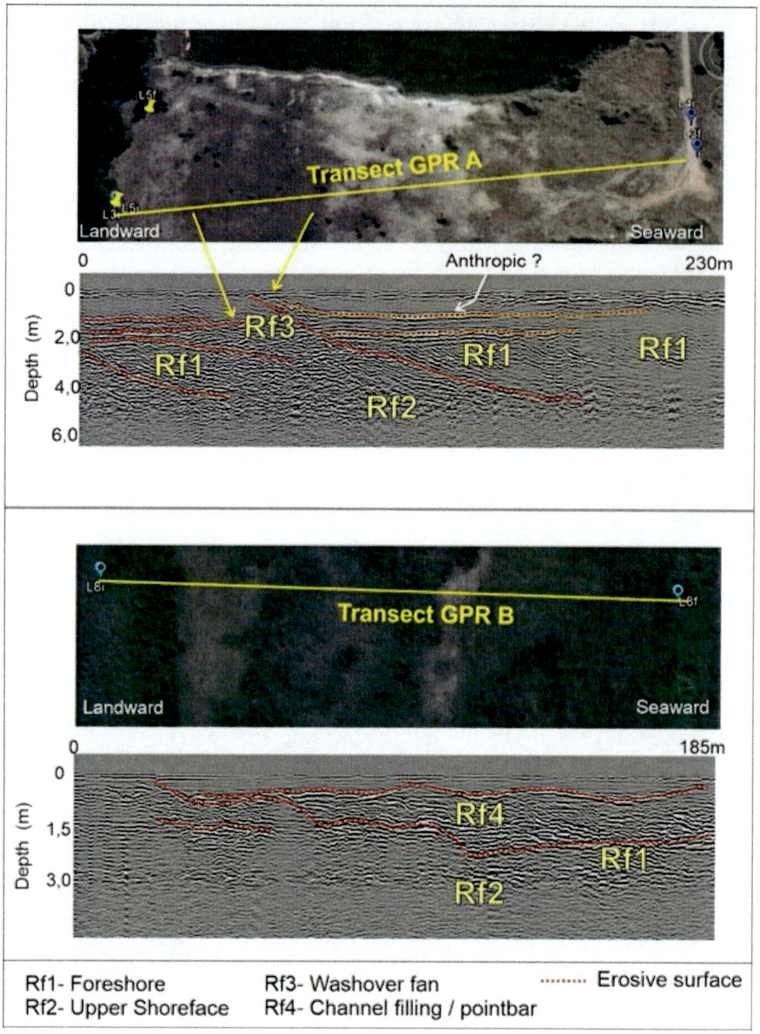

Fig. 3 Interpretation of GPR line A and GPR line B reflectors

(d) Rf 4—Channel filling/point-bar: It presents a sub-parallel to inclined geometry, with dip predominantly toward the sea. It presents limits marked by erosive surfaces that, in the case of the lower limit, appear indicating a paleochannel. This configuration was interpreted by fluvial channel filling by sediments derived from fluvial and marine processes, probably in the form of point bars (inlets), as identified in GPR line B.

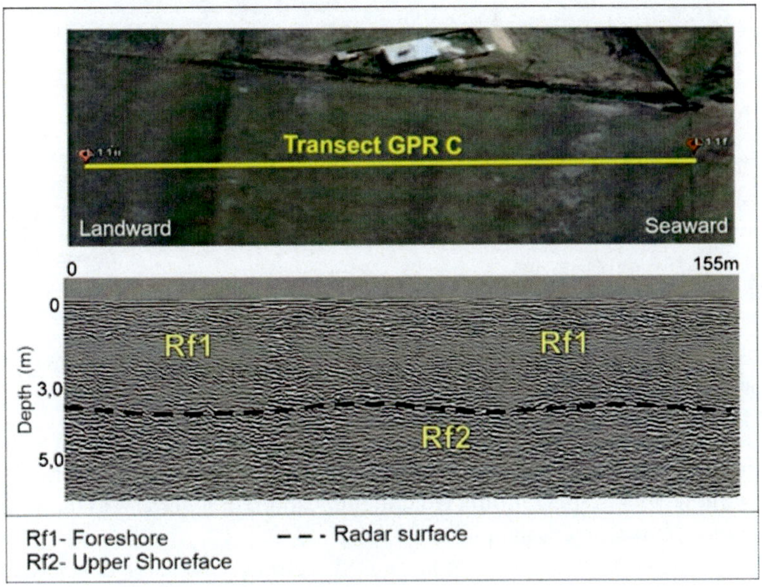

Fig. 4 Interpretation of the GPR line reflectors C

4.2 Geochronology of Beach Ridges and Progradation Rates in the Holocene

The results of the OSL samples indicated ages from middle to late Holocene, whose values ranged between 6.880 ± 630 years and 1.940 ± 180 years (Table 1). The samples positioning in the contact between Precambrian and Quaternary deposits (see Figs. 1 and 2) presented results of 4.140 ± 490; 5.800 ± 750 (OSL 1 A and

Table 1 Numerical ages obtained by the OSL method from samples collected in the coastal plain between the São João and Una rivers

Code	Sample	Depth (m)	Height (m)	Annual dose (μGy/y)	P (Gy)	Age (Ka)
5256	OSL 01 A	1.20	2.9	710 ± 47	2.9	4.140 ± 490
5257	OSL 01 B	2.30	1.8	625 ± 45	3.6	5.800 ± 750
5258	OSL 02	1.50	1.7	780 ± 50	1.8	2.390 ± 210
5259	OSL 03	1.50	1.5	550 ± 40	2.8	5.080 ± 460
5260	OSL 04	1.20	2.9	640 ± 45	2.5	3.880 ± 400
5261	OSL 05	2.00	0.7	760 ± 40	1.5	1.940 ± 180
5262	OSL 06	1.40	2.5	760 ± 50	5.2	6.880 ± 630
5263	OSL 07 A	1.20	1.9	1.000 ± 50	4.0	4.000 ± 330
5264	OSL 07 B	2.20	0.9	800 ± 45	3.7	4.550 ± 325

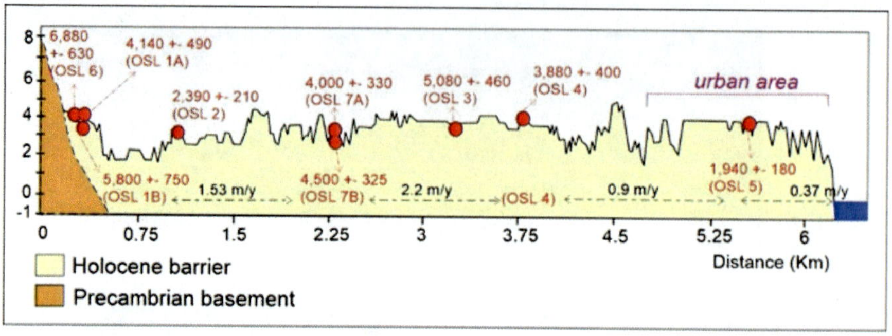

Fig. 5 Spatial distribution of OSL ages along the coastal plain between the São João and Una rivers, and evolutionary rates calculated by ages obtained

B, respectively); and 6.880 ± 630 years (OSL 6). Figure 5 shows the distribution, along with the topographic profile from the Precambrian basement to the shoreline, covering the Holocene coastal plain.

The geochronological distribution of OSL samples indicates that OSL 6 and OSL 1B mark the first Holocene deposits and the evolution of the coastal plain. In sequence, the OSL 2 (2.390 ± 210) presents the second most modern result, showing an unexpected result, compare to previous age, and the next deposition (see discussion for a better explanation). The sequence ages define from the geochronology continue from 4.500 ± 325 obtained on OSL 7B, 5.080 ± 460 identified by OSL 3, and 3.880 ± 440 on OSL 4. In the last sample, close to the beach, the age obtained was 1.940 ± 180 (OSL 5). With the geochronological results, we calculated the coastal plain prograding rate, by the comparison of the space between samples, and the ages. We notice that the coastal plain evolution presents 1.53 m/year between OSL 1B and OSL 7B. These rates increase from OSL 7B to OSL 4, where we calculate 2.2 m/yr. From OSL 4 to OSL5 we found a decrease in the velocity, changing to 0.37 m/yr (Table 2).

5 Discussions

5.1 *Relationship Between OSL Ages and Mean Sea Level Behavior in the Holocene*

Beach ridges are considered *Quaternary coastal geoarchives* that showed the wave process according to the mean sea level (**e.g.**, Tamura 2012; Dougherty 2014; Brill et al. 2015). Specifically, in the case of strandplains or regressive plains, the wave process forms a successive ridge deposit, by incorporating sub-horizontal and shore-oriented sandy materials (foreshore). This mechanism not only proves the incorporation of successive beach environment, and abandoned of the previous deposits

Table 2 Measured concentrations of the radioactive isotopes ^{232}Th, ^{238}U + ^{235}U, ^{40}K were used to calculate the annual dose

Code	Sample		Th (ppm)	U (ppm)	K (%)	Humidity (%)
5256	OSL 01 A		2.459 ± 0.215	0.196 ± 0.085	0.347 ± 0.053	3.2
5257	OSL 01 B		2.809 ± 0.221	0.115 ± 0.083	0.285 ± 0.051	3.8
5258	OSL 02		3.069 ± 0.245	0.327 ± 0.095	0.352 ± 0.057	2.9
5259	OSL 03		2.875 ± 0.213	0.281 ± 0.079	0.170 ± 0.047	16
5260	OSL 04		2.436 ± 0.213	0.487 ± 0.088	0.234 ± 0.052	11.9
5261	OSL 05		3.531 ± 0.235	0.498 ± 0.083	0.322 ± 0.049	14.8
5262	OSL 06		3.662 ± 0.256	0.185 ± 0.089	0.325 ± 0.055	3.6
5263	OSL 07 A		2.728 ± 0.244	0.201 ± 0.098	0.669 ± 0.062	8.2
5264	OSL 07 B		3.411 ± 0.255	0.186 ± 0.093	0.487 ± 0.058	17.9

(Tamura 2012), but favor the exposure of these sediments to the sun-light (Brill et al. 2015). In this sense, these features are able to be used as sea-level indicators, considering the actual beach as modern analog (Rocha et al. 2019) and provide confident OSL results (Brill et al. 2015).

In Fig. 6a, we plotted the OSL results distributed along sea-level curves from the most representative Brazilian coastline (Angulo et al. 2006), or local (Jesus et al. 2017). Considering the margin of accuracy of the foreshore deposits, the ages obtained are relatively well adjusted. In our case, the OSL samples were collected along the foreshore deposits described by RF1 and RF2, which indicates that these samples are confident to use for coastal evolution geochronology, even compared with the two curves. To compare the pattern observed for foreshores Rfs, we analyze an actual beach profile. The Rf is comparable, in-depth, as the foreshore reaches around 3.5 m from the berm crest to sea level, and fits with the depth obtains in GPR lines (Fig. 6b).

The age of sample OSL 1B is well adjusted to the Holocene Transgressive Maximum (HTM), which according to Angulo et al. (2006) would have occurred between 5,000 and 5,800 cal years AP. Considering the curve of Jesus et al. (2017), more specific for Cape Buzios area, the age also appears well adjusted using the altimetry and the age (Fig. 6a). In both curves, the MTH the mean sea level would be around 2.4 m above the current level. After MHT, the sea-level curves present a slow fall, and our ages are in according with this pattern (Fig. 6a).

5.2 Discussions on the Radarfacies and Internal Sedimentary Architecture of Holocene Beach Ridges

The identification of four Rfs, distributed in the three GPR transects, shows the regressive process related to the evolution of the coastal plain, where is possible to

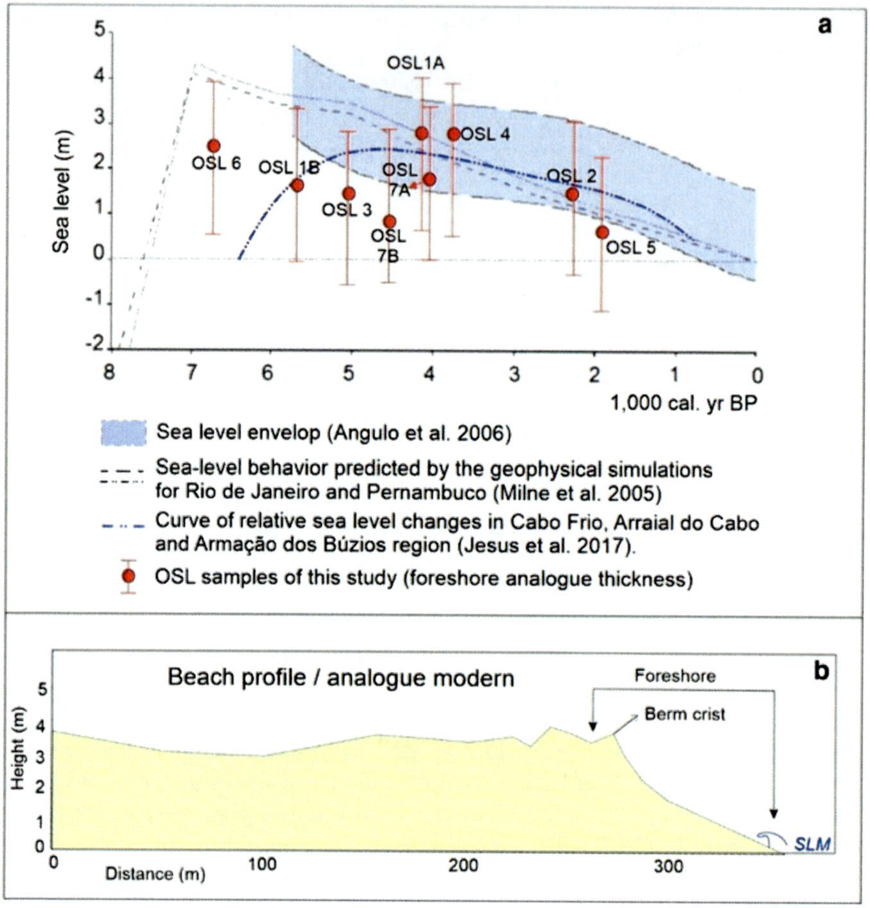

Fig. 6 **a** Overlay of the OSL samples collected in the present study with the sea-level change curve envelope for Brazilian coast established by Angulo et al. (2006); and with the curve for Cabo Búzios area established by Jesus et al. (2017). **b** Current beach topographic profile collected from the study area, to serve as a modern geomorphological paleoindicator analog

attribute fluvial/estuarine and marine processes, by the identification of foreshore and fluvial patterns. In adjustment with the geochronological data, different rates of progradation were established in addition with sea-level behavior. The GPR A (Fig. 3), positioned in the contact between Precambrian and Holocene coastal plain, shows a predominant foreshore pattern defined by Rf 1, but with an upper limit truncated by deposits described by overwash process (Rf 3). Considering that in this line is the OSL 1B sample, dated at 5.800 ± 750 years, which corresponds to the Holocene Transgressive Maximum (MTH) (Angulo et al. 2006), this depositional architecture may represent the transition from a transgressive to regressive pattern. This type of radarfacies presents the same structure as identified by Silva et al.

(2014b) and Nascimento et al. (2018) that probably indicate the changes between the maximum sea level, forming a transgressive barrier to regressive conditioned by sea-level falling.

In the central part of Rio de Janeiro coast (Maricá coastal plain), Silva et al. (2014b) and Silvestre et al. (2015) identified dip reflectors in inland direction, which the authors interpreted as records of the retrogradational phase of the barrier, corresponding to the MTH. In the coastal plain of the Itabapoana River, on the southern coast of Espírito Santo, Nascimento et al. (2018) also identified from GPR reflectors, the transition zone between the retrogradational and progradational pattern, characterized by reflectors with dip direction toward the continent, close to reflectors with dip direction toward the sea. The authors also associated this change with the transition from MTH to the sea-level falling.

After the MHT, our GPR results show a typical progradational pattern, which would result from the post-MHT phase, which is well characterized in the GPR C, where Rf 1 lies over the reflectors of Rf 2, indicating the migration of the beach deposits (*Foreshore*) over shoreface environment, indicating a regressive behavior. This pattern is compared to Niedoroda et al. (1985) and Holz (2012) and is morphologically characterized by a sequence of beach ridges. Considering the altimetric information obtained with geodetic GPS, coupled with MDE data provided by *Google Earth Pro* platform, the topography of the coastal plain indicates that this progradation probably occurred due to normal regression, i.e., when the sediment input rate is higher than the creation of accommodation space (Holz 2012). In this case, even the relative mean sea level has been falling, especially since the last 5500 years (Angulo et al. 2006; Jesus et al. 2017), the directly fluvial sediment input seems to present an important role on the evolution of regressive coastal plain.

The influence of fluvial sedimentation on the evolution of the coastal plain is also corroborated by the internal sedimentary architecture, as identified in GPR B (Fig. 3). In our interpretation, the beach deposits (Rf 1) observed in GPR B are interdigitated by a fluvial radarfacies, interpreted as probably paleochannel, formed parallel to the shoreline, which is subsequently filled by point-bar deposits (Rf 4). Using a modern analog (Fig. 7), this process was actually identified in the Una River. In 2014, the Una River channel present a morphology parallel to the shoreline before inflected to the ocean, where it is even possible to observe the formation of point bars near the concave margin. In 2016, probably after a climate and/or oceanographic event characterized by heavy rainfall and/or high-energy waves, the Una River suffered an avulsion process and abandoned part of the channel on the coastal plain. In 2017, it is possible to identify the reworking of these fluvial deposits abandoned by the action of storm waves and the partial covering in 2018 by beach sediments.

According to Barboza et al. (2014), the association of marine and fluvial processes can generate an interdigitation between coastal and fluvial sediments, and be preserved in the internal sedimentary architecture. Similar reflectors were also identified by Silva et al. (2014a) and Bogo et al. (2015) as examples of regressive barriers on the southern coast of Brazil. In the same geological/geomorphological context, Hein et al. (2016) identified in Tijucas Plain, Santa Catarina, by sediment analyses in

Fig. 7 *Google Earth* images showing the current example of interaction between coastal and fluvial processes, where it is possible to observe the Una River sectioning the coastal plain, with its channel flowing parallel to the coast; subsequently occurring processes of channel avulsion, abandonment, and reworking of the fluvial deposits by wave action

boreholes, marine materials identified by decreasing the diameter of sands that characterized a typical regressive Holocene plain, over fine (stilts and clays), associated by fluvial/estuarine environments.

Despite the absence of borehole data, the morphological characteristic of the strandplain, surrounded by gentle depressions associated with fluvial materials, over shoreface, characteristic from Rf 2, as observed on GPR A and B, is interpreted as beach ridges interdigitated with fluvial/estuarine pattern. Even trying to use a low-frequency antenna (200 MHz), we do not succeed to obtain deeper reflector, probably because of the attenuation of signal, caused by fine sediments on the shoreface. In fact, this can be attributed by the modern fluvial sedimentation on the shoreface, according to surficial sediment map.

6 Conclusions

The coastal plain between cape Búzios and Rio das Ostras is formed by fluvial and marine deposits, surrounded by Precambrian and Cretaceous geology. The coastal plain, divided by São João river present in the southern part, series of beach ridges and fluvial sediments. The geochronological data obtained along the strandplain showed that the transition from a retrogradational to progradational pattern, with an age of 5.800 ± 750 years, which was associated with the Holocene Transgressive Maximum (THM). The OSL ages varied between 6.880 ± 630 years and 1.940 ± 180 years and reveals variations in the coastal evolution rates during the Upper Holocene, whose rates were faster from 4.500 ± 325 years (2.2 m/yr) and slower from 1.940 ± 180 years (0.37 m/yr).

After MHT, the sea level tends to fall till actual level. This condition favors the formation of beach ridges that occurred under normal regression conditions, indicating that the sedimentary input rate seems to be a determining factor for the progradation, corroborating the importance of fluvial sedimentary input in the construction of the plain. This fluvial-estuarine influence was also identified from reflectors showing the filling of a paleochannel sectioning the coastal barrier.

In possible future scenarios marked by projections of sea-level rise, climate change and anthropic interventions in drainage basins that directly contribute to the coastal system and investigations about the fluvial-estuarine influence on the coastal sedimentation are relevant in the temporal context of the Anthropocene.

Acknowledgements This study was financed with the resources of the Division of Territorial Management (DIGATE) belonging to the Department of Hydrology and Territorial Management (DHT) of the Company of Mineral Resources Research—CPRM Geological Survey of Brazil. Maria Emília thanks the support ofthe following CPRM employees Maria Adelaide Mansini Maia, Maria Angélica Barreto, Marcelo Dantas, Alexandre Lago, Pedro Augusto dos Santos Pfaltgraff, Flasleandro Vieira Oliveira and Marcelo de Queiroz JorgeThanks also for the support de Helga Balbi, Suellen Golvea, Ivandro Patrick, Jonas Fish, Rodolfo José Angulo and Fabio Vinicius Feitosa de Albuquerque. Guilherme Fernandez thanks CNPq for research scholarship. Thais Rocha, Israeli Mathias, and Ivandro Patrick performed the field and laboratory activities with equipment available at the Physical Geography Laboratory (LAGEF) of the Universidade Federal Fluminense, funded by different funding agencies with CAPES and CNPq.

References

Angulo RJ, Lessa GC, Souza MC (2006) A critical review of mid- to late-Holocene sea-level fluctuations on the eastern Brazilian coastline. Qua Sci Rev 25(5–6):486–506. https://doi.org/10.1016/j.quascirev.2005.03.008

Barboza EG, Dillenburg SR, Ritter MN et al (2021) Holocene sea-level changes in southern Brazil based on high-resolution radar stratigraphy. Geosciences 11(8):326. https://doi.org/10.3390/geosciences11080326

Barboza EG, Dillenburg SR, Rosa MLCC et al (2009) Ground-penetrating radar profiles of two Holocene regressive barriers in southern Brazil. J Coast Res (SI 56):579–583

Barboza EG, Rosa MLCC, Dillenburg SR et al (2014) Stratigraphic analysis applied on the recognition of the interface between marine & fluvial depositional systems. J Coast Res (SI 70):687–692

Barboza EG, Rosa MLCC, Hesp PA et al (2011) Evolution of the Holocene Coastal Barrier of Pelotas Basin (Southern Brazil) a new approach with GPR data. J Coast Res (SI 64):646–650

Bogo M Souza MC, Angulo RJ et al (2015) Depositional architecture of the Holocene barrier in the southern portion of São Francisco do Sul Island, SC, Brazil. Pesqui Geociênc 42(3):281–295

Brill D, Jankaew K, Brückner H (2015) Holocene evolution of Phra Thong's beach-ridge plain (Thailand)—Chronology, processes and driving factors. Geomorphology 245, 117–134. https://doi.org/10.1016/j.geomorph.2015.05.035

Davis Jr RA, Hayes MO (1984) What is a wave-dominated coast?. Developments in Sedimentology 39:313–329. https://doi.org/10.1016/S0070-4571(08)70152-3

Dillenburg SR, Barboza EG, Hesp PA et al (2011) Ground Penetrating Radar (GPR) and Standard Penetration Test (SPT) records of a regressive barrier in southern Brazil. J Coast Res (SI 64):651–655

Dillenburg SR, Hesp PA (eds) (2009) Geology and geomorphology of Holocene coastal barriers of Brazil. Lecture notes in Earth Sciences, vol 107. Springer, Heidelberg

Dougherty, AJ (2014) Extracting a record of Holocene storm erosion and deposition preserved in the morphostratigraphy of a prograded coastal barrier. Cont. Shelf Res. 86, 116–131. https://doi.org/10.1016/j.csr.2013.10.014

Fernandez GB, Muehe D (1995) Recent sedimentary cover and bathymetry of the inner continental shelf between Macaé and Cabo Búzios, RJ. In: 6º Simpósio Nacional de Geografia Física Aplicada, Goiânia, 1995. UFG, Goiânia, pp 196–203

Fernandez GB, Rocha TB (2015) Holocene coastal barriers: geomorphology and depositional architecture on the coast of Rio de Janeiro. Rev Bras Geomorfol 16(2):301–319

Fernandez GB, Rocha TB, Barboza EG, Dillenburg, SR, Rosa, MLC, Angulo, RJ, Souza, MC, Oliveira, LHS, Dominguez, JML (2019). Natural Landscapes along Braziliann Coastline. In: Salgado A., Santos L., Paisani J. (eds) The Physical Geography of Brazil. Geography of the Physical Environment. Springer, Cham. https://doi.org/10.1007/978-3-030-04333-9_10

Figueiredo MS, Brill D, Rocha TB et al (2021) Late Holocene evolution of São Tomé cape (Rio de Janeiro, Brazil): insights from geomorphological, geophysical and geochronological data. Quaternary International 602:15–29. https://doi.org/10.1016/j.quaint.2021.04.001

Figueiredo MS, Rocha TB, Fernandez GB (2018) Geomorphology and internal depositional architecture of the Holocene coastal barrier of Massambaba, coast of Rio de Janeiro state. Rev Bras Geomorfol 19(3):447–464

Hein CJ, FitzGerald DM, Souza LHP et al (2016) Complex coastal change in response to autogenic basin infilling: an example from a sub-tropical Holocene strandplain. Sedimentology 63(6):1362–1395. https://doi.org/10.1111/sed.12265

Hesp PA, Dillenburg SR, Barboza EG et al (2005) Beach ridges, foredunes or transgressive dunefields? Definitions and an examination of the Torres to Tramandaí barrier system, Southern Brazil. An Acad Bras Ciênc 77(3):493–508

Holz M (2012) Sequence Stratigraphy: history, principles and applications. Editora Interciência, Rio de Janeiro

Jesus PB, Dias FF, Muniz RA et al (2017) Holocene paleo-sea level in southeastern Brazil: an approach based on vermetids shells. Journal of Sedimentary Environments 2(1):35–48. https://doi.org/10.12957/jse.2017.28158

Lessa GC, Angulo RJ, Gianinni PC et al (2000) Stratigraphy and Holocene evolution of a regressive barrier insouth Brazil. Mar Geol 165(1–4):87–108. https://doi.org/10.1016/S0025-3227(99)00130-9

Martin L, Suguio K, Flexor JM (1993) Sea level fluctuations during the upper Quaternary and the geological evolution of Brazilian "deltas". IG-USP Bulletin. Special Publication, 15: 01–186. https://doi.org/10.11606/issn.2317-8078

Murray, AS, Wintle, G (2003) The single aliquot regenerative dose protocol: Potential for improvements in reliability. In: Radiation Measurements. Pergamon, pp. 377–381. https://doi.org/10.1016/S1350-4487(03)00053-2

Nascimento FJS, Barboza EG, Rocha TB et al (2018) Chronostratigraphic analysis of coastal ridges in the coastal plain of the Itabapoana River mouth (Espírito Santo, Brazil). Rev Bras Geomorfol 19(3):503–523

Neal A (2004) Ground-penetrating radar and its use in sedimentology: principles, problems and progress. Earth Sci Rev 66(3–4):261–330

Niedoroda AW, Swift DJ, Hopkins TS (1985) The shoreface. In: Davis Jr RA (ed) Coastal sedimentary environments, 2nd edn. Springer, New York, p 533–624

Nielsen L, Clemmensen LB (2009) Sea-level markers identified in ground-penetrating radar data collected across a modern beach ridge system in amicrotidal regime. Terra Nova 21(6):474–479. https://doi.org/10.1111/j.1365-3121.2009.00904.x

Otvos EG (2012) Coastal barriers: nomenclature, processes, and classification issues. Geomorphology 139–140:39–52. https://doi.org/10.1016/j.geomorph.2011.10.037

Otvos EG (2019) Coastal barriers-fresh look at origins, nomenclature and classification issues. Geomorphology 355:107000. https://doi.org/10.1016/j.geomorph.2019.107000

Rocha TB, Fernandez GB, Peixoto MNO (2013) Applications of ground-penetrating radar to investigate the Quaternary evolution of the south part of the Paraiba do Sul river delta (Rio de Janeiro, Brazil). J Coast Res (SI 65):570–575. https://doi.org/10.2112/SI65-097.1

Rocha TB, Vasconcelos SC, Pereira TG et al (2019) Optically stimulated luminescence (OSL) dating of the beach ridges in the Paraíba do Sul River delta (RJ): considerations about the geomorphological evolution between Upper Pleistocene and Holocene. Rev Bras Geomorfol (20):563–580

Roy PS, Cowell PJ, Ferland MA et al (1994) Wave-dominated coasts. In: Carter RWG, Woodroffe CD, van de Plassche O (ed) Coastal evolution: Late Quaternary shoreline morphodynamics. Cambridge University Press, Cambridge, p 121–186

Silva AB, Barboza EG, Rosa MLCC et al (2014a) Meandering fluvial system influencing the evolution of a Holocene regressive barrier in southern Brazil. J Coast Res (SI 70):205–210. https://doi.org/10.2112/SI70-035.1

Silva ALC, Silva MA, Gambôa LAP, Rodrigues AR (2014b) Sedimentary architecture and depositional evolution of the Quaternary coastal plain of Maricá, Rio de Janeiro, Brazil. Brazilian Journal of Geology, 44(2): 191–206

Silvestre CP, Silva ALC, Silva MAM, Rodrigues AR (2015) Investigation of the internal structure and evolution of the Holocene barrier of Maricá (Rio de Janeiro). Revista Brasileira de Geofísica, 33(3): 461–475

Souza MC, Angulo RJ, Assine ML et al (2012) Sequence of facies at a Holocene storm-dominated regressive barrier at Praia de Leste, southern Brazil. Mar Geol 291–294:49–62. https://doi.org/10.1016/j.margeo.2011.10.009

Tamura T (2012) Beach ridges and prograded beach deposits as paleoenvironmental records. Earth Sci Rev 114(3–4):279–297

Geoenvironmental Analysis Under the Perspective of Geographic Information System (GIS) and Landscape Archaeology: Guarani and Kaingang Sites in the Anhumas Stream, Lower Paranapanema Region, SP

Larissa Figueiredo Daves and Neide Barrocá Faccio

Abstract The relationship of society with the environment in past times shows the relevance of interdisciplinary studies between Geography and Archaeology. In this case, we discuss the landscape of Guarani and Kaingang archaeological sites located near the banks of Ribeirão Anhumas, Municipality of Narandiba, SP. Cartography and the logic used in GIS for spatial analysis were carried out with emphasis on location, distribution of archaeological remains, as well as to demonstrate the form of implantation of the indigenous site Guarani or Kaingang in the relief, thus contributing to the analysis of geoenvironmental contexts in interdisciplinary research, especially by the triad—Geography, Landscape Archaeology, and Cartography, through the study of material culture and the form of implantation of archaeological sites.

1 Introduction

In the twenty-first century, contemporary geographers and archaeologists are approaching, especially in research into the organization of past occupations (KORMIKIARI 2000). Boado (1993) emphasizes that archaeology, when used as a research strategy, includes the study of all social and historical processes in its spatial dimension or, rather, intends to interpret areas of archaeological sites and the objects that specify them, either by recording the archaeological culture or material of a spatial matrix and then converting space into the object of archaeological research.

L. Figueiredo Daves (✉)
Human Geography, University of São Paulo—FFLCH-USP, São Paulo, Brazil
e-mail: larissadaves@usp.br

N. Barrocá Faccio
Department of Planning, Urbanism and Environment of Paulista State University—FCT/UNESP, São Paulo, Brazil

G. Barbosa dos Santos et al. (eds.), *Geomorphology of Brazil: Complexity, Interscale and Landscape*, Springer Proceedings in Earth and Environmental Sciences,
https://doi.org/10.1007/978-3-031-05178-4_10

Thus, the new technologies that enable professionals in the area to make maps linked to technologies with the use of software in order to represent the geographic space.

> (…) society, when appropriating nature, imprints on this objectivity an order that is expressed by geographical principles. And the appropriated nature becomes geographical environment, from then on the relation becomes society/geographical environment. In fact, the process of subjectivation/objectivation in the construction of the geographical environment is realized through the geographical principles as a dimension of existence, both of the subject and the object, embodying a process of totalization. Here is the geographic as an expression of the existence of totality. And between the geography of man and the environment are built the mutual geographical determinations in the objectification/subjectivation relationship (Martins 2016, p. 62).

Archaeological studies in Pontal do Paranapanema are scarce in relation to the establishment of Guarani, small-sized, and Kaingang sites, which are always small compared to the Guarani. According to Faccio (1998) the archaeological records are identified in different contexts and extensions. The denomination of small sites refers to archaeological sites of smaller extension, with low density of artifacts and distant from navigable rivers, located near springs, streams, and/or creeks. The small and medium-sized sites located in Pontal do Paranapanema is close to small watercourses (streams and creeks).

The pottery groups of the Pontal do Paranapanema, such as the Guarani, inhabited the terraces and mid-slope areas, passing through river channels and tributaries where they entered the Paranapanema River area. The possibility of contact between Guarani and Kaingang is evidenced by the analysis of pottery from Guarani sites, mainly in the Lower Paranapanema Valley area (Faccio 2011). The Kaingang of São Paulo, together with the Kaingang of the state of Paraná, constitute the largest southern Jê group. In the state of São Paulo, they occupied valleys and spurs in the interior of the state, bordering the Tietê, Peixe, Feio-Aguapeí, Paranapanema Rivers, and their tributaries. In southern Brazil, the Kaingang inhabited the valleys of the Tibaji, Ivaí, Iguaçú and Uruguai rivers (Pinheiro 1999).

The regional farming systems of the Paranapanema were dismantled by the various fronts of the Iberian invasion in the sixteenth century (Morais 2002). Regarding the regional farming systems in the Paranapanema River area, formed by sedentary communities originating from the Southwest and the South, they migrated through the Paranapanema and its tributaries along the left bank of the river (Morais 2002). Archaeological data reveal that these migratory fronts were occupied around the beginning of the Christian era, marked mainly by archaeological records of remnants of the villages of the regional Guarani system (Pallestrini and Morais 1982).

This paper presents the preliminary results of the spatial analysis of archaeological sites located near the banks of the Ribeirão Anhumas, from the description of their geoenvironmental components. They are indigenous Guarani and Kaingang archaeological sites: Santa Cruz do Anhumas II, Santa Cruz do Anhumas III, Santa Cruz do Anhumas IV, Santa Cruz do Anhumas V, São Saprino, Santa Helena, Tatu Galinha, Córrego da Boa Vista I and Córrego da Boa Vista II.[1]

[1] The Preventive Archaeology research carried out in the sugarcane plantation area of COCAL, Narandiba Unit, under the coordination of the archaeologist and Professor Neide Barrocá Faccio,

2 Study Area

The pottery groups of the Pontal do Paranapanema, such as the Guarani, inhabited the terraces and mid-slope areas, passing through river channels and tributaries where they entered the Paranapanema River area. The possibility of contact between Guarani and Kaingang is evidenced from the analysis of pottery from Guarani sites mainly in the Lower Paranapanema Valley area (Faccio 2011). The municipalities of Narandiba, Anhumas, and Taciba are located in the western part of the State of São Paulo, bordered to the south by the Paranapanema River and to the west by the Paraná River.

The Pontal do Paranapanema was occupied by groups of migrants from Minas Gerais and the Northeast Region of the country, between the late nineteenth century and early twentieth century, when the coffee cultivation consolidated in other regions of the State of São Paulo expanded, which resulted in an extraordinary demand for "land" by farmers (Abreu 1972). This occupation process, according to Abreu (1972), Monbeig (1984), and led to an intense process of environmental degradation (deforestation and soil degradation), followed by the extermination of the traditional indigenous and caboclo populations.

The delimitation for the region known as Pontal do Paranapanema is called Water Resources Management Unit (UGRHI-22) and is considered the 10th-Administrative Region of the State of São Paulo, according to the governmental administrative delimitation (Fig. 1).

According to Faccio (2011) in the Paranapanema Project (ProjPar) area, Kaingang archaeological sites were evidenced in the middle valley area; however, in the Baixo Paranapanema Paulista area, although Ethno-History reports point to the presence of Kaingang Indians and Guarani Sites have presented ceramics with brunidura (blackening technique of ceramics, recognized as Kaingang). Robrahn (1988) and Chmyz conducted an archaeological assessment in the middle Ribeira de Iguape area and detected Kaingang sites with dates between 600 and 270 BP.

Araújo (2001) reported that in the Alto Taquari basin, a tributary of the Paranapanema River, near the city of Itapeva, 60 km from the Paraná border and only 40 km upstream from the Middle Ribeira sites, he found 39 Kaingang-related archaeological sites, including open-air ceramic sites, in shelters, mounds, and underground houses, which confirm the preference of this indigenous group for areas of high relief.

The geomorphology of the Paranapanema River basin favored the occupation of these human groups, as it presents in its relief an abundance of resources such as silexite rocks, silicified sandstone, basalt, and also clay, thus making it easier for the indigenous groups to obtain raw materials for the manufacture of lithic and ceramic materials. The distinction of the occupation mode of indigenous groups in the Paranapanema River basin was influenced by the relief morphology and settlement pattern since in the upper basin of the Paranapanema River, the implantation of these

---

coordinator of the Laboratory of Guarani Archaeology and Landscape Studies (LAG) and of the Regional Archaeology Museum José Luiz de Morais (MAR)—FCT/ UNESP, Presidente Prudente Campus.

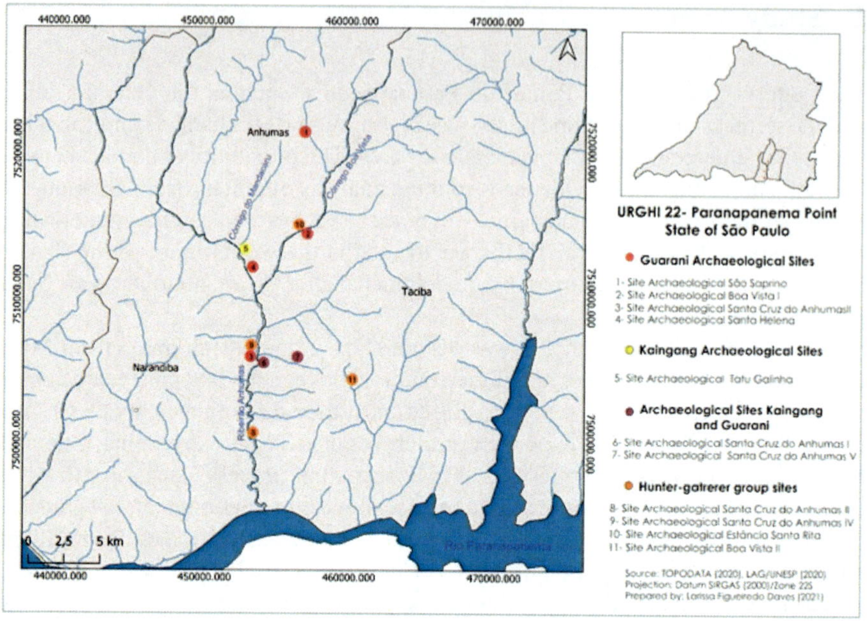

Fig. 1 Location of the archaeological sites located near the banks of Ribeirão Anhumas, Pontal do Paranapanema region (UGRHI-22). Municipalities in the study area: Narandiba, Anhumas, Taciba, State of São Paulo. *Source* Google Earth Pro Landsat Image. IBGE (2021). **Organization**: Larissa Figueiredo Daves

archaeological sites occurred in the hills interspersed by small drainage channels (Upper Paranapanema). In the middle basin, on the other hand, the settlement pattern continues to be amplified by a network of camps and chipping workshops along the large hydrographic channels (Middle Paranapanema).

3 Methodology

3.1 Theoretical Assumptions: The Concept of Landscape and Landscape Archaeology

Landscape Archaeology is an interdisciplinary knowledge with a vast amount of theories and methodological approaches to other sciences. When used as a research strategy, Landscape Archaeology includes the study of social and historical processes, in order to interpret landscapes and archaeological remains in their temporality in the spatial dimension.

The space is produced, resulting from this interaction and uses of nature, mediated by technique (artifacts) and also production of a landscape. Thus, the man does not occupy the space, he produces the space, and the landscape is the result of this dynamic interaction, as Bertrand (2009) puts it.

> The postulate that bases the analysis of the landscape then can only be social: "it is the production system in the broad sense, i.e., producing material and cultural goods, which, within a defined social group and in a given space, draws the material and cultural content of a landscape". (Bertrand and Bertrand 2009, p. 221).

The investigation of the settlement pattern makes it possible to relate geographical characteristics, such as topographic use and use of vegetation types; the social structure, inferring about its organization in the territory and, finally, the changes through time, with a view to relating the characteristics of the materials for comparations (Vogt 1956). Thus, we have a method the Landscape Archaeology, in order to analyze the elements of the physical and cultural environment in an integrated way.

> Because it understands the landscape in a dynamic way and is related to the integrative and polysemic debate, bertrand's proposal (2009) is used as the main approach, in which he states that "Landscape Archaeology should be apprehended as an attempt to refind the vestige of the historical relations established between society and nature." (BERTRAND 2009, p. 171).

As an example of this debate, we have the GIS (Geographic Information Systems) that presents extremely relevant potential for spatial analysis, researchers use such tool for cartographic production in social and/or environmental studies, in addition to computational systems for management of spatial data (Ferreira 2006).

GIS can be defined in two categories—1st GIS is a field of research that contains a **broad set of spatial analysis issues and** is known as the science of geographic information; 2nd **GIS is a toolbox of multiple technical uses** -set of techniques ancillary to the sciences in general (Goodchild 1992). In the first definition, GIS is geared to the paradigms of geographic information science—*GISystem* by *GIScience*. At this level, it presents functions of map analysis with theoretical-methodological basis, which were established by spatial analysis. At the second level, the GIS is seen as a set of techniques at the service of a science, either by the image processing system or by the system and access to database (Ferreira 2006).

The interdisciplinarity allows showing that the landscapes examined by the collaborators form a subjective perspective, locally situated, as something that not only shapes but is shaped by human experience (Bender 1993). The interpretation of archaeological data and interpretation of the landscape by spatial logic is fundamental to the study of human groups who inhabited the territory in past times, such perspective refers us to the "spatial turn," definition used in the human sciences for understanding in the function of images in the context of knowledge by imagery language (Fonseca 2007).

It is noted the intense use of GIS in archaeological research. According to Llobera (2003) the theme of GIS in archaeology has the purpose of representing the occupation of human groups in past times, as stated in the article "Exploring the topography of the mind: GIS, social space and archaeology". It is verified that from the visibility in the landscape and the topographic space it is possible to find evidence of the past occupation. Llorera (2003), and Gaffney and Stancic (1995) emphasize that the use of GIS for landscape analysis in archaeological sites makes it possible to represent the memory of occupation in a given place in the past, from evidence rescued in the current landscape.

3.2 *Methodological Procedures*

To analyze the Kaingang and Guarani settlement pattern in Pontal do Paranapanema on a local scale, the methodology proposed by Morais et al. (2000), called Regional Occupation System, is used as a starting point. Comparative studies are used to identify the settlement pattern. The intrasite spatial analysis has several objectives, both at the inferential level, which takes into consideration the reconstruction and explanation of past behaviors and unobservable activities and at the operational level of the relations between archaeological observations. In this context, the landscape analysis and the settlement pattern are fundamental for the discussion of the context of the archaeological sites, both in the cultural scope and in its relation to the physical aspects of the area (Carr 1984).

The data systematization was performed in three steps: a) Obtaining and manipulating SRTM data, by the pre-processed image with spatial resolution of 30 m from the letter 22S52RSN extracted by INPE—TOPODATA Project, for slope and altitude analysis. Such maps were prepared through the raster model, in order to perform the mapping of archaeological sites and their situation in the relief, in order to discuss the settlement pattern; b) The analysis of the spatial distribution of archaeological remains was performed based on the maps of vector model, initially, the data were entered into the software Q.GIS 3.18.2, classified by type and amount of archaeological material (ceramics and chipped lithic), with their respective coordinates in KML format—UTM coordinates (file compatible with GoogleEarth Pro).

The concentrations of the vestiges were represented by means of the graduated method in proportional circles (percentage %). c). The mapping was performed through the analysis of panoramic photographs and satellite images (Google Earth Pro- Image 2020 CNES/Airbus and Image Landsat 8). In the realization of the mapping was extracted geomorphological features (drainage, floodplains, and terraces, characterization of valley bottoms, and areas with presence of geoindicators) for a digital planialtimetric base, through juxtaposition of these features on a topographic sheet (from the IBGE, title of Esperança do Norte- SF-22-Y-B-VI-1, scale: 1:50 000, year 1973), according to drainage network—(Scale-temporal map of the Paranapanema River watercourse: years 1973 and 2021).

After the elaboration of vector data according to the concentration of traces in the perimeter of each archaeological site, we perform the treatment from symbols (triangles, pentagons, arrows) for representation of chipped lithics, polished lytics, and indigenous migration of each ethnic group with ethnohistorical survey basis according to the ethnologist Curt Nimuendajú (1943),[2] in order to show migratory flows of the Guarani and Kaingang in the twentieth century with the archaeological data of each tradition (Tupiguarani, Jê).

The fieldwork was conducted between 21 and 25 September 2020, by means of exploratory research in the study area, as well as landscape analysis of physical-geographic characteristics (relief, soil, vegetation, and hydrology). The methodology used was based on the levels of treatment proposed by Ab'sáber (1969), with emphasis on topographic compartmentalization, geomorphological features, and archaeological remains.

4 Results and Discussions

Cartography assists the representation of space in Landscape Archaeology through the preparation of maps, from the use of aerial photos, topographic charts (altimetry and planimetry of the terrain and contour lines), being fundamental to topographic representation, one of the ways of apprehending reality, because it is a quantitative apprehension of reality: the morphology through the slopes, altitudes, lengths and breaks of a ramp, slope (Crosby 1999).

The remote sensing through the analysis of satellite images, besides the elaboration of slope maps, in this case in studies of slope morphology and landscape dynamics (Casseti 1989; Ab'Sáber 1969). For possible discussion regarding the material culture and dispersion of archaeological materials on the perimeter of archaeological site, having probable interpretation of the settlement pattern and/or occupation system of these human groups, in time and space, together with data obtained by the dating of archaeological remains (environmental geoindicators and operating chain) (Faccio 2011; Morais et al. 2000).

Ab'sáber (1969)[3] made use of three levels of treatment for geomorphological research, with respect to the structuring of landscapes in the past and their evolution. According to Casseti, the understanding of the choice of the geographic position of the archaeological site offers auxiliary elements for the analysis of evidence of later colluvial settlements, by the physical–chemical characteristics of the correlative deposits. Thus, the analysis of the surface structure allows the researcher to prove the chronogeomorphological studies, from the dating of the archaeological remains and geoindicators with evidence of habitation in past times.

---

[2] from ethno-historical reports and field research, the ethnologist Curt Nimuendajú (1943) elaborated a map of the linguistic families of the State of São Paulo in the first half of the twentieth century.

[3] The first level studies the **compartmentalization of the regional topography**, with characterization and description, as exact as possible, of the relief forms of each of the studied compartments. The second level of treatment seeks to obtain systematic information about the **superficial structure of** the landscapes referring to all the observed compartments and relief forms. The third level seeks to understand the current morphoclimatic and pedogenetic processes in their full action in order to understand the **physiology of the landscape**.

From this mapping, we had the delimitation of the morphology of the relief and deployment of archaeological sites by spatial analysis and their location on the relief. The visual interpretation was listed by the visualization and quantification, laws of the two natures. In this sense, the cartographic representation is both a representation and a construction. From these elements were extracted geomorphological features (drainage, floodplains and terraces, characterization of valley bottoms and areas with presence of geoindicators) for a digital planialtimetric base, through juxtaposition of these features on a topographic chart, according to the drainage network, in this case, the watercourse of the Paranapanema River, Ribeirão Anhumas, near the area of the archaeological sites.

Thus, we contributed a historical approach to the landscape in order to represent by Cartography the occupation of the sites in question, mainly the pottery groups, in the area of Pontal Paranapanema, besides analyzing the landscape transformation and approaching the scenario in two moments: 1°-occupation of the landscape by the Guarani and Kaingang groups 2°-after the impact on the environment and its change in the landscape.

In this sense, the logic used in GIS for prehistoric spatial analysis on location, distribution of archaeological remains, and the implantation of archaeological sites in the relief show the relevance of landscape studies in archaeology. Thus, the application of GIS in archaeological research allows the analysis of archaeological attributes associated with geographic ones, and "this presents the possibility of tracking distribution and movements, as well as interactions between archaeological cultures" (Csáki and Jerem 1995, p. 85).

The physical-geographical characteristics of the Paranapanema River, especially its morphology and lithology, conditioned the way in which the geographic space was appropriated by pottery groups, mainly by the Guarani group. The area where the archaeological sites are located presents fluvial terraces and sloping surfaces of colluvial-aluvial deposits, predominance of sedimentary rocks of the Bauru Group (Adamantina Formation), and volcanic rocks of the São Bento Group (Serra Geral Formation). The macrorelevo is formed by the Peripheral Paulista Depression (Depression of the Paranapanema) and the morphoculture of the Paulista Western Plateau. The mesorelevo is characterized by broad hills of hilly relief with the presence of soils of the type: Red Latosolo (VL) and Red Podzsolos (RW) (Fig. 2).

The union of such a study allows us to understand the current physiognomy and reflect on its transformation over the history of use of the researched environment. The analysis emphasized the location for delimitation of the morphology and implementation of the archaeological site (relief compartmentalization). Table 1 shows the archaeological materials evidenced during the prospective diagnosis for the Guarani and/or Kaingang occupations in western São Paulo (Table 1 and Fig. 2).

Such physical-geographic characteristics, mainly the morphology and lithology, conditioned the way of appropriation of the geographic space by ceramic groups, mainly the Guarani and Kaingang groups. The concentration of ceramic material is found near the course of this stream, in an area of medium/low slope, with altitudes ranging from 340 to 430 m.

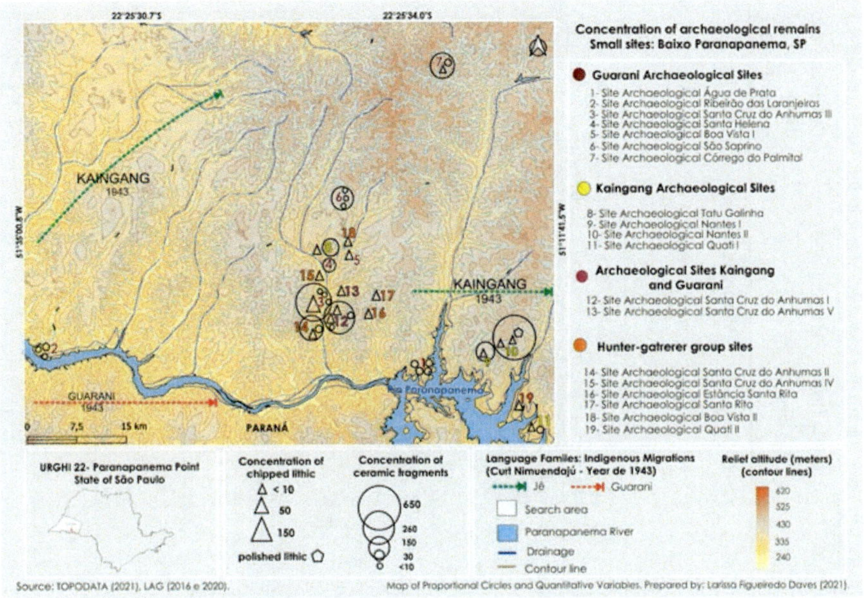

Fig. 2 Concentration of archaeological remains on the perimeter of the sites located near the banks of the Paranapanema River, Paulista side—Ribeirão Anhumas tributaries, Boa Vista Stream, Mandacaru Stream, Laranja Doce River. State of São Paulo. *Source* TOPODATA (2020). LAG/UNESP (2020). **Elaboration**: Daves

The archaeological survey on Kaingang archaeological sites shows the form of occupation on the relief, their preference for hilltop areas, although it presents the existence of sites located on low slopes near valley bottoms, while Guarani sites present location in medium/low slope areas near large navigational rivers (Araújo 2001; Fig. 3).

We found that the Kaingang and Guarani sites with the presence of ceramics are located on soft and/or undulated terrain, in areas of wide hills, while the lithic sites occupy the flat terrain, near the Ribeirão Anhumas fluvial terrace, near possible rocky outcrops with the presence of silicified sandstone pebbles and silexite, while the possible clay sources are located in areas of sand deposition in marginal dikes of the Ribeirão Anhumas.

The archaeological sites Tatu Galinha, Santa Helena, São Saprino, Boa Vista I and II, Santa Cruz do Anhumas I, II, III, IV, and V are located in the municipality of Narandiba, situated in the area of medium/low slope, near the confluence of the Mandacaru Stream with the Anhumas Stream, tributary of the Paranapanema River. The landscape of the Ribeirão das Laranjeiras site shows that it is located on a low slope, approximately 30 m away from the Paranapanema River, with the same characteristic as the Água de Prata site (Taciba municipality) and the Quati II site (Iepê municipality).

Table 1 Archaeological Sites located near the Ribeirão Anhumas, Boa Vista Stream and tributaries of the Paranapanema River. Municipality of Narandiba, SP

Archaeological sites	Lithic materials	Ceramic materials	Lytic polite	Site classification	Occupation in relief	Altitude (m)	Hydrography	Distance from the nearest body of water
Saint Cross of Anhumas I	35	61	1	Tupiguarani/Itararé	Medium/low strand	341	Anhumas Stream	220 m
Saint Cross of Anhumas II	5	–	–	Lithic Site	Medium/low strand	312	Anhumas Stream	380 m
Saint Cross of Anhumas III	–	17	–	Tupiguarani	Medium/low strand	340	Anhumas Stream	290 m
Saint Cross of Anhumas IV	5	–	–	Lithic Site	Low strand	340	Anhumas Stream	1,5 km
Saint Cross of Anhumas V	–	11	–	Tupiguarani/Itararé	Medium/low strand	404	Anhumas Stream	410 m from the spring 3 km from Ribeirão Anhumas
Boa Vista Stream I	–	37	–	Tupiguarani	Low strand	354	Boa Vista Stream	75 m of the Boa Vista stream
Boa Vista Stream II	7	–	–	Lithic Site	Low strand	365	Boa Vista Stream	120 m of the Boa Vista stream
Saint Saprino	–	9	–	Tupiguarani	Medium/low strand	407	Anhumas Stream	140 m from boa vista stream
Chicken Armadillo	8	40	–	Tupiguarani/Itararé	Low strand	360	Anhumas Stream	360 m

(continued)

Table 1 (continued)

Archaeological sites	Lithic materials	Ceramic materials	Lytic polite	Site classification	Occupation in relief	Altitude (m)	Hydrography	Distance from the nearest body of water
Saint Helena	–	10	–	Tupiguarani	High and medium strand	380	Anhumas Stream	630 m

Source Faccio et al. (2016). **Organization**: Larissa F. Daves

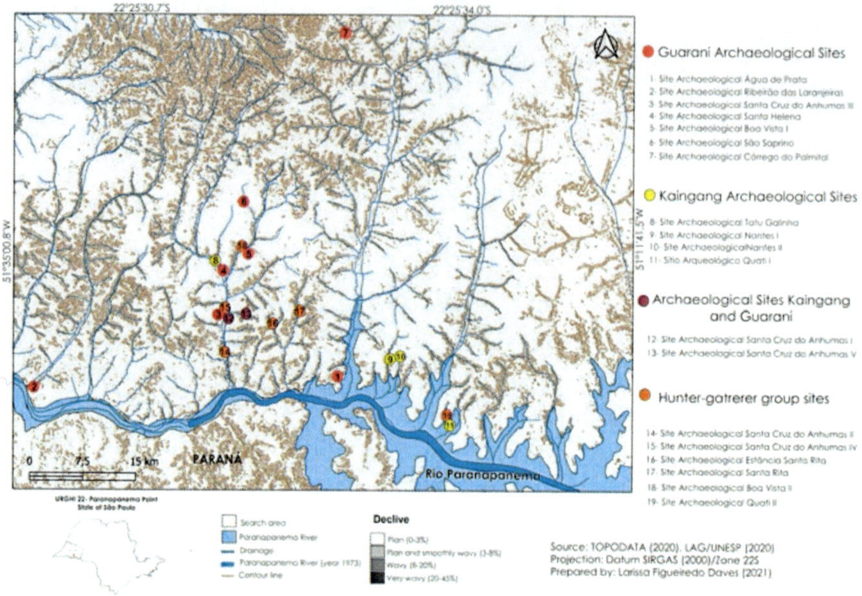

Fig. 3 Occupation in the relief of the lithic, Guarani, and Kaingang sites in the study area, Lower Paranapanema River course. *Source* TOPODATA (2020). Elaboration: Daves

These sites presented the presence of Guarani pottery, and in some cases also Kaingang pottery. The Nantes I and II sites are situated on the middle and lower slopes of a wide hill, approximately 50 m away from the Coroado Stream, also a tributary of the Paranapanema River, located in the Municipality of Nantes. They present blackened black ceramics as a result of the presence of burnishing (Fig. 4).

The archaeological sites located near the banks of the Ribeirão Anhumas make it possible to characterize the indigenous settlement pattern of Guarani and Kaingang groups far from the banks of the Paranapanema River.

In the Lower Paranapanema area, the large Guarani sites are characterized by occupation on hilltops and villages near the large marginal terraces of the Paranapanema River.

The Santa Cruz do Anhumas I archaeological site is located in an area of medium/low slope, 200 m west of the Anhumas stream and 12 km away from the Paranapanema River watercourse. We found the presence of a clay source (fluvial neosolo), being a possible geoindicator (highlighting floodplain and terraces) for pottery making in prehistoric times, located 300 m away to the west of the Santa Cruz do Anhumas I site (Fig. 5).

Analyzing the Paranapanema River watercourse, having as parameter the current context (2021), a new delineation can be noticed, leaving evident the present modification in the drainage with the submersion of the terraces and plains in the low concavity areas, besides the landscape transformation during the years, from 1973

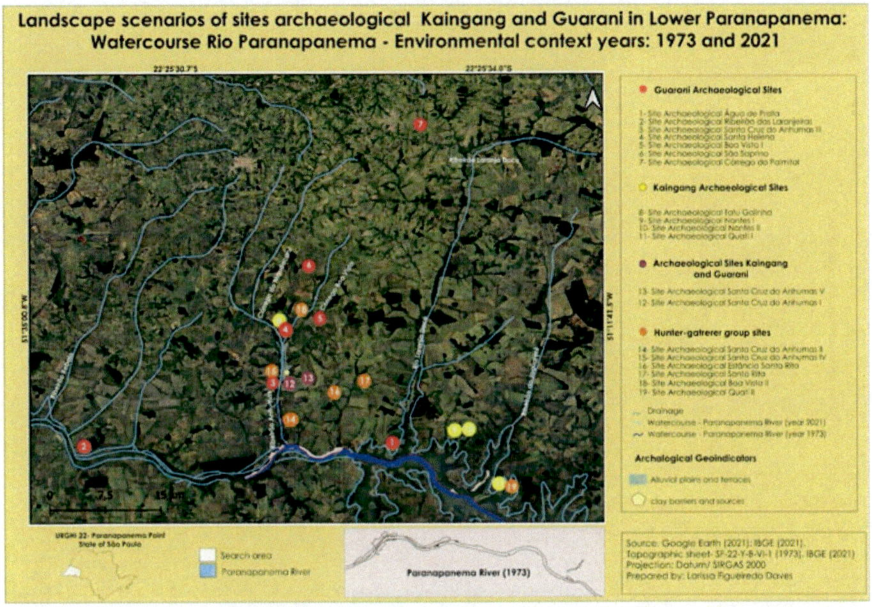

Fig. 4 Landscape of archaeological sites, Guarani and Kaingang course of the Lower Paranapanema River. Time scale 1973 and 2021. *Source* Google Earth (2021); IBGE (2021). Topographic sheet (1973), IBGE

to 2001, with the disappearance of the archaeological geoindicator sites, due to the formation of the Ada Usina Capivara lake, in the municipality of Iepê, SP.

5 Final Considerations

The cartographic representation of the geoenvironmental context of Guarani and Kaingang archaeological sites in the area surrounding the Ribeirão Anhumas allowed us to understand the spatial distribution of the phenomena studied, both in their vertical stratification and in their horizontal structure. From this cartographic representation, it was possible to indicate the dynamics of these indigenous groups, who lived in the period from 370 to 570 AP, which can be identified in the landscape of the present and in the logic of the location of the remains and materials collected from the past.

We conclude that from the landscape analysis, the settlement pattern of the archaeological sites located in Ribeirão Anhumas can be characterized by the presence of ceramics—Guarani, and Kaingang—in areas of gently undulating and/or undulating relief in areas of wide hills. The lithic sites occupy the flat relief near the low slope, on the fluvial terrace of the floodplain of this tributary of the Paranapanema River.

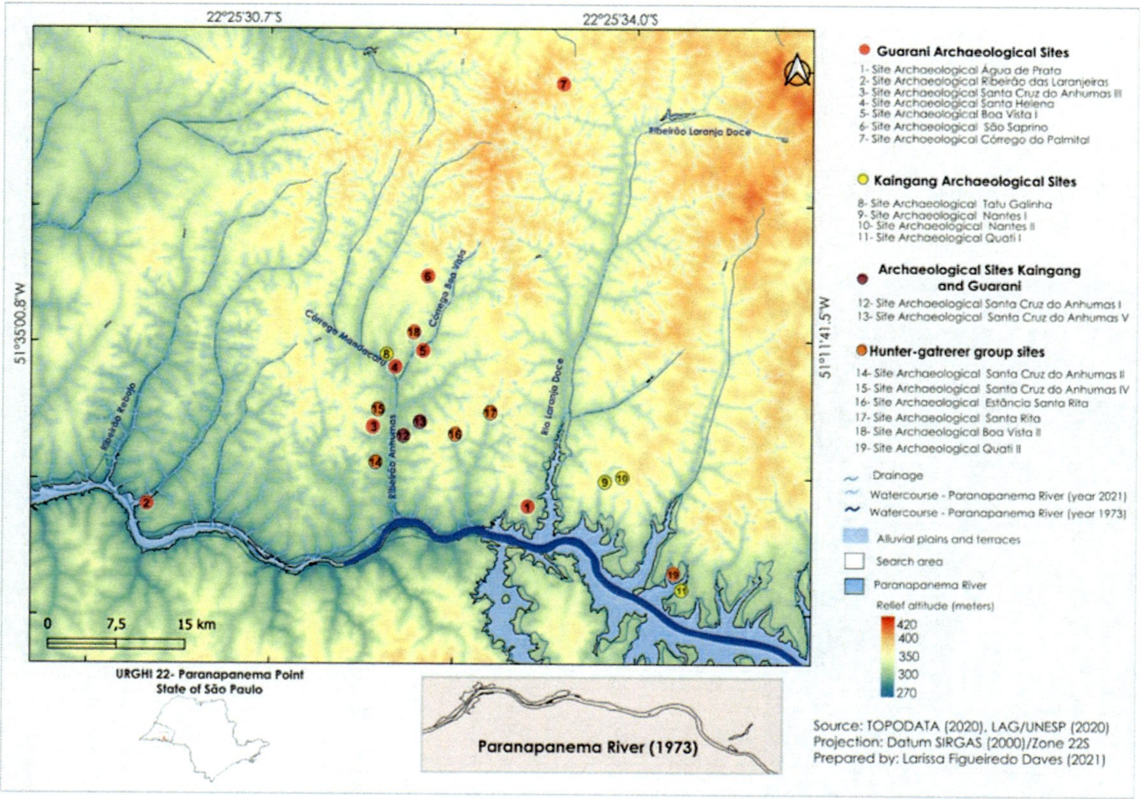

Fig. 5 Hypsometry map of the Guarani and Kaingang archaeological sites of the Lower Paranapanema River. Time scale 1973 and 2021. *Source* Google Earth (2021), IBGE (2021). Topographic sheet (1973), IBGE

The relevance of the analysis of geoenvironmental aspects in interdisciplinary research is noted, especially by the triad—Geography, Archaeology, and GIS, through the study of the material culture of archaeological sites and landscape analysis. This involves the mapping and elaboration of cartographic products about their occupation in the relief, in order to highlight the reasons that led these groups to establish settlement in this particular area.

Acknowledgements To the Laboratory of Guarani Archaeology and Landscape Studies (LAG) and the José Luiz de Morais Museum of Regional Archaeology (MAR) -FCT/UNESP, Presidente Campus. To Capes for funding the doctoral research (Graduate Program in Human Geography—FFLCH/USP).

References

Abreu, DS (1972) Formação histórica de uma cidade pioneira paulista: Presidente Prudente. Presidente Prudente. Ph.D. thesis. FFCLPP.

Anschuetz, KF; Wilshuen, RH.; Scheik, CL (2001) An Archaeology of Landscapes: Perspectives and Directions. Journal of Archaeological Research, Vol. 9, No. 2, 2001. https://doi.org/https://doi.org/10.1023/A:1016621326415

Ab'sáber, NA (1969) Um conceito de geomorfologia a serviço das pesquisas sobre o Quaternário. In: Geomorfologia, São Paulo, no. 18, 14p.

Araújo, AG (2001) Theory and Method in Regional Archaeology: a case study in Alto do Paranapanema, State of São Paulo. Ph.D. thesis. Faculty of Philosophy, Literature and Human Sciences, University of São Paulo.

Bertrand, G.; Bertrand, C (2009) **Uma Geografia Transversal e de Travessias:** o meio ambiente através dos territórios e das temporalidades. Editora Massoni, Maringá, PR.

Bender, B (1993) Cognitive Archaeology and Cultural Materialism. Cambridge Archaeological Journal, vol. 3, 1993, pp. 258-259.

Boado, FC.; Rodriguez, B. N.; Landín, D. C.; Vázquez, MD. et al (1991) **Arqueología del Paisaje.** El área Bocelo-Furelos entre los tiempos Paleolíticos y Medievales (Campañas de 1987, 1988 y 1989), vol. 6, p. 35.

Boado, CF (1993) Construcción social del espacio y reconstrucción arqueológica del paisaje. **Boletín de Antropología Americana**, 24, p. 5-29.

Casseti, V (1989) Elementos da Geomorfologia aplicada à Arqueologia. Revista do ICHL-Universidade Federal de Goiás.

Carr, C (1984) The Nature of Organization of Intrasite Archaeological Records and Spatial Analytic Approaches to their Investigation. **Advances in Archaeological Method and Theory**, 7 New York, Academic Press: 103–222. https://doi.org/10.1016/B978-0-12-003107-8.50008-6

Chmyz, I. et al (1968) Notas sobre a arqueologia do Vale do Rio Itararé. Revista do Centro de Ensino e Pesquisas Arqueológicas, Curitiba, 1:7-24.

Crosby, A (1999) The measurement of reality: quantification and Western society 12501600. São Paulo: UNESP Publisher.

Csáki, E.; Jerem, FR (1995) Data recording and GIS applications in landscape and intra-site analysis: case studies in progress at the Archaelogical Institute of the Hungarian Academy of Sciences. In: **Archaeology and Geographic Information Systems:** a European perspective. Lock, G.; Stancic, Z London: Taylor & Francis, p. 392.

Faccio, NB (1998) Estudo do Sítio Arqueológico Alvim no Contexto do Projeto Paranapanema. Master's Dissertation. São Paulo: FFLCH/USP, 1992. FACCIO, N. B. Arqueologia dos Cenários das Ocupações Horticulturas da Capivara, Baixo Paranapanema - SP. São Paulo: FFLCH/USP.

Faccio, NB (2011) **Arqueologia Guarani na Área do Projeto Paranapanema:** estudo dos sítios de Iepê, SP. Volume I. Tese de Livre Docência - Museu de Arqueologia e Etnografia, Programa de Pós-Graduação em Arqueologia - Universidade de São Paulo, São Paulo.

Faccio, NB, et al (2016 and 2021) Field report of the sugarcane plantation area of Usina Cocal. Museu de Arqueologia Regional/FCT/UNESP, Presidente Prudente, SP.

Ferreira, MC (2006) Considerações teórico-metodológicas sobre as origens e a inserção do Sistema de Informação Geográfica na Geografia. In: **Contribuições à história e a epistemologia da Geografia**. Bertrand Brazil, Rio de Janeiro, p. 101-125.

Fonseca, FP.; Oliva, JT (2012) Space and Cartography: Theory of Space and Assessments of Cartography and Pictorial Landscapes. **Territorium Terram Magazine**. v.1, p.24 - 45.

Fonseca, FP (2007) O potencial analógico da Cartografia. **Boletim Paulista de Geografia**, v.87, p.85 - 110.

Gaffney, V; Stancic, Z (1995). **GIS approaches to regional analysis:** A case study of the island of Hvar, Yugoslavia: Znanstveni institut Filozofske fakultete, University of Ljubijana.

Goodchild, MF (1992) Geographical information science. **International Journal of Geographical Information Science**. New York: Taylos & Francis, v. 6, n. 1, p. 31–45.

Hollenback, KL (2010) Landscapes. In: SCHIFFER, M. B. Behavioral Archaeology. London: Equinox, p.186–193.

Kormikiari, CN (2000) Landscape archaeology. **ResearchGate electronic journal**, p. 1–20.

Leite, JF (1981) A ocupação do Pontal do Paranapanema. 256 folhas. Ph.D. thesis - Institute of Planning and Environmental Studies of the Universidade Estadual Paulista - Campus de Presidente Prudente.

Lévy, J (2008) A cartographic turn? In: Social Cartographies and Territory. Henri Acselrad (organizer). Rio de Janeiro: Federal University of Rio de Janeiro, Institute of Research and Urban and Regional Planning.

Lévy, J (2003) Anamorphose. In: LÉVY, Jacques; LUSSAULT Michel (Org.). Dicitionnaire de la Géographie et de l'espace des sociétés. Paris: Belin, p. 74.

Llobera, M (2003). "Understanding movement: a pilot model towards the sociology of movement". In: G. LOCK (Ed.). Beyond the Map: Archaeology and spatial Technologies. Amsterdam: IOS Press, 65–84.

Martins, E (2016) Is geographical thought geography in thought? **GEOgraphia** - Ano. 18 - Nº 37. https://doi.org/10.22409/GEOgraphia2016.1837.a13758

Morais, JL.; Faccio, NB.; Piedade, SM (2000, 2002 and 2003) Salvamento arqueológico da UHE PIRAJU - ARQPIRAJU. Technical report of preventive archeology, vol. 1, 2 and 3. São Paulo.

Monbeig, P (1984) Pioneiros e fazendeiros de São Paulo. Translation Ary França and Raul de Andrade e Silva. São Paulo: Hucitec/Polis.

Moreira, R (2009) Velhos temas, novas formas. In: Mendonça, FA.; Kozel, S. (Org.). Elements of Epistemology of Contemporary Geography. Curitiba, PR: Editora UFPR, v. 2. P. 111–120.

Pallestrini, L.; Morais, JL (1982) Arqueologia pré-histórica brasileira. Editora Universidade de São Paulo - Museu Paulista Fundo de Pesquisas. 2nd edition, São Paulo,.

Passos, MM (2003) Biogeografia e Paisagem. 2ª ed: UNESP - PPGE, 2nd Edition.

Pinheiro, NS (1999) Vanuíre - conquista, colonização e indigenismo: Pontal do Paranapanema, 1912 - 1967. Ph.D. thesis. Programa de Pós-Graduação em História. Assis, 292 p.: il.

Robrahn, EM (1988). A Ocupação Pré-Colonial do Vale do Ribeira de Iguape, SP: Os Grupos Ceramistas do Médio Curso. Master's Dissertation. São Paulo, University of São Paulo.

Suertegaray, DM (2009). Physical Geography(?) environmental geography(?) or geography and environment (?). In: MENDONÇA, F. A.; KOZEL, S. (Org.). **Elements of Epistemology of Contemporary Geography**. Curitiba, PR: Editora UFPR, v. 2. P. 111–120.

Vitte, AC (2007) The development of the landscape concept and its insertion in Physical Geography. Revista Mercator, UFC, vol. 6, nº11, p. 59-70.

Vogt, EZ (1956) An appraisal of Prehistoric Settlement Patterns in the Nwe World. See Ref. 99, p. 173-82.

Cerro do Jarau and the Importance of Its Preservation as Records of the History of the Land and Its Current Scenic Beauty

Roberto Verdum and Lucimar de Fatima dos Santos Vieira

Abstract The Cerro do Jarau, which is located the southwest of Rio Grande do Sul, in the Pampa region, is the sixth impact crater (astrobleme) identified in Brazil, being considered a set of extreme quality, from the landscape point of view, with its uniqueness and its local and regional representativeness. In this sense, the main objective of the research is to identify the basic elements of landscape qualification of the space, from the point of view of perception, as a spatial and social element of reference for residents and passersby, especially for its tourist interest. As procedures, the methods that identify the landscapes, considered of great aesthetic value, as a consequence of the junction of significant visual properties, such as differentiated forms, exuberant colors, and elements of great proportions, among others, are adopted. These combinations form spectacle landscapes, the case of Cerro do Jarau, a privileged landscape, by tourism activities and interest, as geopatrimonial and regional references, because it has elements with outstanding geological and geomorphological particularities, which are of easy attribution to tourism and identity interests.

1 Introduction

For common sense, the term landscape suggests two distinct understandings: the objective and the representation. The idea that landscape is based on what the vision can reach—spatial scale—makes us build the notion of a more or less ordered mosaic of shapes and colors. In terms of temporal scale, we notice that the spatial cut, given by the vision, changes, that is, the landscape is endowed with a dynamic. All landscapes, which are transformed over time, can be objects of study, both from the isolated elements that compose it and their totality. However, this temporal dynamic suggests that each landscape contains an essentially unique structure and functioning, characteristics that lend a specific character to each landscape. Thus, studying the

R. Verdum (✉) · L. de Fatima dos Santos Vieira
Federal University of Rio Grande Do Sul (URFGS), Porto Alegre, Brazil
e-mail: verdum@ufrgs.br

G. Barbosa dos Santos et al. (eds.), *Geomorphology of Brazil: Complexity, Interscale and Landscape*, Springer Proceedings in Earth and Environmental Sciences,
https://doi.org/10.1007/978-3-031-05178-4_11

relationship between nature and society, having the landscape as a category of analysis, is extremely important because, through it, it is possible to understand, in part, the complexity of geographic space at a given moment, or over time, and its importance as a geo-historical reference of a human group in a given space.

By the studies carried out, Grehs (1969) and Lisboa and Schuck (1988), researchers at the Federal University of Rio Grande do Sul (UFRGS), based on the analysis of aerial photographs, satellite images and geomorphology of the region, proposed that the Cerro do Jarau, located in the municipality of Quaraí, in the southwest of Rio Grande do Sul, would have been formed by the impact of a meteorite (Fig. 1). Later, research conducted by the geologists Crósta et al. (2010), from the Institute of Geosciences of the State University of Campinas (Unicamp), found evidence that these elevations were formed as a result of the impact of a meteorite, which fell in the region millions of years ago, opening a large crater. Microscopic analysis of the rocks confirmed that they could only have been formed at extremely high temperatures and pressures, such as those generated by the fall of a celestial body. Over millions of years, wind, rain, and the movement of the planet's surface eroded the edges of the Cerro do Jarau, raising the geomorphological feature to altitudes of around 200 m at its vertical end, whose constituent rocks form a ring 3.5 km in diameter, marking the most central region of the crater, where the crash possibly occurred.

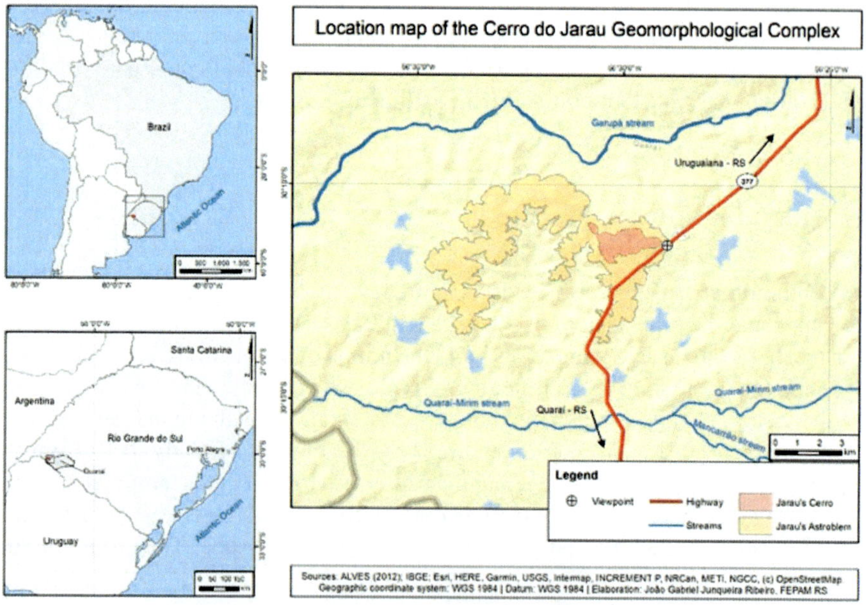

Fig. 1 Location of Cerro do Jarau, in the municipality of Quaraí, Rio Grande do Sul, Brazil. *Source* prepared by João Gabriel Junqueira Ribeiro, based on Alves (2012)

2 Presentation and Relevance of the Study Area

According to Sánchez et al. (2014), the evaluation of the structure of the rocks of Cerro do Jarau indicates two pieces of evidence of the fall of a celestial body in the area in question (Fig. 2). The first is the location of so-called impact breccias, rocks

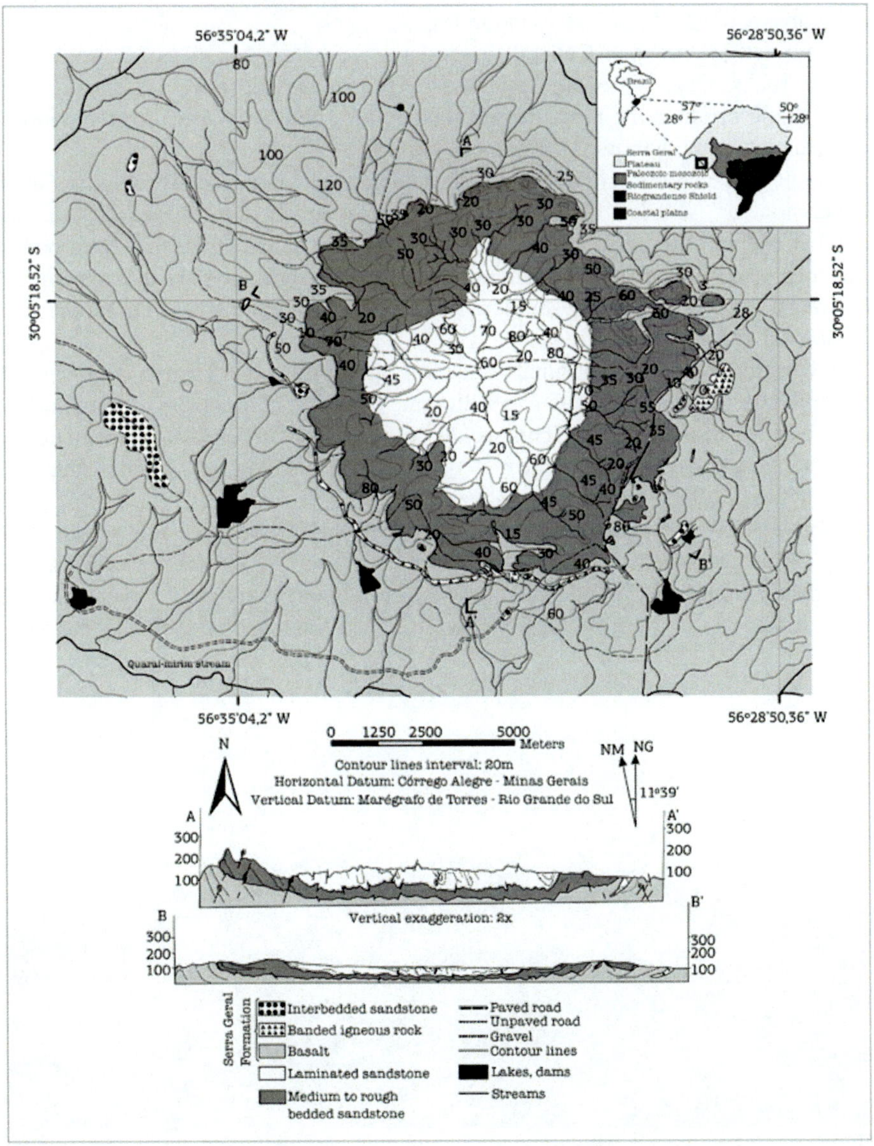

Fig. 2 Geological map of Cerro do Jarau. *Source* Adapted from Sánchez et al. 2014

formed by fragments of other rocks. The second and more conclusive evidence is that the quartz grains of the rocks suffered a phenomenon known as planar fracturing, as the samples analyzed present parallel traces of vitrified material, different from the natural structure of the quartz crystals. These grains are only formed in deeper regions of the planet, such as the mantle, located between 30 km and 2.9 thousand kilometers below the surface, where the temperature is thousands of degrees Celsius and the pressure is hundreds of thousands of atmospheres, enabling the formation of structures equivalent to those found in impact craters.

However, the rocks of Cerro do Jarau have characteristics of surface rocks, not mantle rocks. Only the energy released by the shock of a body such as a meteorite would produce the pressure and temperature necessary to cause this kind of deformation in the quartz of the planet's surface.

As mentioned, the Cerro do Jarau is the sixth impact crater—or astrobleme, Greek expression for "scar left by a star"—identified in Brazil (Crósta et al. 2010). The number is small but tends to rise, with time, as knowledge about the space bodies that hit Brazil in the distant past should increase. Geologists believe that the number of known astroblemes in the southern hemisphere is small because comprehensive geological surveys are lacking.

It is estimated that the original crater was approximately 13 km in diameter, but the difficulty in determining its size with some precision comes from the fact that the edge is quite eroded (Fig. 3). This would be a piece of fundamental information to accurately calculate the size of the meteorite that fell in the region, which is supposed to be between 600 and 700 m in diameter.

Another priority question is to find out when the impact occurred, which is no simple question to answer. To determine the crater's age, it will be necessary to find rock samples that had melted at the exact moment of impact, to measure their isotope ratio of the chemical element argon. The problem is that the rocks melted at

Fig. 3 Current morphology of Cerro do Jarau and its crater rim structure, eroded over time. *Source* Roberto Verdum's collection on November 27, 2019

the moment of impact may be very similar to those that make up most of the terrain of Cerro do Jarau, basically, basalt, igneous rock, formed at high temperatures, like those inside volcanoes. Moreover, considering the correction of the diameter of about 13 km of the crater opened by the impact and the action of weathering, the surface of occurrence of such rock fragments can be millimeters long.

According to Crósta et al. (2010), the maximum age of the youngest rocks affected by the impact (basalts) is around 135 million years, but as the crater edges are quite eroded, they are thought to be tens to a hundred million years old. This dating is important because it may reveal another story hidden in the geological record, given that an impact of this scale may have strongly affected life in the southern part of the South American continent, causing considerable local extinctions.

In addition, Jarau may also reveal more about Earth's past, since the collision of meteorites with basaltic rocks possibly causes specific transformations, which would allow us to differentiate their evolution from those of other types of rock—and reveal details of how other rocky planets were formed, such as Mars and Venus, where there is a lot of basalts.

2.1 Methodological and Operational Procedures

For the definition of the geographic space of reference of the study, using the landscape as a category of analysis, two levels of information were chosen:

(a) The landscape units, defined by FEPAM, for the environmental licensing of wind turbines;
(b) The municipal territory is defined as a reference for the request for licensing of wind farms, with FEPAM, by the entrepreneurs.

The proposed steps for the study of landscape perception indicators are as follows:

(a) A bibliographic survey of the methods, related to the study of the landscape, through the perceptual landscape approach;
(b) Bibliographic and visual surveys of studies on the deployment of wind turbines in the world, and adoption of methods for evaluating the indicators of perception given their installation;
(c) Elaboration of the research instrument, for the definition of landscape perception indicators;
(d) Search for iconscapes (identities) on the *websites* of municipalities potentially favorable to the installation of wind turbines, in this case, Quaraí;
(e) Research on *Google Maps* images, related to photographic records of landscapes of aesthetic and heritage interest, made by tourists, in municipalities potentially favorable to the installation of wind turbines.

Thus, to achieve the proposed objectives, the landscape perception methodology was developed, based on geographical, historical, and ecological recognition of the landscape. In this sense, levels of analysis were established, which refer to:

(a) The protection of the landscape as regards its natural and heritage features;
(b) Human perceptions, valuing individual and collective identities, related to the landscape, as elements or sets that people identify as references, through observation, characterization, and differentiation of landscapes (identity landscapes or icons);
(c) The publicization of municipal territories, by understanding the relations of social groups with their living spaces, i.e., the local landscape commons, which typify or function as an identity, brand, or attraction of a (municipal) territory;
(d) Differentiation of landscapes according to the temporal scale.

3 Presentation and Discussion of the Results

For Vieira (2014, p. 15), as "[…] object of contemplation, the landscape is usually linked to the memory of a place of great scenic beauty, concerning which one has, in memory, the record of some pleasant experience", while the scenic beauty is "[….] is characterized by being the central place of the observer's gaze when reading a landscape, that is, it is the scenario with formal and structural aesthetic properties marked by harmony, proportion, brightness and balance" (Vieira and Verdum 2017, p. 155).

The dichotomous classification of beautiful/weak is the simplest way to assess a landscape. However, there are other aspects to evaluate the quality of a landscape, such as integrity, diversity, uniqueness, and representativeness. The basic aspects of landscape perception consist of the spatial element (the landscape), the social element (the observer), and the subjective element (perception).

The landscape can be divided into three planes (Fig. 4), according to the elements captured by the viewer's vision and the distance of the elements arranged in space,

Fig. 4 Landscape plans. *Source* Roberto Verdum's collection on November 27, 2019

concerning the observer: the foreground, which is the zone of details, is located a few meters away from the observer; the landscape itself, in which the details are not distinguished, but the shapes of the elements of the landscape, observed at a distance of up to one kilometer; and the background, in which the eye no longer accurately distinguishes the characteristics of the elements, capturing only volumes, located more than one kilometer away (Brandão 2018).

Landscape plans are important and should be considered in the evaluation of a landscape and its constituent elements, especially if the evaluation is intended to identify its tourism potential, because the landscape is the product of tourism and there must be harmony between the three plans, forming a balanced and pleasant to the eye. This look, which is given, is from certain points of observation, which are as important as the landscape itself.

Therefore, landscape plans influence the intrinsic visual quality intrinsic to the landscape, as well as the visual quality of the immediate surroundings and the scenic background. The most important elements motivating the observer's perception of the landscape and consequently determining its visual quality are geomorphology, vegetation, the presence of water or rocky outcrops, and the altitude of the horizon.

In addition to landscape planes, shape, line, color, texture and scale and spatial configuration are important in determining the visual quality of the landscape (Kroeff and Verdum 2011; Vieira 2014), as well as the following properties:

- Diversity: expresses the landscape variety of a given territorial space. It is assumed, then, that a varied landscape contains more value than a homogeneous landscape, for having differentiated parts, with distinct visual elements and absence of monotony;
- Naturalness: degree of approximation of current conditions, verified in the landscape, to its natural form, free of human actions. The closer to this condition, the greater the naturalness;
- Singularity: natural or man-made occurrences in the landscape become points of visual attraction because of their uniqueness, scarcity, strength, traditional value, or historical interest;
- Topographic complexity: degree of movement or irregularity of the relief. The more irregular, with greater differences in level and with more distinct cardinal orientations of the slopes, the greater visual value the landscape has;
- Surface and water's edge: these are the natural forms of surface water, such as the sea, lagoons, and rivers. In turn, the water's edge is the boundary between the water surfaces and the other components, such as land, vegetation, and sky;
- Human actions: are responsible for the introduction of structures and artificial elements of superficial character (urban settlements, industrial complexes, crops), of linear character (roads, transmission lines), and punctual character (buildings, bridges, towers). Human actions modify the natural characteristics of the landscape (PIRES, 1996 apud Kroeff and Verdum 2011, p. 25).

Based on the forms and characteristics cited, the landscapes of great aesthetic value are a consequence of the junction of significant visual properties, such as differentiated forms, exuberant colors, and elements of great proportions, among others. These combinations form the landscapes privileged by tourist activity. Concerning the Cerro do Jarau, this has specific elements, which stand out and are easily visualized and appreciated.

It is also noteworthy that the perception of the visual quality of tourist areas is related to the natural potentialities, especially those that are prominent in the landscape, as is the case of this hill.

The *Atlas of the Scenic Beauties of the Pampa Landscapes: look, read, reflect and understand to value the landscape—Cuesta do Haedo region* (Vieira et al. 2018) presents some speeches, collected in interviews:

> "[...] by the contrast in the landscape, when seeing it." "[...] it is a form of relief that differs from the flatness of the fields. It brings to the residents a different dimension of nature. It has a cultural expression, a place of stories, legends, and films". "[...] by the presence of ornamental species, the view of the surrounding landscape and the traditional management of herds by the gaucho". "[...] for its imposing morphology, its ecological composition, and its historical references that have even made it a regional cultural icon". "[...] for its rare beauty, pristine environments, endemic/rare species, among others".

Concerning the concept of landscape, expressed by people, who register this morphology of landscape exceptionality, through photography, and by the municipality, which defines it as an outstanding set on its *website*,[1] it is emphasized that this is associated with the elements of nature, which are considered beautiful and pleasant: the green (field and bush), the coxilhas, and the animals in the field.

Additionally, this landscape is said to be notable for its natural beauty and its historical value, as this element is highlighted as a municipality symbol, through poems, and represents a monument of interest to be preserved as geopatrimonial (Borba 2014). The Cerro is also depicted in one of the oldest legends of the literature of the state of Rio Grande *do* Sul: *A Salamanca do Jarau*, by João Simões Lopes Neto, written in 1913 (Vieira et al. 2018). In this sense, the sensory evaluation of the Cerro do Jarau landscape, from the publicization of this landscape, by the municipal government, and the individual records, found on the satellite images of *Google Earth*, can be considered a rating of 5, on a scale between 1 and 5.

Among the main economic activities developed in the municipality, agriculture and livestock are recognized as activities that do not alter the landscape, both in the past and in the present, being part of the "natural context". However, the new projects, located near the hill, linked to the production of wind energy, have tensioned local actors, tourists, researchers from various fields of knowledge and the environmental licensing body of Rio Grande do Sul, the Henrique Luiz Roessler State Foundation for Environmental Protection (FEPAM), since such elements will be able to alter the landscape and interfere with its recognition as heritage.

[1] http://www.quarai.rs.gov.br/CONHECENDO_fotos_de_quarai.htm.

4 Conclusions

Through the studies carried out in the late 1980s, when the interest and need to know the geological-geomorphological genesis of the Cerro do Jarau, the recognition of this space as an icon of the local and regional landscapes was further increased, already recorded in the literature and the landscape matrix of the surrounding residents and passersby. The thesis that the morphology of this hill results from the impact of a meteorite reaffirms its originality and the interest in dating more precisely the episode is still considered an unknown in the field of science.

Today, this geomorphological feature, whose highest altitudes, which are around 200 m, mark a landscape referenced by the elements of nature that surround it. In this sense, the visual quality of this landscape can be specifically related to its naturalistic value (landscape unit, whose ecosystem conservation status has remarkable animal species or even natural singularities, related to geological-paleontological-geomorphological factors). From this, regarding the points of view of perception and human expressions, which refer to it and revere it, the Cerro do Jarau is undoubtedly an identity icon for its visual, ecological, and cultural qualities.

Therefore, if wind farm installation projects advance in the area surrounding this hill, it will be essential to consider this landscape as of aesthetic interest, as well as cultural heritage and geopatrimonial. In this case, it is essential and mandatory to propose scenarios that establish in detail the landscape of the future, inserting the wind turbines in the landscape, so that the population can have the ability to build a reference of the new landscape to be produced. In doing so, one can obtain a notion of the scalar dimension of the new elements that will be inserted into the landscape (wind turbines) and that are not necessarily (re)known by most of the people around the hill or even by passersby. There will certainly be changes in the shape of the landscape and its functionality, as well as restrictions and precautions regarding access to the hill and its surroundings from the time the wind farms are in operation.

Acknowledgements This article is part of the research on institutional relations between the PAGUS Landscape Laboratory, Department of Geography, the PPG in Geography (POSGEA) – from the Institute of Geosciences of the UFRGS –, and the State Foundation of Environmental Protection Henrique Luiz Roessler (FEPAM-RS), in the contexts of university research and extension.

References

Borba A (2014) Perspectives for research and action in geoconservation at the Federal University of Santa Maria (UFSM) with a focus on less developed areas of southern Brazil. Revista Ciência e Natura, 36.

Brandão GS (2018) Geotourism potential of the Caraá municipality: inventory of geodiversity sites as a subsidy for the development of geotourism. Dissertation (Master's Degree) Graduate Program in Geography, UFRGS, Porto Alegre.

Crósta AP, Lourenco FS, Priebe, GH (2010) Cerro do Jarau, Rio Grande do Sul: a possible new impact structure in southern Brazil. Special Paper of the Geological Society of America 465:173–190. https://doi.org/10.1130/2010.2465(12)

Grehs SA (1969) Aspectos Geológicos e geomorfológicos do Cerro do Jarau, Rio Grande do Sul. In: Proceedings of the 23rd Brazilian Congress of Geology of the SBG, pp 265 –272.

Katsaprakakis DA (2012) A review of the environmental and human impacts from wind parks. A case study for the Prefecture of Lasithi, Crete. Renewable and Sustainable Energy Reviews, 16(5):2850 –2863.

Kroeff L, Verdum R (2011) Identification of potential areas to the mapping of ecotourism trails in the Ecoparque property, in Canela/RS. Revista Brasileira de Geomorfologia, 12(3):131 –136. https://doi.org/10.20502/rbg.v12i0.266

Lisboa NA et al (1987) Reconhecimento geológico da região do Jarau, Quaraí, RS. In: Atas do 3 Simpósio Sul-Brasileiro de Geologia da SBG, 1:319 –332.

Lisboa NA, Schuck MTGO (1988) Caracterização de formas e padrões estruturais no Grupo São Bento da Bacia do Paraná no Rio Grande do Sul em imagens orbitais e suborbitais. In: Anais do Simpósio Brasileiro de Sensoriamento Remoto da SBG, 2:323 –333.

Sánchez JP et al (2014) Stratigraphy and structure of Cerro do Jarau: new proposal. Brazilian Journal of Geology, 44(2):265–276.

Siefert CAC, Santos I (2016) Assessment of the visual impact of wind farms on the quality and aesthetics of the landscape around protected areas: case study of the Guartelá State Park, PR. Ra'eGa Magazine, Curitiba, 38:221 –244.

Vieira LF dos S (2014) A valoração da beleza cênica da paisagem do bioma Pampa do Rio Grande do Sul: proposição conceptual e metodológica. Tese (Doutorado) – Programa de Pós-Graduação em Geografia, UFRGS, Porto Alegre.

Vieira LF dos S et al (2018) Atlas of the Scenic Beauties of the Pampa Landscapes: look, read, reflect and understand to enhance the landscape - *Cuesta do Haedo* Region. IGEO/UFRGS, Porto Alegre, 2018. Available at: https://lume.ufrgs.br/handle/10183/180921

Vieira LF dos S, Verdum R (2017) The landscape as a reading of scenic beauty, organization and use of rural space in the Pampa. In: Medeiros RMV, Lindner M (org.) Dinâmica do espaço agrário: velhos e novos territórios. Evangraf, Porto Alegre, 2017.

Technogenic Modifications in River Channels Associated with Urbanization—Ribeirão Brandão Basin, Middle Paraíba Do Sul River Valley, Southeastern Brazil

Lucas Cesar Figueiredo Hoepfner de Almeida, Eduardo Vieira de Mello, and Maria Naíse de Oliveira Peixoto

Abstract The study presented in this chapter aimed to identify and map technogenic modifications in the margins of river channels that compose the hydrographic network of the Paraíba do Sul River, one of the most important drainage basins of the Brazilian Southeast, seeking to relate them with the regional urbanization process. We used it as analysis cutout of the watershed of Ribeirão Brandão, which encompasses some of the growth axes of the industrial city of Volta Redonda (RJ), and, as a basis for identifying the types of margins of the main collecting river channels of this watershed, the classification of (Wheaton et al., Geomorphology 248:273–295, 2015), employing Google Street View and Google Earth Pro images for recognition of natural and built features that exert confinement of rivers. The results obtained show that: (i) in urbanized areas the main confining elements are landfills and buildings, often implemented in floodplains, plains, and terraces, modifying them significantly; (ii) in rural areas or areas with low densification of the urban network, slopes are the predominant confining features; (iii) the classification was effective in the spatialization of technogenic modifications on the banks of the channels, when combined with the recognition of Quaternary depositional features, contributing to the mapping of the technogenic relief.

L. C. F. H. de Almeida (✉)
Graduate Program in Geography (PPGG)/UFRJ, Center for Quaternary and Technogenetic Studies (NEQUAT)-IGEO/UFRJ, Rio de Janeiro, Brazil
e-mail: lucas.hoepfner@gmail.com

E. V. de Mello
Colégio Pedro II, Rio de Janeiro, Brazil

Center for Quaternary and Technogenetic Studies (NEQUAT)-IGEO/UFRJ, Rio de Janeiro, Brazil

M. N. de Oliveira Peixoto
Department of Geography, Graduate Program in Geography (PPGG)/UFRJ, Center for Quaternary and Technogenetic Studies (NEQUAT)-IGEO/UFRJ, Rio de Janeiro, Brazil
e-mail: marianaise@igeo.ufrj.br

G. Barbosa dos Santos et al. (eds.), *Geomorphology of Brazil: Complexity, Interscale and Landscape*, Springer Proceedings in Earth and Environmental Sciences,
https://doi.org/10.1007/978-3-031-05178-4_12

1 Introduction

The evolution of drainage systems during the last thousands of years in the columinous geomorphological compartments of the Atlantic Plateau of southeastern Brazil has been investigated by means of stratigraphic and geomorphological studies conducted by the Center for Quaternary and Technogenic Studies (NEQUAT) of the Federal University of Rio de Janeiro (UFRJ). Tecnogeno (NEQUAT) of the Federal University of Rio de Janeiro (UFRJ), using and developing and applying methodologies for reconstructing the sequences of erosional/depositional events, elaborating typologies, and identifying differentiated behaviors in the fluvial systems and drainage headwaters (Moura et al. 1991; Moura e Mello 1991; Mello et al. 2005; Mello 2006; Peixoto et al. 2010; among others). Based on the information produced, the most recent research has been dedicated to identifying and characterizing morphological and functional patterns in river systems, analyzing the role of urban and rural spatial transformations in the geomorphological dynamics of river and slope environments.

The modifications in rivers due to urban expansion and changes in land use in rural areas have been increasing in Brazil since the second half of the twentieth century, associated with the great change that occurred in the spatial concentration of the population, as pointed out by Pelech and Peixoto (2020), which made the country leave a markedly rural occupation pattern for a predominantly urban one. The urban expansion was mostly detached from environmental concerns in the various Brazilian regions, producing intense transformations both in the hydrological system (through soil sealing and construction of new drainage networks) and in the morphology and structure of the banks and beds, affecting the longitudinal profile and the shape in plan of the rivers, the hydrological connectivity between channel and plain and the riparian vegetation (Pelech and Peixoto 2020). The city of Volta Redonda (RJ), as a *company-town* founded in the 1940s with the installation of the Companhia Siderúrgica Nacional (CSN), experienced an accelerated urban growth marked, in a special way, by the different phases of the structuring of the productive networks of the base industry in southeastern Brazil, which allows identifying determining relations with technogenesis (Castro and Peixoto 2015; Mello, unpublished).

In this context, the previous studies carried out in Volta Redonda sought to elaborate a typology of rivers that would allow the identification of different morphological and evolutionary patterns and anthropogenic interventions, using as bases the River Styles® methodology proposed by Brierley and Fryirs (2000; 2005) and Brierley et al. (2002) integrated with the morphostratigraphic approach developed in the NEQUAT researches (Mello 2006; Peixoto et al. 2010). Subsequently, with the works of Del Pozo (2011), Ribeiro (2016), Oliveira (2017), Almeida (2021), and Mello (unpublished), new approaches for mapping and treatment of technogenic information have been applied and analyzed, and it is in this line of research that the present work is inserted. These authors identified, from cartographic bases and digital images, several types of alterations in the river channels by means of technical actions—such as concreting the gutters or banks and straightening the course of rivers, or even works that made the channels underground—which express the close relationship between geotechnogenesis in river environments and the metabolism of the city.

Thus, this study aims to contribute to the development of procedures for mapping and analysis of the interventions operated on the banks of river channels in urban and rural areas in an integrated way to the recognition of features linked to the geomorphological structure of river systems, from the application of the taxonomy of river features presented by Wheaton et al. (2015), aiming to contribute to investigations on the technogenic relief in the region and in the country.

2 Study Area

Volta Redonda is located in the Tectonic Depression of the Paraíba do Sul River, a compartment inserted in domains of crystalline plateaus that border the southeastern coast of Brazil and limited to the south by the Serra do Mar, and to the north by the Serra da Mantiqueira (Fig. 1). The hydrographic basin of Ribeirão Brandão, chosen as study section, is one of the main tributaries of the Paraíba do Sul River in its Middle Valley, covering an extensive area of the southern portion of Volta Redonda and part of the neighboring municipalities (Barra Mansa and Pinheiral—RJ).

The hydrographic basin of Ribeirão Brandão river comprises geomorphological compartments of hills and hills developed, in the upper and middle courses, over gneisses and granitoids of the Precambrian crystalline basement and, in its middle and lower courses, also over Neogene sedimentary sequences that fill the Casa de Pedra Graben and the Volta Redonda Sedimentary Basin (units 6 and 7, respectively, in the geological framework of Fig. 1), limited by normal faults (Silva 2001). The variation in the dissection and morphology of the slopes and river valleys and in the orientation of the hydrographic networks express the structural controls of the Precambrian substratum and Cenozoic tectonics.

The hill domain where the Ribeirão Brandão basin is located has been remodeled by erosive and depositional processes during the late Quaternary, which is associated with the evolution of fluvial systems and drainage headwaters in amphitheater (the regional drainage networks) as well as the long-term evolution of the relief in southeastern Brazil, marked by neotectonic, lithostructural, and climatic controls. It has as characteristic Quaternary depositional features the different levels of fluvial terraces, the complexes of colluvium ramps, alluvium-coluvium ramps, and the present floodplains, sheltering distinct depositional materials and soils that have direct relation with the landscape development along the last thousands of years (Moura and Mello 1991; Moura et al. 1991). In this context, the main geomorphological and stratigraphic landmark corresponds to the Holocene entrenchment of the valleys and recesses of the drainage headwaters in amphitheater by the Manso Alloformation, associated with the formation of alluvium-coluvium ramps in topographic continuity with the highest regionally recognized terrace level. Subsequent phases of

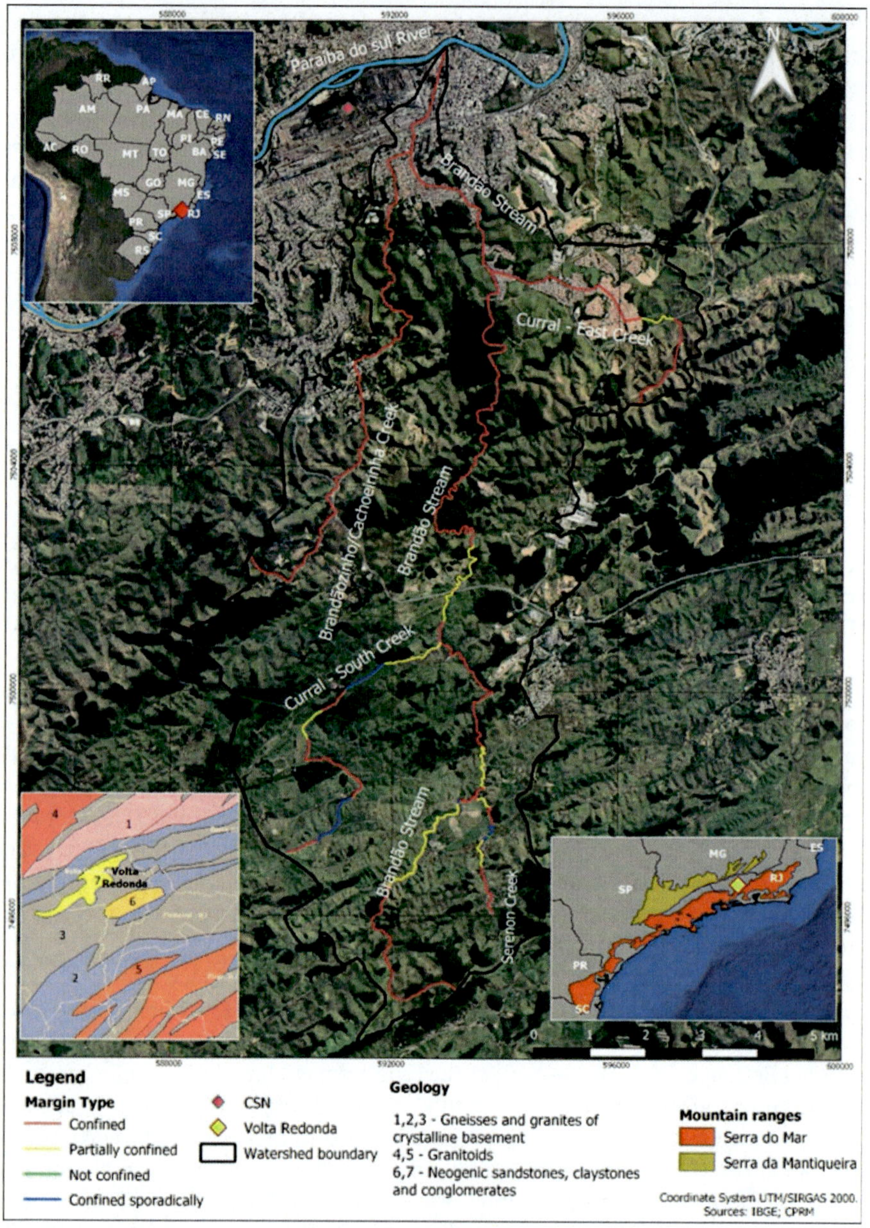

Fig. 1 Localization of Volta Redonda on Rio de Janeiro State (RJ), SE Brazilian Plateau, and Brandão Stream drainage basin showing confinement types of the main river and tributaries

fluvial entrenchment and hydrographic re-hierarchization by accelerated channelized erosive processes (gullies) have affected the watersheds and their tributaries differently, resulting in distinct patterns of preservation of the entrenchment features and generation of low terraces and fluvial plains, as can be identified in the mappings of Del Pozo (2011) and Ribeiro (2016), for example.

The hydrographic basin of Ribeirão Brandão embraces distinct hillside compartments with urban and rural uses, having previous studies of mapping and classification of features indicative of technogenic alterations in the drainage networks, detailed in several stretches of channels (Mello 2006; Peixoto et al. 2010). The Ribeirão Brandão has as main tributaries, from upstream to downstream (Fig. 1):

- the Serenon stream, located in the headwaters of the basin, and the Curral stream, the main tributary in the upstream section of the basin (called "Curral-Sul Stream" in this study);
- the Córrego do Curral tributary in the central stretch of the basin (here called "Córrego do Curral-Leste", to distinguish it from its homonym) and the Córrego Brandãozinho or Cachoeirinha, the latter an extensive tributary that drains a large part of the western sector of the basin, in its middle and lower courses.

The main collector of the basin has a clear south-north orientation, crossing more dissected and elevated hill and hillside compartments in the headwaters and in its middle-lower course, which contrast with more depressed compartments of gentle hills and plains in the middle course and in the confluence zone with the Paraíba do Sul River (Fig. 1). The tributary sub-basins of Ribeirão Brandão are predominantly elongated, following the design and general orientation of the hill geomorphological compartments, and in several stretches, one can observe lattice drainage patterns, with higher drainage density (Mello 2006). The basin shows clear asymmetry, with the western tributaries more extensive than the eastern ones, observing also the structural control in the organization of the drainage network mainly in the middle and upper courses, by the orientation and sharp inflections of the main collector and the Curral-Sul stream, clearly controlled by the NE-SW regional lineaments. The predominant NW–SE orientation of the 1st and 2nd order channels in narrow and parallel aligned valleys is also evidence of this structural control. As described by Almeida (2021), the main collecting river channels of the Ribeirão Brandão basin present margins predominantly limited by slopes or fluvial terraces, being thus mostly confined (48.34 km) or partially confined (10.13 km), with few unconfined segments (0.1 km) or sporadically confined (3.29 km).

The Middle Valley of the Paraíba do Sul River region where the basin under study is located is characterized by a tropical climate, with average annual precipitation of around 1,300 mm. The largest precipitation volumes and the highest temperatures occur in the summer (December to February) when the average monthly rainfall totals range from 600 to 800 mm, which represents approximately 42.7–46.7% of the total annual precipitation volume. In autumn, the average monthly rainfall varies between 130 and 230 mm (about 10.1–13.4% of the total annual precipitation), and in winter (June to August), this average remains between 100 and 170 mm, (7.7–9.7% of the annual rainfall). Spring corresponds to the second wettest season, with the average monthly rainfall totals ranging between 450 and 550 mm (31.5–36.1% of the total annual precipitation), according to Costa et al. (2012). In the case of Ribeirão Brandão, the rainfall regime, in association with the effects of changes in land use, has resulted in frequent floods, especially in the summer months, generating significant impacts in urban areas.

3 Methodology

3.1 The Study of Rivers in the Technogenic Approach

The basis of the studies on the processes and products related to the human agency in the region under study is the geotechnogenic approach, executed through the identification of the sets of technogenic processes in their specific geographic, territorial and historical contexts—the technogenic systems. The geotechnogenesis comprises: the modifications in the relief, resulting in the creation or remodeling of technogenic reliefs; the changes in the physiology of landscapes, with the creation, induction, intensification, or modification of the behavior of the processes of external dynamics; the generation of technogenic surface deposits (Oliveira et al. 2005, 2018; Pelech and Peixoto 2020).

In this approach, human technical action constitutes a modifying agent of geomorphological landscapes, being associated with changes in land use and land cover. The individualization of geotechnical processes and forms, as well as their correlated deposits and sedimentary facies that record the processes and stages of land use, allow the sequencing of anthropogenic environmental transformations (Peloggia et al. 2014; Oliveira et al. 2018). For the study of river systems, identifying the modifications in relief forms and spatializing them in drainage basins are fundamental steps of analysis, especially considering that rivers, due to their high geomorphological sensitivity, are important archives of recent environmental history.

In this sense, the methodology adopted for the identification of types of river channel margins and technogenic interventions was the taxonomy of fluvial geomorphological features developed by Wheaton et al. (2015). This methodology takes into consideration, the classification and taxonomy of fluvial forms, the river banks, structural elements, and geomorphic units, which requires, at first, a high level of detail analysis with high-resolution images. Wheaton et al. (2015) recommend, for the channel scale, that cartographic scales from 1:100–1:2000 be employed in the identification and representation of these components, and that structural elements, margins, and geomorphic units be discrete and represented by means of symbols.

The use of the taxonomy of river forms elaborated by Wheaton and collaborators in the present study is also based on the methodological framework of River Styles®, defined by a set of geomorphological units in a river section that establishes an integrated framework of the biophysical processes along the hydrographic basin, being identified by the character and behavior of the channel (Brierley and Fryirs

2000). The character—defined by the channel planform, geometry, the assembly of geomorphological units, the type of vegetation cover and detrital material—and the behavior—comprising the hydraulic characteristics, the channel-plain connections, the sediment regime and the propensity to geomorphological transformations—allow identifying discontinuities and local controls that result in the diversity of behaviors of rivers and river networks (Brierley et al. 2002; Fryirs 2003). The use of this framework in conjunction with the morphostratigraphic approach allows the identification and mapping of Quaternary depositional features of regional expression and individualizes different patterns of fluvial behavior (Peixoto et al. 2010; Del Pozo 2011).

Technological advances, especially in the last decade, have allowed the use of new remote sensing imaging techniques in geomorphological studies and mapping of fluvial environments in different climatic and tectonic contexts. In this scenario, the taxonomy proposed by Wheaton et al. (2015) stands out for providing a multilevel classification that can be applied in a wide diversity of environments and with the available inputs, based on principles of natural fluvial dynamics and allowing the incorporation of the diversity of technogenic features.

3.2 *Methods and Materials*

The taxonomy of Wheaton et al. (2015) defines different stages of analysis for river channels, which, for the purposes of the present study, were adapted to the purpose and the available inputs. Thus, the first step of the mapping carried out for the Ribeirão Brandão watershed was the identification of the different types of banks (channel, valley bottom, and valley) used to define the configuration of the valley. Within this step it is necessary to identify, firstly, if the river channel bank has characteristics that are indicative of direct technogenic interventions or not, examples of these interventions being the different types of built features such as embankments, fences, dikes, railways, roads, walls or other structures that limit the channel bank. According to Wheaton et al. (2015), channel banks with these types of intervention are present in many rivers, occupying, fragmenting, and dissecting many valley bottoms. River channel banks that do not contain features directly produced by human agency (termed "natural" by the authors) may also be associated with confining features such as terraces and slopes.

It is important to highlight the distinction between river valley margins and river valley bottom margins, which also differ from river channel margins, according to the authors: the valley margins correspond to the boundary between the slope environment and the valley bottom, while the valley bottom margins delimit the scope of only the active river channels and the contemporaneous floodplains (plains); the river channel margins correspond, in turn, exactly to the limits of the river channel, that is, they are defined by the boundary between the channel flow and the area reached when it overflows into the floodplains or plains during episodes of higher flow.

Thus, from the identification of the different margins and their layout, the confining conditions of the fluvial channels are firstly recognized, with the configurations of confined, partially confined and laterally unconfined valleys also referenced in the methodology of fluvial styles (Brierley and Fryirs 2005). According to Wheaton et al. (2015), a confining bank is defined as any channel section that is confined by a valley bank, valley bottom bank, or anthropogenic bank. Confining margins are those that are acting to restrict the lateral setting of a channel, and thus channel margins can be confined by slopes when they are contiguous to them (i.e., by the valley margin) or by other confining features, such as terraces and alluvial fans. In the latter case, these geomorphological features compose the valley, but not the valley bottom, within the conceptualization presented by the authors.

In the analysis carried out, Google Earth Pro images were adopted as the main basis for the mapping of technogenic fluvial features, both for presenting a satisfactory resolution to identify the elements focused on in the classification used, as well as for providing recent and good quality coverage (the one used dates from 23/07/2020). The Google Earth images were processed using the Quantum GIS *software*, defining the cartographic scale 1:1000 as a reference for the mapping, because even though it did not allow for detailed visualization of the interior of the channel, it satisfactorily identified the features present on its banks, as well as on the banks of the valley and valley floor, the focus of this study.

To identify and classify the existence and types of technogenic interventions, we also used the Google Street View tool, obtaining visualizations closer to those seen in the field. Thus, based on the more detailed observation of locations chosen as viewing points on Google Street View, the elements that confine the banks were identified, and with the images from the Google Earth platform, the mapping of their distribution along the main collectors defined for analysis was performed, identifying the degree of confinement. Subsequently, we mapped the confining features of the banks of the channels in the stretches not visualized in Google Street View, based on Google Earth images and using visual records of documents and studies produced previously.

To identify the type of bank confinement and the confining elements, it was necessary to recognize, whenever possible, the Quaternary depositional features present in the channel surroundings and their effect on the arrangement of the valley, valley floor, and channel margins. This identification was made by means of Google Street View, three-dimensional images of the relief by Google Earth Pro, and also by the analysis of contour lines.

4 Technogenic Modifications in River Channels

With an area of approximately 75 km^2 and a perimeter of approximately 55 km, the hydrographic basin of the Ribeirão Brandão river has its outlet in the Paraíba do Sul River, near the Companhia Siderúrgica Nacional (CSN) plant, encompassing in this area sections of the central neighborhoods of the city (Fig. 1). The Brandão river is 31.03 km long and its main tributaries are 12.99 km long—Brandãozinho stream/Cachoeirinha stream, which stands out as its largest tributary—and 8.02, 6.41, and 3.39 km long—Curral-Sul stream, Curral-East stream and Serenon stream—the latter, as already mentioned, located in the basin head. The collector and its tributaries go through neighborhoods located in the southeast expansion front of the city of Volta Redonda, in the Gráben Casa de Pedra (case of the Córrego do Curral-Leste), and in the connection axis with the Presidente Dutra highway (case of the Brandãozinho/Cachoeirinha stream, in its lower course). The tributaries of the middle and upper courses run through areas with predominantly rural use, in which pastures dominate with the occurrence of small forest fragments.

The initial identification of river channel bank types, according to the Wheaton et al. (2015) classification, allowed the recognition of three main patterns, named "Anthropogenic" bank, "Partially Anthropogenic" bank, and "Natural" bank. These classes were individualized from the presence or absence of technogenic confining elements, recognized in the visual interpretation of the images. Constructed features and terrain, such as embankments and building walls, for example, define margins classified as "Anthropogenic", while situations of confinement by hill slope segments or by higher fluvial terrace levels were considered within the "Natural" margin category. It was considered here, as a fundamental criterion, the significant preservation of Quaternary morphology, and therefore, disregarding other aspects acting in the modification of surface coverings by geotechno-genesis. The situations of coexistence of technogenic confining elements in a margin and natural on the opposite margin were classified as "Partially Anthropogenic". As a result, we identified 44.13 km of channel margins classified as "Natural", predominant in rural areas; 12.05 km of "Anthropogenic" channel margins, predominating markedly in urban areas; and 5.23 km of "Partially Anthropogenic" margins, generally associated with stretches in which built elements, such as roads, interfere in the natural confinement of rivers (Table 1).

Based on the recognition of differentiated structures on the banks of the channels and in the context of the river segments, a typology of channel banks was elaborated, resulting in six types that seek to synthesize the information regarding the elements that border the banks of the collecting channels of the Brandão river basin (Table 2).

It is evident from the data presented in Table 3 and the map is shown in Fig. 3 that the most common type of confinement of the banks of the main collectors of Ribeirão Brandão is produced by the slopes of the hills and hills, a situation that predominates in rural areas or where the urban network has a low degree of densification, and where the floodplain cannot be seen or does not exist.

Table 1 Length of channel margins types of Ribeirão Brandão and main tributaries, Middle Paraíba do Sul River Valley-Brazil

Margin type	Length (km)	Length (%)
Anthropogenic	12.5	20.23
Natural	44.1	71.36
Partially anthropogenic	5.2	8.41

Table 2 Confinement features of river channel margins* in Ribeirão Brandão drainage basin, Middle Paraíba do Sul River Valley-Brazil

1. Landfills and buildings Channel banks confined by landfills with buildings or roads, often making it impossible to view the floodplain. They occur more frequently downstream of the Ribeirão Brandão basin, where the relief is softer, and with a high urban density	
2. Low terraces (sporadic) Channel banks with continuous floodplains bordered by low terraces in a few stretches. Their occurrence is restricted to upstream tributaries, in places where channels still present a meandrical pattern	
3. Hillslopes **4. Hillslopes and roads or buildings** River channel banks juxtaposed to the edges of the valley, with no floodplains. They occur in stretches with narrow valley bottoms, located mainly in the upper and middle courses of the basin. They may also be confined by highways, roads, or buildings, a situation found throughout the entire hydrographic basin	
5. Low terraces and hillslopes **6. Low terraces, hillslopes, or roads** Conditions with a discontinuous floodplain where the channel's sinuosity is controlled by the valley floor margin; control can also be carried out by roads or highways. Its occurrence is restricted to the upper course of the watershed, where rural use predominates	

Table 3 Length of confining features of the fluvial channel margins for Ribeirão Brandão and main tributaries, Middle Paraíba do Sul River Valley-Brazil

Types of confining features	Length (Km)	Length (%)
Landfills and buildings	12.692	20.51
Low terraces (sporadic)	3.2398	5.33
Hillslopes	31.913	51.58
Hillslopes and roads or buildings	3.739	6.04
Low terraces and hillslopes	9.513	15.38
Low terraces, hillslopes, or roads	0.616	1.00
Not identified	0.095	0.15

In the more urbanized stretches, the main element of confinement of the banks is the embankments, which are often made in floodplain and river plain areas, modifying them significantly.

In the map of Fig. 2, sections A and B are highlighted where Google Street View was used to visualize them in greater detail in order to serve as a reference for the areas without photo coverage from roads, where we could only perform a visual analysis in plan. In section A it is observed the predominance of channel margins is limited by embankments and buildings in the urban area, and by slopes or roads in the rural area, situations exemplified in types "1" and "4" of Table 2. The urban density of this area of the city implies a large number of interventions in the drainage networks, and most of the stretches of river channels with banks classified as "Anthropogenic" are inserted in this location. The stretch highlighted in B, upstream of the basin, encompasses rural areas with lower density of direct interventions in the river channels, observing the predominance of "Natural" and "Partially Anthropogenic" banks (types "2," "4," and "6" of Table 2).

The channel margin features identified in the present study show strong correspondence with the typology proposed by Mello (2006). Mello (2006) proposed nine types of fluvial channels in the Ribeirão Brandão basin: silted, clogged, erosive, forested, impermeable, incised, non-incised, rectified, rocky, and underground, identified based on the morphostratigraphic approach combined with the fluvial styles methodology. The channels classified by Mello (2006) as Rectified, Impermeabilized, and Underground correspond, in the current mapping, to those that present banks confined by embankments or constructions installed on the floodplain since the rectifications of the rivers are associated with engineering works carried out in various stretches of the basin in order to reduce flooding in urban areas and/or near highways. The channels that suffer the impermeabilization of their gutter and banks, in general, are also rectified, and it is frequent in urban areas with more dense occupation, the confinement in underground galleries, which make the rivers no longer visible.

On the other hand, the channels classified as Incised type by Mello (2006), which result from the Holocene or current fit of the drainage in valley bottoms filled by the Manso Aloformation, present, mostly, margins confined by slopes, terraces, or roads, and the inexistence of significant plains are explained by this behavior of

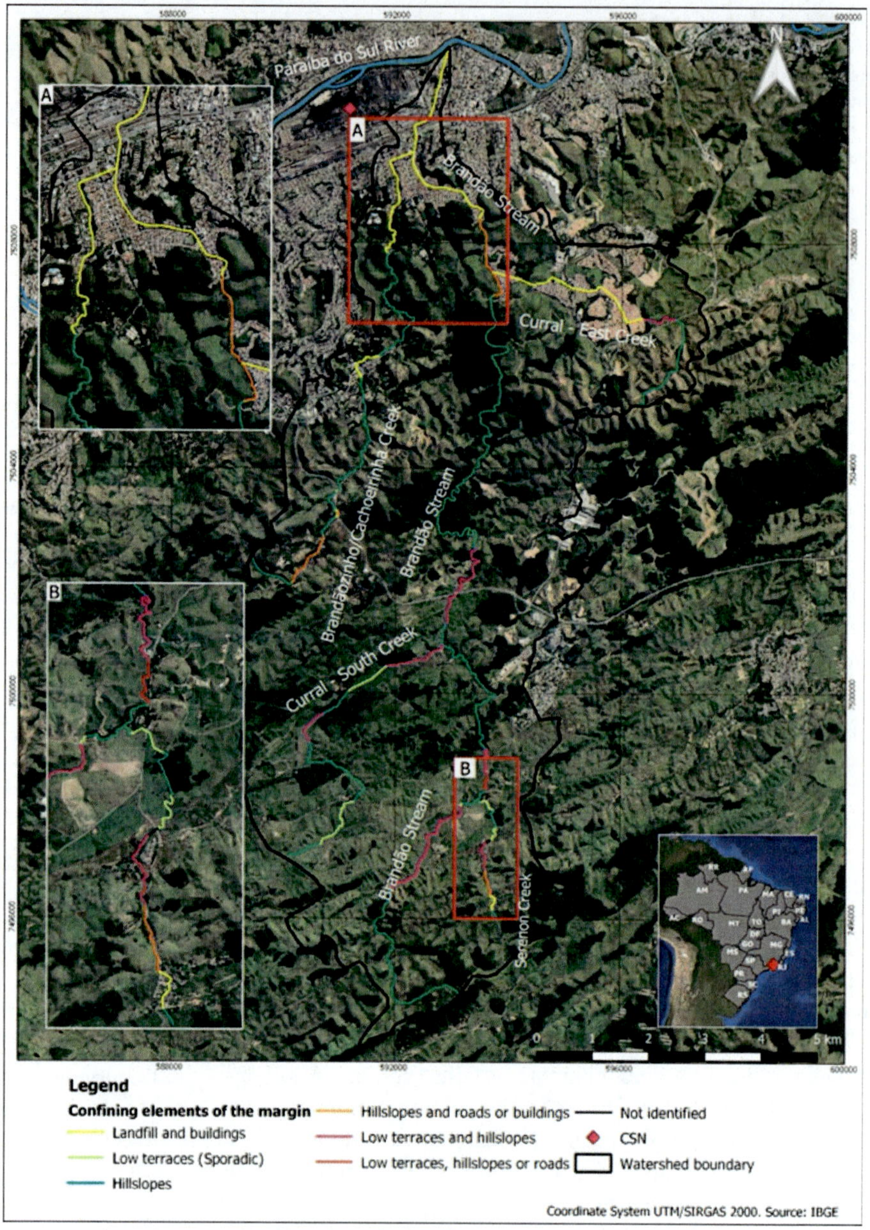

Fig. 2 Spatial distribution of confining features of fluvial channel margins of Brandão Stream and main tributaries, Middle Paraíba do Sul River Valley-Brazil

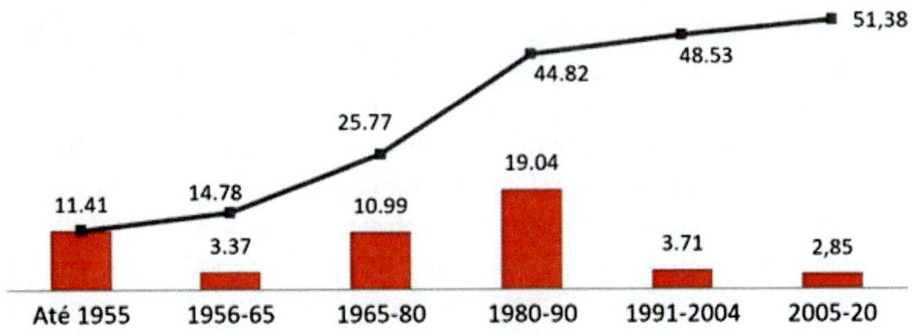

Fig. 3 Technogenic changes in river channels of Volta Redonda urban area between 1955 and 2020, in kilometers per time interval (red vertical bars) and accumulated values (black line)

fluvial incision. The non-incised type channels recognized by Mello (2006), in turn, which are generally found in wider valleys, show correspondence with the stretches of channels with unconfined margins, which present continuous floodplains limited or not by low terraces and with or without the presence of meandering channels, and can also correspond to channels with partially confined margins with discontinuous floodplains and sinuosity controlled by the valley or by roads.

It is verified, on the other hand, that the introduction of technogenic elements in the identification criteria of channel confinement resulted in a significant change in the interpretation of the general picture of fluvial confinement since most of the margins are shown to be confined by both Quaternary relief features and built structures associated with the urbanization process of the city of Volta Redonda and the region where it is inserted.

In the context of the regional evolutionary history, the intense dissection of the crystalline and sedimentary substrate by the drainage systems in the hillsides where the basin is located resulted in a topography with few stretches of significant valley widening, restricting the formation of broad floodplains. The fitting of the drainage network after the deposition of the Manso Aloformation, resulting in the formation of the highest fluvial terrace or, sometimes, also in low terraces, configure confining elements of the current fluvial channels, according to the classification used. In the lower course of the basin, on the other hand, although plains and terraces associated with the Paraíba do Sul River predominate, Ribeirão Brandão has its banks confined by landforms and buildings, showing a condition of significant restriction by the presence of technogenic features and deposits.

Another important aspect to be highlighted with respect to the methodology employed refers to the interpretation of the fluvial terraces in the identification of the type of confinement of the fluvial channels. In the case of the low terraces, we should consider the possibility of an induced technogenic origin, as shown by studies and records of technogenic fluvial deposits in the region (Mello et al. 1995), which would indicate a situation of confinement associated with anthropogenic channel margins. Thus, there is a need to improve this criterion, starting with specific investigations on the origin and evolution of the fluvial sedimentation levels (especially the low terraces).

Mello (2006; unpublished), when mapping the extension of technogenic modifications in the river channels in Volta Redonda, identifies several pulses of modification in the rivers and streams of the city, linked to different phases of production of the urban-industrial space (Fig. 3). In the first phase, which goes from the beginning of the 1940s until 1965, the construction of Companhia Siderúrgica Nacional (CSN) caused great population attraction to the region and was a determining factor for the Brazilian industrial advance. In the urbanization process that occurred in Volta Redonda, the planned occupation areas were intended for workers, technicians, and managers of the steel mill, excluding a large mass of temporary workers who began to occupy areas without infrastructure adjacent to the projected urban core.

In the second phase, from 1965 to the early 1990s, CSN began to incorporate other urban areas with a view to building and selling housing units, but still not meeting the high demand from population groups attracted by job opportunities and not absorbed by the company, which began to form "nuclei of possession" and irregular subdivisions. This is the phase of greatest territorial expansion of the city and, consequently, of acceleration in the changes in the river channels, both in the planned areas and in those of disorderly occupation. In the third phase, after the early 1990s, marked by the privatization of CSN and reduction of jobs, the pace of production of urban space slowed down, and the pace of expansion of the modified river channels was also reduced, although new works on the banks and the rectification of previously altered channels have been frequent.

These aspects demonstrate differences in the forms and technical sets of intervention in the river channels, with the types of confinement and modification of rivers and streams being associated with the urban expansion fronts and the spatially selective processes of implementation of the city's basic infrastructure, as well as with the urbanization models adopted.

5 Finnal Considerations

The application of the methodology proposed by Wheaton et al. (2015) in the definition of river channel bank types for the main collectors of the Ribeirão Brandão watershed presented positive overall results for the mapping of technogenic fluvial features. The use of Google Earth Pro and Google Street View images enabled the visualization of the characteristics of the surroundings of the different stretches of the analyzed fluvial channels, allowing the classification of the channel, both regarding its level of confinement and its confining elements (either constructed or associated with regionally recognized Quaternary depositional features) to be satisfactorily performed. The limitations of Google Street View due to the fact that the photographs are taken from roads led us to select stretches for detailed analysis, being an important complement to the visual plan analysis.

Thus, we verified that the spatialization and analysis of technogenic modifications in the margins of river channels based on the taxonomy of Wheaton et al.

(2015) can contribute to the advancement of research on the mapping of technogenic relief in the crystalline plateaus of southeastern Brazil. Considering that the employed classification is also integrated with the framework of fluvial styles, we understand that the improvement of the procedures and their application in other geomorphological domains will contribute to the identification of patterns of fluvial behavior and geomorphological conditions of rivers in different urban and rural environments of Brazil. This knowledge, together with the evolution of basins and drainage systems, should subsidize the construction of scenarios and trajectories of technogenic geomorphological changes in already urbanized areas and in the urbanization expansion fronts, both in the southeast and in other Brazilian regions.

The study carried out in the Ribeirão Brandão watershed allows confirming, still, the relevance of the geotechnical approach for the fluvial geomorphology. Although in the urban evolution of Volta Redonda the role of CSN (together with the state power) as hegemonic modeling agent of the urban space is determinant for the rhythm of rivers confinement (Mello, unpublished), the processes of selectivity and spatial marginalization documented there are present in several Brazilian cities and may have effects on the speed and spatiality of the modifications in the river channels. Thus, the mapping of technogenic modifications in the river networks must be anchored in the investigation of the processes and forms of materialization of the urban metabolism.

References

Almeida, LCFH (2021) Alterações **Geomorfológicas em Canais Fluviais - Marcas da Geotecnogênese na Bacia do Ribeirão Brandão, Médio Vale do Rio Paraíba do Sul (RJ)**. Monografia (Graduação em Geografia), Universidade Federal do Rio de Janeiro, Rio de Janeiro, 74p.

Brierley GJ, Fryirs K (2000) River Styles, a Geomorphic Approach to Catchment Characterization: Implications for River Rehabilitation in Bega Catchment, New South Wales, Australia. **Environmental Management,** 25(6): 661-679 https://doi.org/10.1007/s002670010052

Brierley GJ, Fryirs K; Outthet, D; Massey, C (2002) Application of the River Styles framework as the basis for river management in New South Wales, Australia. **Applied Geography**, 22: 91-122 https://doi.org/10.1016/S0143-6228(01)00016-9

Brierley GJ, Fryirs K (2005) **Geomorphology and River Management: applications of the River Styles framework.** Blackwell Publishing. https://doi.org/10.1111/j.1541-0064.2007.00168_1.x

Castro CM, Peixoto MNO (2015) Urbanization, vulnerability and environmental risks associated with drainage systems: an analysis from Volta Redonda (RJ). In: Fernandes U. et al. (Org.). **Velhos Saberes, Novas Abordagens: a Geografia à Luz da Contemporaneidade.** 1st ed. Gramma, Rio de Janeiro, p181–200

Costa AO, Salgado CM, Dinali, YT (2012) Characterization of precipitation in the middle valley of the Paraíba do Sul Fluminense River (RJ). **Revista Geonorte** (2)5:1000-1013

Del Pozo IFS (2011) **Modeling Landscape Connectivity Based on the Spatial Analysis of Geomorphological Variables and Land Cover in Watersheds - Municipalities of Volta Redonda and Pinheiral (RJ).** Dissertation (Master in Geography), Federal University of Rio de Janeiro, Rio de Janeiro, 134p

Fryirs K (2003) Guiding Principles of Assessing the Geomorphic Condition of Rivers: Application of a Framework in Bega catchment, South Coast, NSW, Australia. **Catena** 53:17- 52. https://doi.org/10.1016/S0341-8162(02)00199-6

Mello EV (2006) **Alterações Tecnogênicas em Sistemas Fluviais no Município de Volta Redonda, Médio Vale do Rio Paraíba do Sul Fluminense**. Dissertation (Master in Geography) Universidade Federal do Rio de Janeiro, Rio de Janeiro, 168p

Mello EV (unpublished) **Volta Redonda and CSN, symbols of the Anthropocene in Brazil.** Thesis (Doctorate in Geography) Federal University of Rio de Janeiro, Rio de Janeiro

Mello CL, Moura JRS, Carmo IO, Silva TM, Peixoto MNO (1995) Sedimentation Events during the Holocene in the Middle Paraíba do Sul River Valley (SP/RJ): Allostratigraphy and Radiocarbon Dating. In: Congresso da Associação Brasileira de Estudos do Quaternário, Niterói. **Proceedings**... ABEQUA, 5:193–197

Moura JRS, Mello CL (1991) Allostratigraphic classification of the Upper Quaternary in the Bananal Region (SP). **Rev. bras. Geociências** 21(3):236-254

Moura JRS, Peixoto MNO, Silva TM (1991) Geometry of Relief and Quaternary Stratigraphy as a Basis to the Typology of Headwaters in Amphitheater - Middle Paraíba do Sul River Valley. **Rev. bras. Geociências** 21(3):255-265.

Oliveira SA (2017) **Vulnerabilidade Ambiental na "Cidade do Aço": a Geotecnogênese na Construção de Paisagens de Perigo no Setor Leste de Volta Redonda (RJ)**. Dissertation (Master in Geography), Federal University of Rio de Janeiro, Rio de Janeiro, 213p.

Oliveira AMS, Brannstrom C, Nolasco MC, Peloggia AUG, Peixoto, MNO, Coltrinari L (2005) Technogen: records of the geological action of Man. In: Souza CRG, Suguio K, Oliveira AMS, Oliveira PE (Eds.) **Quaternary of Brazil**. Holos, Ribeirão Preto, p 363–378.

Oliveira AMS, Peloggia AUG, Oliviera AA (2018) Technogenic-Antropocene. In: Oliveira, AMS, Monticeli JJ **Engineering and Environmental Geology**. ABGE, São Paulo

Peixoto MNO, Cesário FV, Pimentel ML, Mello EV; Oliveira DFP (2010) Identification of river styles in drainage basins inserted in hill compartments - Middle Paraíba do Sul River Valley (RJ). **Geography Journal** 27 (3):347-362

Pelech AS, Peixoto MNO (2020) Urban Rivers: contribution of Anthropogeomorphology and Technogenic Terrain studies. **Brazilian Journal of Geography** 65:2–22 https://doi.org/10.21579/issn.2526-0375_2020_n1_2-22

Peloggia AUG, Oliveira AMS, Oliveira AA, Silva ECN, Nunes JOR (2014) Technogenic geodiversity: a proposal on the classification of artificial ground. **Quaternary and Environmental Geosciences**, 05(1):28-40

Ribeiro AS (2016) **Caracterização geomorfológica, sedimentológica e da vegetação de canais embrejados na bacia do Córrego Santa Rita, município de Volta Redonda (RJ).** Dissertation (Master in Geography), Federal University of Rio de Janeiro, Rio de Janeiro,148p

Silva LC (2001) **Geologia do Estado do Rio de Janeiro: texto explicativo do mapa geológico do Estado do Rio de Janeiro** In: Silva, LC; Cunha, CS (org.) Programa Levantamentos Geológicos Básicos do Brasil. 2nd ed, CPRM, Brasília

Wheaton JM, Fryirs K, Brierley GJ, Bangen SG, Bouwes N, O'Brien G (2015) Geomorphic mapping and taxonomy of fluvial landforms. **Geomorphology** 248: 273-295 https://doi.org/10.1016/j.geomorph.2015.07.010.

Geoeducation and Geoculture: Concepts, Characteristics, and Contributions to Geoconservation in Brazil

Marcelo Martins de Moura Fé, Thaís de Oliveira Guimarães, Cristina Rodrigues Holanda, Marcos Antonio Leite do Nascimento, João Victor Mariano da Silva, and Raquel Landim Nascimento

Abstract Geodiversity is notable for the diversity of the abiotic factors that compose it and for its importance for the occurrence and maintenance of natural and ecosystem processes fundamental to life on planet Earth. Considering the socio-economic risks of degradation, the need to develop projects, activities, and actions for (geo)conservation is peremptory. In this context, this work aims to present concepts, characteristics, and contributions of geoeducation and geoculture, strategies aimed at the knowledge, appreciation, and application of geoconservation. Methodologically, the research was based on a qualitative approach, with descriptive purposes; segmented in a careful and systematic bibliographic survey, readings and analysis of the theoretical framework, and several field surveys. As results are presented conceptualizations, assumptions, parameters, and bases are characteristics that substantiate possible contributions through their applications. It is expected that these strategies are consolidated and effective, contributing to the expansion of knowledge, appreciation, and conservation of geodiversity in various contexts and scales.

M. M. de Moura Fé (✉)
Regional University of Cariri (URCA), Crato, Brazil
e-mail: marcelo.mourafe@urca.br

T. de Oliveira Guimarães
University of Pernambuco (UPE/Petrolina), Recife, Brazil
e-mail: thais.guimaraes@upe.br

C. R. Holanda
Culture of the Government of the State of Ceará, Ceará, Brazil

M. A. L. do Nascimento
Federal University of Rio Grande do Norte (UFRN), Rio de Janeiro, Brazil
e-mail: marcos.leite@ufrn.br

J. V. M. da Silva
Sustainable Regional Development (Proder/UFCA), Ceara, Brazil

R. L. Nascimento
Geography (MAG/UVA), Rio de Janeiro, Brazil

G. Barbosa dos Santos et al. (eds.), *Geomorphology of Brazil: Complexity, Interscale and Landscape*, Springer Proceedings in Earth and Environmental Sciences,
https://doi.org/10.1007/978-3-031-05178-4_13

1 Introduction

Geodiversity is notable for the heterogeneity of the abiotic factors that compose it, having in geology its basic aspects, including mineralogy, petrography, stratigraphy, and paleontology, among other branches. Derived from and intrinsically related to the geological elements are other abiotic factors, namely: geomorphology with its forms and processes, compartmentalized and classifiable through the geographical concept of scale; pedology, whose characteristics are directly derived from rocks and may have in the elements time and climate considerable means of classification and analysis; and hydrology, compartmentalized, finally, by segmentation between surface water and groundwater.

Despite such diversity and its importance for the occurrence and maintenance of natural and ecosystem processes, fundamental to life on the planet, geodiversity, as a scientific, educational, and/or heritage object, and considering the socio-economic risks of its degradation, demands more studies about its genesis, evolution, and interrelationships, with biodiversity, for example, as indicated by Uceda (1996; 2000), Erikstad (2014), Crofts (2014), Liccardo and Piekarz (2017), or with culture (Delphim 2009; Ruchkys and Machado 2010; Alvarenga et al. 2018; Lima and Carvalho 2020); besides generalizations and regional and local particularities, among other aspects.

In this context, it is peremptory the need that such studies may consider, approach, and/or reach, whenever possible, directly or indirectly, the nearby localities, where elements of geodiversity are even more remarkable, the geological sites or geosites. When referring to "localities" it refers, also and necessarily, to people (or communities, encompassing people and places) living in these places, who need to understand the importance, care, and participate in the management of these geosites, which, in turn, are assets that belong to them, in the most different ways—materially, historically and culturally, for example, among other aspects.

But how can this happen? Through geoconservation, can be understood as a set of activities designed and developed for the conservation of geodiversity (Brilha 2005) and all its aspects. In the set of geoconservationist activities, there are geotourism, geoeducation, and geoculture.

The first branch is more widely discussed, developed, and applied (Hose 1995; 2000; Moreira 2011; Ollier 2012; Urquí 2012; Arouca Declaration 2011; Dowling 2013; Guimarães et al. 2017; Liccardo and Piekarz 2017), the second and third branches are in process of theoretical and applied construction, being thought and directed to an essential public to any geoconservation proposal: the student public, the communities close to the geosites and, also, the various actors associated with these segments.

In this context, this manuscript aims to present concepts, characteristics, and contributions of geoeducation and geoculture, strategies aimed at the knowledge, appreciation, and application of geoconservation, especially in Brazil.

In the country, the geoconservation strategies have been developed on several fronts, to mention the geoeducation and geotourism strategies, through independent actions and/or from the geopark territories, in the Project, Aspirant, and UNESCO Global Geopark (UGGp) levels, to mention the until then, the only one belonging to the International Geosciences and Geoparks Programme (IGGP), the Araripe UGGp.

It is also important to highlight that in November 2021 two projects of aspiring geoparks received the visit of UNESCO evaluators, in the search for the entrance to the IGGP, they were, the Caminhos dos Cânions do Sul aspiring UGGp (aUGGp), with 07 municipalities distributed in the states of Santa Catarina and Rio Grande do Sul, and the Seridó aUGGp, territory composed by 06 municipalities, in the state of Rio Grande do Norte -, respectively in the south and northeast regions of the country. In this sense, the visit of the evaluators is a decisive step for the approval of these territories and with this, it is expected that, soon, the country may have two more Geoparks in the program, thus strengthening the Brazilian geoconservation. In sixth statutory meeting held online between the 8th and 11th of December 2021, the UNESCO Global Geoparks Council proposed to forward the nomination of 8 new UNESCO Global Geoparks to the Executive Board of UNESCO, for its endorsement during the 2022 Spring session. Of these 8 are on the Caminhos dos Cânions do Sul and Seridó aUGGp list.

Still with regard to geoconservation in the country, a recent and unprecedented publication in the scope of Environmental Law, entitled, "New Directions in Environmental Law: a look at Geodiversity", presented to society and the academic community, through 24 chapters, geoconservation experiences in various territories of Brazil and other countries in South America and Europe (Sousa-Fernandes et al. 2021).

In the publication Guimarães et al. (2021) presented, for example, a discussion about the geoconservation, highlighting three examples carried out in the country: the strategies developed in the Araripe UGGp, the geoeducation projects applied in formal and non-formal spaces through the extensionist activities linked to Higher Education Institutions and the initiatives taken based on the current legislation, through tombamentos, which have been carried out in the state of Paraná. According to the authors, until December 2019, six geosites were declared protected, two of geological interest, three paleontological, and one geomorphological, namely the Witmarsum Glacial Striae—given its didactic, scientific, and cultural values, in the district of Witmarsum—Palmeira/PR.

2 Methodological Roadmap

Seeking to achieve this objective, the nature of the research was based on a qualitative approach which, according to Gil (1996), aims at the understanding or interpretation of processes in a complex and contextualized manner and is characterized as an open and flexible plan, with descriptive purposes.

Regarding the research techniques, the methodological procedures used were based initially on a careful and systematic bibliographic survey. At this stage, we investigated materials published in relevant national and international journals, books, and legal titles in force in Brazil and in Ceará, with a survey of the main theoretical and methodological references.

In the bibliographical survey specific consultation was made to journals in the areas of Geosciences, Geography, and Environmental Sciences, mainly, while most of the materials are available on the Google Scholar, SciELO, and CAPES Periodicals platforms. The Brazilian Digital Library of Theses and Dissertations (BDTD-IBICT) was also consulted.

The conceptual framework and the theoretical and methodological framework learned were problematized under different realities and natural contexts of geodiversity, from the realization of field surveys, made by the authors at different times and with different objectives about the natural and cultural heritage of European countries, cradle of the concept of geoparks, and especially in Brazil and its Northeast region, with emphasis on the states of Ceará and Rio Grande do Norte, especially in the regions of Cariri cearense, Ibiapaba (south and northwest of Ceará), Seridó potiguar (RN), including also part of the southern coast of the state of Pernambuco.

The development of the methodological roadmap presented provided the preparation, discussion, and analysis of data and significant information for the scope and discussion of the results, presented in the sequence, in three interrelated topics.

3 The "1st C": Concepts

Analytical focus of geoeducation and geoculture, from the conceptual point of view, geodiversity can be understood as the result of the interaction of several factors, such as rocks, climate, living beings, among others, enabling the appearance of distinct landscapes worldwide (Gray 2004; Brilha 2005), thus integrating geological diversity (minerals, rocks, fossils); geomorphological (landscapes, relief forms, and geoforms); pedological (regoliths, paleosols/fossil soils, current soils) and hydrological (surface and groundwater); (Moura-Fé et al. 2021a, b), in addition to the processes that originated them (Bétard et al. 2011) and currently shape them.

As scientific witness of the events that marked the evolutionary history of the Earth, the geodiversity should be conserved as a fundamental part of the natural heritage and used for scientific, educational, cultural, and geotourism purposes (Godoy et al. 2013). For this, it is important the classification of geological sites (geosites), is based on seven fundamental values: intrinsic, cultural, aesthetic, economic, functional, scientific, and didactic (Gray 2004; Nascimento et al. 2008; Mochiutti et al. 2012).

Despite its importance, however, it is known that geodiversity, as a representative of the Earth's abiotic elements, is impossible to be fully conserved, since society needs natural resources in their daily lives. However, there are elements that stand out in this larger scale, composed of all geodiversity (Guimarães 2016). In this sense, when evaluated and verified through appropriate methodologies, the superlative value of a certain element of the geodiversity, is classified as "Places of Geological Interest—LIGs" (Uceda 1996; García-Cortés and Urquí 2009) or geosites (Brilha 2005, 2016).

Due to the economic value, especially, there are many threats to geodiversity, in which the society is the main modifying and degrading agent (Gray 2005). In an attempt to reverse this vulnerability picture, strategies have been created aiming the conservation of the main elements of the geodiversity, or geoconservation, whose main objective is the conservation of geosites as basic units of the natural heritage, whose implementation requires the creation of a careful methodological systematization and divided into: specific inventory, evaluation (quantification and legal protection, as tombamento, registration or seal of cultural landscape), conservation, enhancement, dissemination and monitoring procedures (Brilha 2005; Lima 2008; Henriques et al. 2011).

Figure 1 presents the correlation between the concepts of heritage, natural heritage, and culture, their segments, with emphasis on geoconservation and their most widespread strategies and development.

One way to promote geoconservation, even if in several cases, it is not the central objective, is through the adoption of principles, objectives, strategies, and practices of Environmental Education (EE), despite its historical proximity with biodiversity (Reigota 2009) and the fact that EE is a field under construction of its theoretical and methodological aspects (Torres et al. 2014).

Understanding the EE as an important ally in environmental conservation, given its broad, significant, and diverse theoretical and methodological framework, analyzed more specifically in the next item, it was seen the possibility of working with its applicability in geodiversity, within more specific parameters and considering the intrinsic demands to geoconservation.

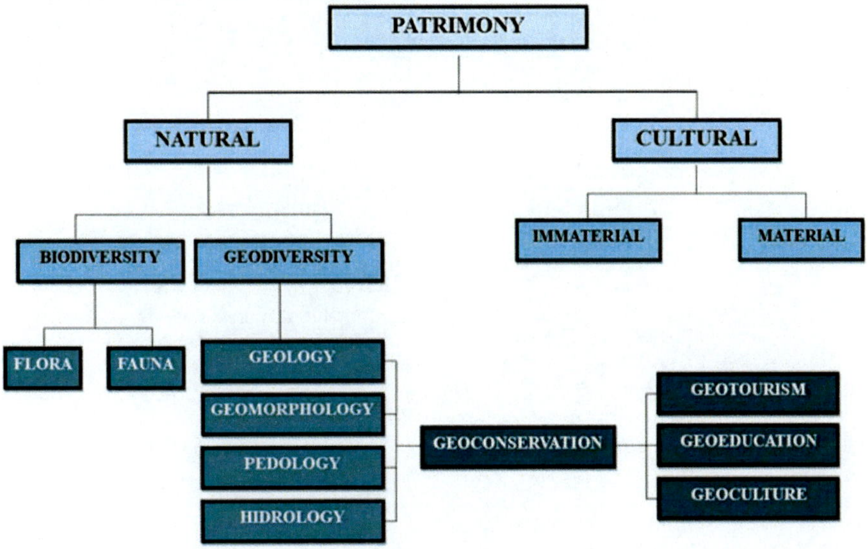

Fig. 1 Conceptual correlation between heritage, geodiversity, and geoconservation. Elaboration: Authors (2021)

Thus, Moura-Fé et al. (2016) proposed, considering the importance of geodiversity and the broad possibility of insertion of EE, the scientific concept of geoeducation, a geoconservationist strategy understood as a specific branch of Environmental Education to be applied in geoconservation of natural heritage, to be treated, fostered and developed in formal and/or non-formal education settings (Fig. 2) (Moura-Fé et al. 2017).

From then on, the geoeducation presented itself as a concept with methodological applicability in development, which can be applied in any place endowed with geodiversity, initially unbound from the compulsory nature of the curricula and school parameters, which brings the possibility of involving a larger audience, variable in age range, level of knowledge of environmental issues, in scope of approach, which may include primarily local aspects, among others.

For Zafeiropoulos et al. (2021), the Geoeducation can be understood as a broader component of environmental education, which aims to promote the geological

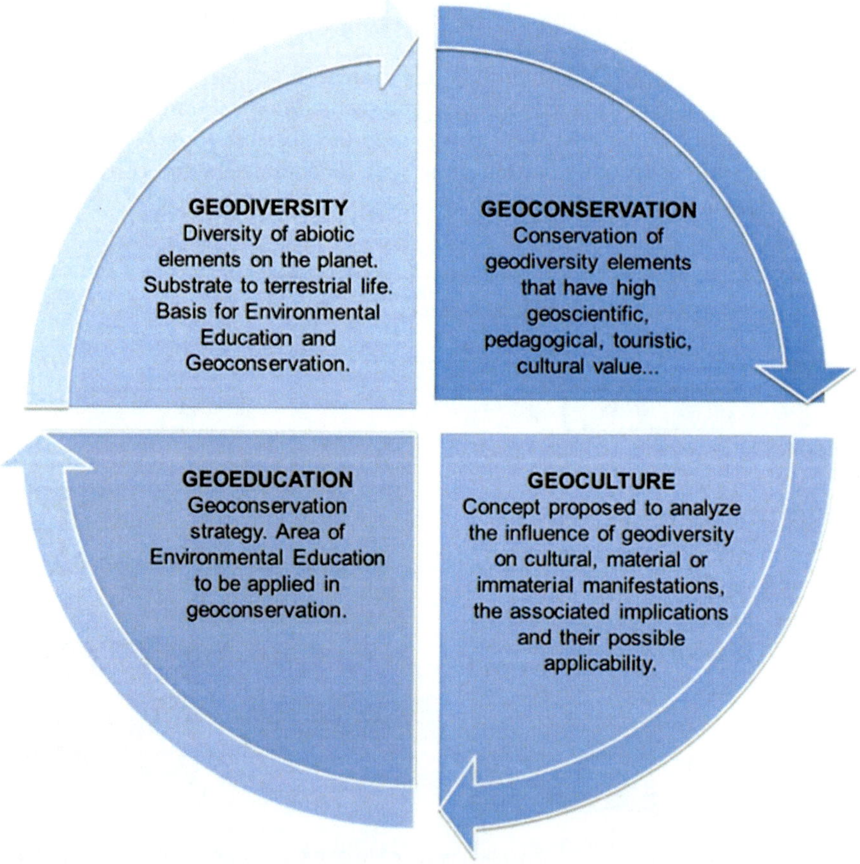

Fig. 2 Conceptual map with key concepts related to geoeducation. Elaboration: Authors (2021)

heritage of a place and its geoconservation. The authors point out that the initiatives aimed at the geological heritage protection since the 1970s and the strategies of Environmental Education, do not deepen or are directly aimed at the issues of geodiversity, geoheritage, and geoethics, for example. In this sense, the importance of promoting and encouraging the geoeducation, this yes, mainly focused on the abiotic elements of the planet, as well as its processes and dynamics. To this end, the aforementioned authors highlight the geotourism as an agent promoter of geoeducation and consequent geoconservation, through, among others, educational and recreational activities (Zafeiropoulos et al. 2021).

In turn, as a geoconservationist proposal with another scope and spectrum of approaches, geoculture was conceptualized as:

> A scientific branch based on the geographic theoretical framework (in the concept of landscape and cultural geography), which should make the analysis of geodiversity and its segments: geoconservation, geotourism, and geoeducation, with emphasis on the influence that the rocks, minerals, fossils, reliefs, geoforms, and soils had and have on cultural manifestations, whether material or immaterial, the associated implications and their possible applicability (Moura-Fé et al. 2017, p. 3074).

Establishing an applicability essay, Silva and Moura-Fé (2018) present a script with indication of geoculture analysis categories, considering the cultural heritage (material and immaterial) and the geodiversity (geology, geomorphology, pedology, and hydrology), with a first stage of identification and inventory; a second stage of specific analysis of the relation between geodiversity and culture; presented in the territory of Araripe UGGp.

Pereira and Cunha (2021) indicate the importance of geotourism as a practice for the connection between the public of the geoheritage and the cultural heritage of a certain visited area, which can happen through geointerpretation strategies designed for the visiting public. Such interaction would still serve as a tool for education of the local population and tourists and for the dissemination of the Geosciences beyond the classrooms, preserving and managing, both the geoheritage and the associated cultural heritage.

4 The "2nd C": Characteristics

The ideal support for geoeducation is in Environmental Education (EE) and its theoretical and methodological assumptions, developed by several researchers in recent decades; and in its already established legal bases (Moura-Fé et al. 2017), present in the current and associated legislation (Soares et al. 2018).

In general, the EA has as mission to form active citizens, presenting itself as a concrete possibility to be present in all spaces of coexistence, including geosites and surrounding communities, for example. Considering also the historically constructed goal of developing proposals by which society can maintain a relationship with the environment, a complex goal, but successful in several case studies throughout the country, the EE presents itself as an educational process articulated, interdisciplinary, which seeks to promote social change, so eminently critical and innovative. In summary, it presents significant elements for the development of geoeducation.

The National Policy on Environmental Education (PNEA) (Brazil 1999) stands out as the main legal reference for EE throughout the country, which was defined by means of 7 (seven) lines of action (Loureiro 2010):

(1) EE in formal education (build capacity of formal, supplementary, and vocational education systems);
(2) Education in the environmental management process;
(3) Conducting specific environmental education campaigns for natural resource users;
(4) Cooperation with those who work in the media and with social communicators;
(5) Articulation and integration of communities in favor of EE;
(6) Intra- and inter-institutional articulation;
(7) Creation of a network of centers specializing in AE, integrating universities, professional schools, and documentation centers, in all the states of the federation.

In turn, illustrating the important role of specific legal diplomas on broader scales (state and municipal), whose creation is part of PNEA (BRAZIL 1999; art. 16), was presented the State Plan for Environmental Education of the state of Ceará (PEACE) (Ceará 2011), a political-pedagogical instrument that presents guidelines, lines of action, objective and conceptions of EE, inserted in the National Environmental Policy, which is subdivided into 6 (six) subprograms (Nikokavouras and Matos 2012):

(1) Capacity Building in Environmental Education;
(2) Education in formal education;
(3) EE and mechanisms for community articulation and mobilization;
(4) EE and local natural resource management mechanisms;
(5) AE, communication and art;
(6) Study and research in Environmental Education.

It is worth mentioning that further studies may adopt the municipal sphere of legislation, expanding the scale of analysis and, therefore, being able to insert a larger contingent of local elements of geodiversity in the context of geoeducational proposals, an important aspect. The review of these parameters should also consider possible changes in the legal diplomas.

Moving on to the second branch addressed in this paper, a key concept for geoculture is that of landscape. The notion of landscape is already present in the memory of mankind even before the elaboration of the concept, whose embryonic idea already existed based on the observation of the environment. The idea linked to this concept, according to Maximiano (2004), seems to have emerged with Humboldt in the eighteenth century, who, in his analyses, started from the observation of vegetation to characterize the space, checking landscape differences of vegetation to apply an explanatory and comparative method (Wulf 2016).

Simply put, the concept of landscape, according to Castro (2007), is always linked to the visual language of the human being, i.e., to the sense of looking and identifying the elements that compose and are found in space. On the other hand, other authors, among them Ferrara (2012), deepen this concept by placing that the landscape is configured by what is beyond what the vision can reach, standing out for having a "semiotic" understood as a wider version of the visible, i.e., all the historical and cultural transformations that that particular location has undergone until today.

Geographic landscape, more specifically, was elaborated to address the set of natural and cultural forms associated with a given area, where the phenomena (or processes) that compose an area are not simply gathered but are associated and interdependent (Sauer 1998).

Thus, specifically regarding the anthropic and social dimension, the modification of a given area by society and its appropriation for its use are important for the occurrence of the succession of these landscapes as a succession of cultures. Thus, the natural landscape is submitted to a social transformation, the last and, in Sauer's (1998) view, the most important factor.

This succession of landscapes seen as a succession of cultures, as well as the analysis of each of them, is a fundamental approach to think how nature, notably, geodiversity, has related to the cultural development of people, communities, and societies. That is, the concept of geographic landscape, by proposing the joint analysis between nature and cultural expressions of society, presents a consolidated theoretical basis that can be used for the development of geoculture (Moura-Fé et al. 2021a).

On the other hand, the Cultural Geography presents in its specific theoretical framework, the cultural approach to the relationship between nature and society, emphasizing and aiming to analyze the social marks printed in nature; allowing the counterposition of geoculture, which, in turn, is a form of analysis that seeks to propose ways to verify the influence of nature on culture, its elements and its material and immaterial manifestations (Moura-Fé et al. 2021a).

Thus, in short, the theoretical assumptions of cultural and landscape geography are the conceptual basis for the development of geoculture (Fig. 3).

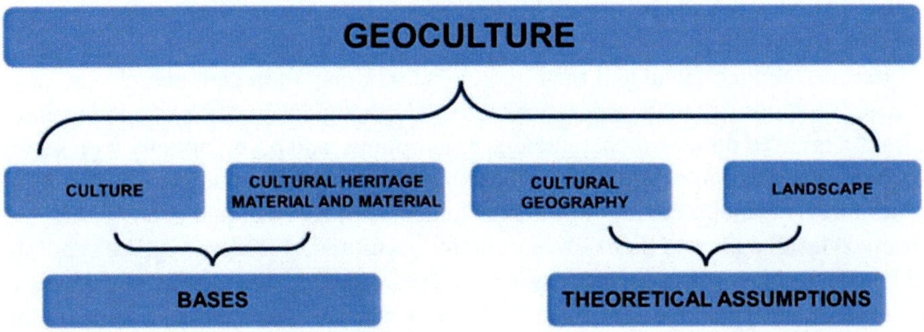

Fig. 3 Parameters, assumptions, and bases of Geoculture. Elaboration: Authors (2021)

5 The Final "C": Contributions

Proposals of theoretical-methodological roadmaps for geoeducation (Moura-Fé et al. 2021b) and geoculture (Moura-Fé et al. 2021a) were elaborated and are being analyzed. While in this item, by way of disclosure and debate, general lines of them will be presented.

The theoretical-methodological roadmap of geoeducation, to be applied in formal and/or non-formal spaces, should take as a starting point, significant elements of local/regional geodiversity, i.e., geosites that can be identified, cultural and referential elements for the communities, regardless of whether they have national or international relevance.

Thus, considering the importance of geosites as starting points, it was decided to adopt the geoconservation strategies initially proposed by Uceda (2000) and then by Brilha (2005, 2016). Thus, the geoeducation itinerary should be preceded by (1) the geosites inventory, (2) the quantification (an important item, although not conditioning for the development of the other steps), (3) the legal protection or conservation and consider, finally, (4) the geosites conservation.

With the approach and development of these 04 (four) geoconservation strategies, there are conditions for the 5th and 6th stages, i.e., valuation and disclosure, in which geoeducation is included. This, which in turn, from what was raised and analyzed so far, is configured as geoconservationist strategy that can be developed, specifically, from the following steps: I—select the geosite; II—compartmentalize the geosite from the dimension of geodiversity and the respective category of analysis; III—make the analysis of the inventory of the geosite, including the information related to the possible realization of quantification, protection or tumbling and analysis of conservation (basic strategies of geoconservation); IV—define the scope of the Geoeducation Work Plan; and V—indicate the strategies of application of the Geoeducation Work Plan.

This theoretical and methodological proposal of geoeducation is composed of conceptual elements, guiding the proposition, and by an application proposal, to be developed in studies directed to specific areas, but that, a priori, can be thought of in the form of work plans, which, in turn, should consider the local and/or regional specificities of geodiversity.

In turn, the theoretical and methodological roadmap of geoculture, similar in its lines of development to the theoretical and methodological roadmap of geoeducation (Fig. 4), is based on specific parameters, assumptions, and bases, notably with regard to their inter-relationships and contributions to geoculture; should take as a starting point for its development the significant elements of local/regional geodiversity, i.e., the geosites that present themselves as identity, cultural, referential and/or symbolic elements for people, localities, communities, regardless of whether the geosites have national or international relevance from the scientific, aesthetic and/or economic point of view.

Thus, the geosites are the verification and analysis field of the term "geo"; and the term "culture" of the geoculture? The elements and cultural manifestations, whether

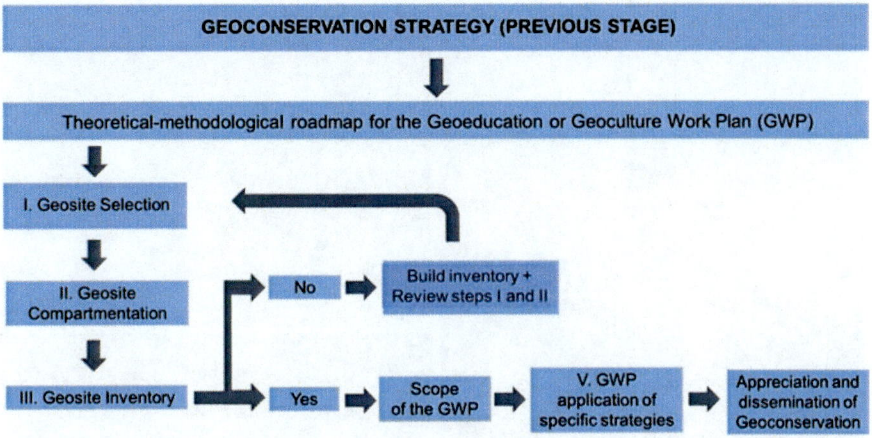

Fig. 4 Correlation of the stages of the Geoculture and Geoeducation theoretical-methodological guide

material and/or immaterial, will be emphasized in analyses made in communities and localities, directly and indirectly, associated, linked, referenced to/with the geosites and their different meanings, even if the people and their cultural manifestations do not recognize/present these places as "geosites", where the geodiversity is the central element of their relations (Fig. 5). Moreover, such possible ignorance by the communities only justifies the gradual adoption of geoconservationist measures in their daily lives.

In a broader scale of analysis, the regions that encompass geosites and localities should also be considered and addressed, that is places and people, real or imaginary (legendary, mythological), making these relationships elements of their identities.

With the approach and development of the four geoconservation strategies mentioned above, there are conditions for the fifth and sixth steps, i.e., valuation and disclosure, in which geoculture is included, which, in turn, from what was raised and analyzed so far, is configured as a geoconservationist strategy that can be developed, specifically, from the following steps:

I. Identify and select the geosite [with the verification of the respective associated locality(ies)].
II. Characterize the geosite from the dimension of geodiversity and the respective category of analysis;
III. Make the analysis of the geosite inventory, including information related to the possible realization of quantification, officinal protection and conservation analysis (basic geoconservation strategies);
IV. Identify and characterize the forms of relationship between the elements of geodiversity of the geosite and the cultural heritage (material and/or immaterial), which will compose the scope of the Geoculture Work Plan (GWP);

Fig. 5 **a** Stone of Sin. Thermoclasty acting on granite from the basement of the Araripe sedimentary basin, whose breach created by the natural process is culturally interpreted by religious devotees of Padre Cícero as proof of the absence of sin by those who pass by. Colina do Horto Geosite—GeoPark Araripe (Juazeiro do Norte/CE). Photo: Marcelo Moura Fé (Mar/2016). **b** Ichnofossils recorded in the silicified sandstones of the Mauriti Formation (Paleozoic age), formed in a lacustrine paleoenvironment and were interpreted by the population as bird tracks, which would have been embedded in the earth's mud after the waters of the biblical flood dried up. Cachoeira de Missão Velha Geosite—Araripe GeoPark (Missão Velha/CE). Photo: Marcelo Moura Fé (Nov/2019). **c–d** Pedra da Santa—Araruna/PB. Granite shelter with the occurrence of diorite enclaves and alveolar cavity, in the shelter there are cave paintings and there is a small sanctuary in honor of Our Lady of Fátima—in the place, geodiversity and religious tourism is mixed. Photo: Thaís Guimarães (Mar/2014)

V. Elaborate the GWP with emphasis on strategies for the application of geoculture in localities/communities for the purposes of knowledge, appreciation, and application of geoconservation and, consequently, of local/regional cultural elements and manifestations.

The theoretical-methodological proposal of geoculture is composed of conceptual elements, guiding for the proposition, and by an application orientation, to be developed in studies directed to specific areas (geosites and localities/communities), but that, a priori, can be thought of in the form of work plans, which, in turn, should consider the local and/or regional specificities of geodiversity and, of course, of the localities, associated communities and their manifestations and specific cultural elements.

6 Final Considerations

The theoretical and methodological proposal of geoeducation, focused on the knowledge, appreciation, and application of geoconservation, gestated over the last years, presents itself as a geoconservationist strategy that can enhance the whole scope present in Environmental Education, directing it to the specificities of geodiversity, its dimensions and respective categories of analysis.

On the other hand, the theoretical and methodological proposal of geoculture, also developed over the last years, presents itself as an alternative for the consolidation of a geoconservationist strategy that seeks to identify, correlate and apply all (or, at least, part) of the diversity of relationships that occurs between geodiversity and the culture present in music, theater, poetry, constructions, sculptures, paintings, archaeological records, but still little captured under the geoscientific view.

What do they both have in common?

As it concerns the methodologies, notably as proposals, in general, the confrontation of these theoretical and methodological roadmaps (which consider theoretical and applicable elements, chosen from studies in various spaces of remarkable geodiversity and experiences of practical order) with the local realities and their respective and diverse specificities (the necessary clash between "theory versus reality"), to be made in more specific studies, should bring elements that can improve the proposals made and thus provide more consolidated forms for the preparation and effective implementation of work plans.

What do both have to offer to Brazilian geoconservation?

The expansion of debates and possibilities:

(1) By bringing environmental education to the core of local geoconservationist demands, with more specific proposals that meet the scales of detail, thus expanding the possibilities of collaboration and contribution to the promotion of geoconservation in communities and localities of the country; and
(2) In addition, by promoting the meeting of culture and geodiversity under a scientific analytical bias, it will also provide the inclusion of various fields of knowledge in the context of reflection, debate, and construction about the scope, importance, and relative urgency of taking geoscientific knowledge to as many spaces as possible.

It is hoped that, finally, the ideas and methods contained in Geoeducation and Geoculture may gain experiences and advances, so that on the "geodiversity table" new chairs are placed, that we have researchers and researchers from the most diverse backgrounds, and that, together, innovations and new contributions come, because the Brazilian geoconservation, in the most diverse places and contexts, needs and deserves.

Acknowledgements Marcelo Martins thanks the support granted by the Research Productivity Scholarship Program, Stimulus to the Interiorization and Technological Innovation (BPI 03/2018) of the Cearense Foundation for the Support of Scientific and Technological Development—FUNCAP, through financial aid (project no. BP3-0139-00130.01.00/18) to research and research productivity scholarship.

Marcos Nascimento thanks CNPq for granting the Research Productivity Scholarship 2.

João Victor and Raquel Landim thank the Scientific Initiation Scholarship Program of URCA (PIBIC-URCA), for granting the scholarships (2018–2019) that contributed in the financial support of part of this research.

The authors would also like to thank the other researchers, scholars and members of the Núcleo de Estudos Integrados em Geomorfologia, Geodiversidade e Patrimônio—NIGEP (URCA/CNPq).

Dedicated to the memory and legacy of Professor Kenitiro Suguio.

References

Alvarenga LJ, Castro PTA, Bernardo JM, Campos IC (2018) Cultural Landscape and Geoconservation: conceptual contributions applied in the Serra da Canastra, Brazil. Anuário do Instituto de Geociências 41(2): 241-251.

Arouca Declaration (2011). International Congress of Geotourism - "Geotourism in Action. Arouca.

Bétard F, Peulvast J-P, Magalhães AO (2011) Biodiversité, géodiversité et enjeux de leur conservation dans les montagnes humides du Nordeste brésilien. BAGF. Géographies: 17–26.

Brazil (1999) Casa Civil. Federal Law nº 9.795, 27 April 1999. Dispõe sobre a educação ambiental, institui a Política Nacional de Educação Ambiental e dá outras providências. Brasília.

Brilha JBR (2005) Geological heritage and geoconservation - nature conservation in its geological aspect. Palimage, Braga.

Brilha J (2016) Inventory and Quantitative Assessment of Geosites and Geodiversity Sites: a review. Geoheritage 8(2): 119-134.

Castro DG (2007) Significados do conceito de paisagem: um debate através da epistemologia da Geografia. Universidade Estadual do Rio de Janeiro.

Ceará (2011). Casa Civil. State Law No. 14.892 of March 31, 2011, which provides for environmental education, establishes the state policy for environmental education and other provisions. Fortaleza: DOE published on April 4, 2011. Series 3, Year III, n 064. Caderno 1/2.

Delphim CF (2009). Cultural Heritage and Geopark. Revista do Instituto de Geociências 5:75-83.

Dowling RK (2013) Global Geotourism - An Emerging Form of Sustainable Tourism. Czech Journal of Tourism 2 (2): 59-79.

Erikstad L (2014) Geodiversity, biodiversity and landscape - key elements in modern nature management strategies. Geophysical Research Abstracts, 16.

Ferrara LD (2012) The mediations of the landscape. Libero, 15(29): 43-50.

García-Cortés A, Urquí LC (2009) Documento Metodológico para la elaboracion del inventario espanñol de lugares de interés geológico (IELIG). Madrid: Instituto Geológico y Minero de España.

Gil AC (1996) Como elaborar projetos de pesquisa. Atlas, São Paulo.

Godoy LH, Sardinha DS, Bertini RJ, Conceição FT, Del Roveri C, Moreira CA (2013) Potencial Geoparque de Uberaba (MG): geodiversidade e geoconservação. Sociedade & Natureza 25(2): 395-410.

Gray M (2004) Geodiversity: valuing and conserving abiotic nature. John Wiley and Sons, Chichester, England.

Gray M (2005) Geodiversity and Geoconservation: what, why, and how? Geodiversity and Geoconservation 4–12.

Guimarães TO (2016) Geological heritage and geoconservation strategies: popularization of the Geosciences and sustainable territorial development for the Southern Coast of Pernambuco (Brazil). Doctoral thesis presented to the Graduate Program in Geosciences - UFPE, 406p.

Guimarães TO, Mariano G, Sá AAA (2017) Geotourism: proposal for territorial enhancement and sustainability alternative to "sun and beach" tourism in the southern coast of Pernambuco, Brazil. Ciência e Sustentabilidade v3(1): 33–57.

Guimarães TO, Moura-Fé MM, Piekarz GF (2021) Geoconservation in Brazil: theory and practices. Geoconservação no Brasil: teoria e práticas. In: Souza-Fernandes LC, Aragão A, Sá AA (ed). Novos Rumos do Direito Ambiental - um olhar para a Geodiversidade. Unicamp, Campinas, p. 251–274.

Henriques MH, Reis RP, Brilha J, Mota T (2011) Geoconservation as an Emerging Geoscience. Geoheritage 3: 117-128.

Hose TA (1995) Selling the Story of Britain's Stone. Environmental Interpretation 10(2): 16-17.

Hose TA (2000) "Geoturismo" europeo. Interpretación geológica y promoción de la conservación geológica para turistas. In: Barretino, D, Winbledon, WAP, Gallego, E, (ed). Patrimonio geológico: conservación y gestión. Madrid: Instituto Tecnológico Geominero de España, p. 212.

Liccardo A, Piekarz GF (2017) Tropeirismo e geodiversidade no Paraná. Text Studio, Ponta Grossa.

Lima FF (2008) Proposta Metodológica para a Inventariação do Património Geológico Brasileiro (Dissertation). School of Sciences of the University of Minho, Braga (POR).

Lima JTM, Carvalho IS (2020) Geological or Cultural Heritage? The Ex Situ Scientific Collections as a Remnant of Nature and Culture. Geoheritage 12(3). https://doi.org/10.1007/s12371-020-00448-5.

Loureiro CFB (2010) Crítica ao teoricismo e ao praticismo na educação ambiental. In: Cabral Neto, A, Macedo Filho, FD, Batista, MSS (Ed). Educação Ambiental – caminhos traçados, debates políticos e práticas escolares. Líber Book, Brasilia, p. 15–32.

Maximiano LA (2004) Considerations on the Concept of Landscape. RA'E GA Journal 8: 83-91.

Moreira JC (2011) Geoturismo e Interpretação Ambiental. Editora da UEPG, Ponta Grossa.

Mochiutti NF, Guimarães GB, Moreira JC, Lima FF, Freitas FI (2012) The values of geodiversity: geosites of the Araripe Geopark/CE. Anuário do Instituto de Geociências 35(1): 173-189.

Moura-Fé MM, Guimarães TO, Holanda CR, Nascimento MAL, Silva JVM (2021a) Geoculture: theoretical-methodological proposal for the knowledge, appreciation and application of geoconservation, in press. Caminhos de Geografia.

Moura-Fé MM, Guimarães TO, Nascimento MAL, Soares LN, Nascimento RL (2021b) Geoeducação: proposta teórico-metodológica para o conhecimento, valorização e aplicação da geoconservação, in press. Caminhos de Geografia.

Moura-Fé MM, Nascimento RL, Soares LN (2017a) Geoeducation: theoretical principles and legal bases. In: Perez Filho, A, Amorim, RR (Ed). Os Desafios da Geografia Física na Fronteira do Conhecimento. Institute of Geosciences - Unicamp, Campinas, p. 3054–3065.

Moura-Fé MM, Pinheiro MVA, Jacó DM, Oliveira BA (2016) Geoeducation: environmental education applied to geoconservation. In: Seabra G (Ed) Educação Ambiental & Biogeografia. Barlavento, Ituiutaba, p. 829–842.

Nascimento MAL, Azevedo, UR, Mantesso-Neto V (2008) Geodiversidade, geoconservação e geoturismo: trinômio importante para a conservação do patrimônio geológico. Rio de Janeiro: SBGeo edition.

Nikokavouras EAQ, Matos KSAL (2012) O Sistema de meio ambiente e a educação ambiental no estado do Ceará. In: Matos KSAL, Sampaio JLF (Ed) Diálogos em Educação Ambiental. UFC, Fortaleza, p. 51–66.

Ollier C (2012) Problems of geotourism and geodiversity. Quaestiones geographicae 31(3).

Pereira LS, Cunha LS (2021) Allying the cultural heritage to geoheritage: a geotouristic coastal itineray in the municipality of Cabedelo, Paraíba (northeast of Brazil). Revista do Departamento de Geografia 41: 1-15.

Reigota M (2009) O que é Educação Ambiental. Brasiliense, São Paulo.

Ruchkys U, Machado MMM (2010) Cultural landscape and landscape potential of the Geopark. Brazilian Journal of Ornamental Horticulture, 16(1):13–16. https://doi.org/10.11606/eISSN.2236-2878.rdg.2021.182159.

Sauer C (1998) The Morphology of Landscape. In: Corrêa RL, Rosendahl Z (Ed) Paisagem, tempo e cultura. Ed UERJ, Rio de Janeiro.

Silva JVM, Moura-Fé MM (2018) Geoculture: the relationship of geodiversity with culture in the "territory GeoPark Araripe". Crato/CE. In: National Symposium of Geomorphology, 12, 2018, Crato. Annals [...]. Crato: URCA.

Soares LN, Nascimento RL, Moura-Fé MM (2018) Proposal of Geoeducation application in Araripe Geopark. Crato/CE. In: Simpósio Nacional de Geomorfologia, 12, 2018, Crato. Annals [...]. Crato: URCA.

Sousa-Fernandes LC, Aragão A, Sá AA (2021) Novos Rumos ao Direito Ambiental: um olhar para a geodiversidade. Editora UNICAMPI, Campinas, 559p.

Torres JR, Ferrari N, Maestrelli SRP (2014) Critical-transformative environmental education in the school context: Freirean theory and practice. In: Loureiro CFB, Torres JR (Ed). Educação Ambiental. Dialogando com Paulo Freire. Cortez, São Paulo, p. 13–80.

Uceda AC (1996) El Patrimonio Geológico. Ideas for its Protection, Conservation and Utilization. Ministerio de obras publicas, transportes y Medio Ambiente (MOPTMA), Dirección General de Información y Evaluación Ambiental. Seriemonografias, Madrid.

Uceda AC (2000) Património geológico; diagnóstico, clasificación y valoración. In: Suárez-Valgrande, J. P. (Ed). Jornadas sobre Patrimonio Geológico y Desarrollo Sostenible, Soria, Serie Monografías, Ministry of Environment, Spain, p. 23–37. 2000.

Urquí LC (2012) Geoconservation. Los Libros de la Catarata, Madrid.

Wulf AA (2016) Invenção da Natureza: a vida e as descobertas de Alexander von Humboldt. Planeta, São Paulo.

Zafeiropoulos G, Drinia H, Antonarakou A, Zouros N (2021) From Geoheritage to Geoeducation, Geoethics and Geotourism: a critical evaluation of the Greek Region. Geosciences 11, 381. https://doi.org/10.3390/geosciences11090381.

Susceptibility to the Development of Debris Flows in the Territory of the Caminhos Dos Cânions Do Sul Geopark in Southern Brazil

Marina Tamaki de Oliveira Sugiyama and Maria Carolina Villaça Gomes

Abstract Several studies have pointed out the strong relationship between morphometric parameters and the triggering of hydrogeomorphological processes such as debris flows. In protected areas, which include elements endowed with scientific, educational, cultural, and touristic interests, their effects, although important for the natural dynamics of the landscape, can lead to momentary decharacterization of this natural heritage, as well as put at risk the visitors of these spaces. In view of the proposal to create the Geopark Caminhos dos Cânions do Sul (RS/SC) in an area historically affected by these processes, and given the regional extension of the territory, this work aims to evaluate from the morphometry the susceptibility to the development of debris flows in the Geopark territory. To this end, 25 watersheds were delimited, for which 12 morphometric parameters were generated, with the result of the susceptibility of the basins confronted with the record of occurrences of mass movements and the location of geosites. The result of the susceptibility presents seven basins of Low, seven of Medium, six of High, and five of Very High Susceptibility. Furthermore, the results point to a high susceptibility to debris flows in basins with the presence of geosites, proving the need for studies on the susceptibility to these processes for the definition of priority areas for the elaboration of preventive plans for risk reduction and disasters.

1 Introduction

Debris runs are mass movements induced by the action of gravity, in which the materials behave as highly viscous fluids (Costa 1984). This process is characterized by high erosion capacity and high speed, which commonly ranges from 5 to 20 m/s, transporting a large volume of material for long distances in a short period (Kanji and Gramani 2001). The material transported consists of a combined mass of solids and fluids, formed by large volumes of fine material, rock blocks, logs, various amounts of water, and other materials present on the slopes.

M. T. de Oliveira Sugiyama (✉) · M. C. Villaça Gomes
Santa Catarina State University (UDESC), Florianópolis, Brazil
e-mail: maritamaki12@gmail.com

G. Barbosa dos Santos et al. (eds.), *Geomorphology of Brazil: Complexity, Interscale and Landscape*, Springer Proceedings in Earth and Environmental Sciences,
https://doi.org/10.1007/978-3-031-05178-4_14

This typology of mass movement is considered one of the most destructive natural processes associated with external dynamics, being responsible for significant transformations in the landscape (Kanji and Gramani 2001). The destructive power of the runs tends to increase with the distance traveled, from its initiation to its deposition, due to the remobilization of pre-existing material that is incorporated along its path. This behavior is linked to the characteristics of the drainage network and the terrain, especially the slope, which contributes to the increase in the speed of this process. In general, debris runs in Brazil are related to slope slides caused by anomalous precipitation that increase the moisture content of the slopes and result in the loss of internal friction of the material at rest (Corrêa 2018). High intensity and short duration rainfall, usually from 60–70 mm/h, is responsible for the triggering of mass movements, with widespread occurrence of landslides and debris flows (Kanji and Gramani 2001).

In addition to high rainfall rates, the occurrence of runs is favored by a set of geological (presence of debris or rocks susceptible to mobilization) and geomorphological factors (slope, generally greater than 25°, and curvature of the slopes). Thus, the debris flows are processes strongly influenced by the basin morphometric characteristics, related to the drainage pattern and the relief configuration. Several authors have used morphometric parameters to study debris flows (Jakob 1996; De Scally et al. 2001; Vieira et al. 1996; Dias et al. 2016; Gomes 2016; Kanji and Gramani 2001; Cabral et al. 2021), demonstrating the importance of morphometry for the triggering of this process.

The analysis of the morphometric characteristics of the drainage pattern and relief, besides expressing the strong relationship with lithology, geological structure, and surface formation of the elements that compose the earth's surface, makes it possible to obtain information regarding the properties of the terrain, such as infiltration, channel flow and surface runoff (Pissarra et al. 2004). The flow of the channels depends on the factors that determine the excess of precipitated water and those that influence the total time that rain takes to travel throughout the basin (Morisawa 1962). The factors that control excess rainfall are closely related to climate, vegetation, infiltration capacity, and surface water storage. Geomorphological factors, such as channel length, basin shape, slope, as well as lithology, and geological structure, influence surface runoff, and river flow.

In addition, the literature shows that most debris runs occur in high slope, small catchments (Slaymarker 1990; Keller 1992; Kanji and Gramani 2001). The slope of a basin is one of the main factors that regulate the time of concentration in streams and the duration of runoff. Besides, basins with higher slopes have greater action of gravity on the geomorphological processes and, consequently, on the transport and deflagration of debris races (Jakob 1996). Slope slopes higher than 30° are considered critical for the generation of runs, although their minimum values vary from 20° to 25°, while in the channels, the slope for movement of the material deposited in the bed varies from 15 to 20° (Kanji and Gramani 2001).

As explained, the morphometric parameters are closely related to the geomorphological processes, however, the parameters alone are not able to provide a conclusive analysis of the processes. Thus, it is the meanings of the correlations of the parameters, morphometric indices, and hydrological characteristics that reveal the action of hydrogeomorphological processes, such as debris flows (Cherem et al. 2020).

In protected areas, which include elements endowed with scientific, educational, cultural, and touristic interests, its effects, although important for the natural dynamics of the landscape, can lead to the momentary decharacterization of this natural heritage, as well as put at risk the visitors of these spaces. In the territory of the Caminho dos Cânions do Sul Geopark (GCCS), located in the states of Rio Grande do Sul and Santa Catarina, the geosites covered are periodically stage of hydrogeomorphologic events of great magnitude, being these places, sometimes, endowed with scientific value due to the occurrence of these events.

Among the events of great magnitude that have occurred in the region, it is worth mentioning those that occurred in December 1995, when the high rainfall associated with the passage of a frontal system and a cyclonic vortex caused widespread erosion processes in the scarps and plateaus of Serra Geral (Valdati 2000). However, according to Pellerin et al. (1997), in spite of being catastrophic, this event cannot be considered totally exceptional, since in 1974 a similar event, of even greater magnitude, was recorded on the slopes of Serra Geral. Furthermore, the deposits found in valley bottoms demonstrate that the phenomenon is recurrent in the geological history of the region (Duarte 1995; Pontelli 2005).

Thus, considering the existence of geosites in an area historically affected by mass movement processes, with records of debris flows, this work aims to evaluate the susceptibility to the development of debris flows in the territory of Caminhos dos Cânions do Sul Geopark (RS/SC), based on the morphometry of basins. The identification of areas with different degrees of susceptibility is essential for the identification of priority areas for the development of preventive plans for risk reduction and disasters.

2 Study Area

With an area of 2829 km^2 the GCCS encompasses seven municipalities belonging to the states of Rio Grande do Sul and Santa Catarina (southern Brazil), being them: Cambará do Sul (RS), Jacinto Machado (SC), Mampituba (RS), Morro Grande (SC), Praia Grande (SC), Timbé do Sul (SC) and Torres (RS) (Fig. 1).

The geological conformation of the study area is the result of the conjunction of numerous geotectonic events that date back to the evolution of the Paraná Basin, which is characterized as an intracratonic volcanosedimentary basin that evolved on the South American Platform. The stratigraphic framework of the Paraná Basin is formed by six Supersequences, and in the territory of the GCCS one can find the records of the Supersequences Gondwana I, with the eolic sandstones, lacustrine pellets, and fluvial deposits of the Rio do Rasto Formation, and Gondwana III, with the eolic sediments of the Botucatu Formation and the volcanic outflows of the Serra Geral Formation.

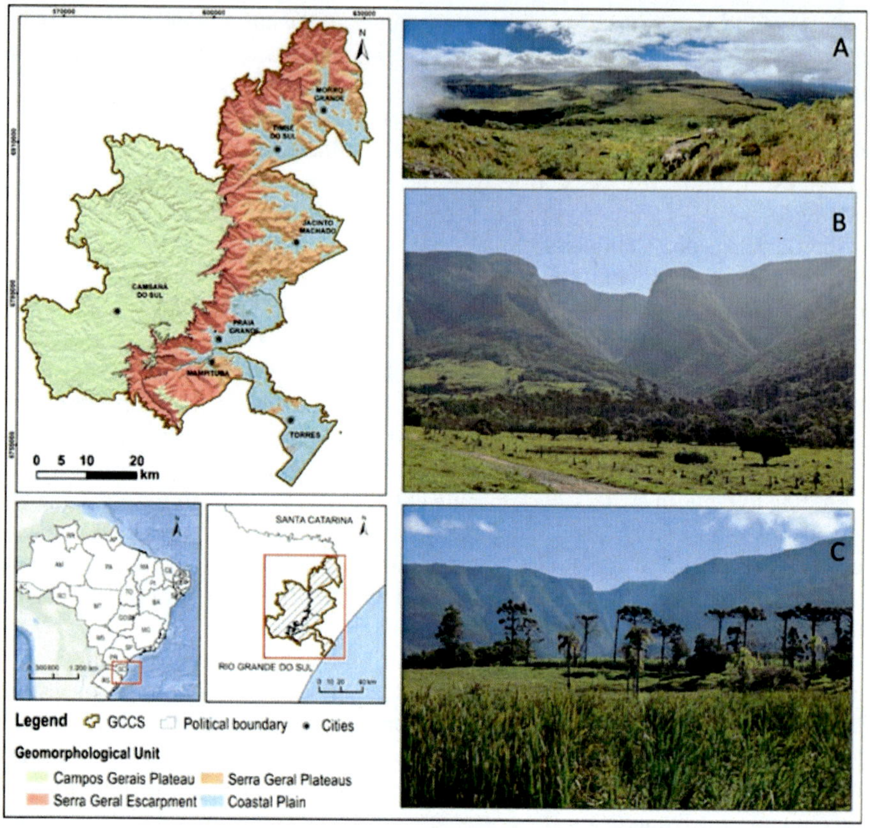

Fig. 1 Location of the study area and geomorphological compartmentalization. **A** Campos Gerais plateau, characterized by its low slope and gentle slope to the west; **B** Escarpments of Serra Geral, which presents a high altimetric gradient, high slope, and deep valleys. **C** In the background, Escarpments and Patamares da Serra Geral and, in front, Coastal Plain

The uplift of the eastern edge of the Plateau, concomitant with the process of Gondwana separation, exposed the entire sedimentary rock package of the Paraná Basin to intense erosive phenomena, forming gradients of more than 1000 m. Thus, in the study area, there are four well-defined geomorphological compartments: the plateau, the mountains, the plateaus, and the plain (Fig. 1).

The Campos Gerais Plateau constitutes the South Brazilian Plateau erected during the Cretaceous and that possesses a soft and continuous inclination of the surface toward west, toward the interior. In this compartment are found the springs of the main rivers of the region, following a clear structural control of faults and diaclases. The relief is marked by flat and gently undulated forms, with low declivity, in altitudes above 900 m. The Serra Geral represents, in reality, the scarps of the plateau and is characterized by the intense dissection of its *front*, many times conditioned by the geological structures of the substratum, presenting deep valleys in "*V*", with abrupt and scarped forms, located in altitudes between 400 and 900 m.

The Serra Geral escarpments present, along the study area, verticalized walls in the form of canyons due to the fluvial incision along the tectonic lineaments, and the occurrence of block falls is common in the contact between the spills of different acidity (Santos, 2021).

At slightly lower altitudes (200–400 m), we find the Patamares of Serra Geral which constitute elongated spikes that project from the escarpments toward the plain as an intermediate relief and testify the past extension of the Plateau and the escarpments. This compartment is largely composed of the sedimentary rocks of Fm. Rio do Rasto and Fm. Botucatu, presented less dissected forms and lower declivity. Finally, the coastal plain, of low declivity and situated in elevations lower than 200 m, is formed by continental and coastal depositional systems. The former, associated with slope processes, includes colluvial, fluvial, and alluvial fan deposition, while the latter system, linked to sea level variations during the Quaternary, encompasses marine, lagoonal, paludial, and eolian deposits (Horn Filho 2003).

The GCCS territory is inserted in two large hydrographic basins: the Mampituba River Basin and the Araranguá River Basin. Most of the headwaters of the Araranguá and Mampituba river basins are located in the steep terrains of the Serra Geral Formation, presenting a clear structural control in the drainage pattern. The headwaters located in the Serra Geral plateau are embedded in sedimentary rocks of the Rio do Rasto Formation and Botucatu Formation, presenting a dendritic drainage pattern. When they leave the steep terrains of the hillsides, the rivers reach the Quaternary sediments of the coastal plain, where the relief is very soft and the drainage is characterized by intertwined channels. Only sparse fragments of the Plateau are inserted in the Araranguá river basin, possibly due to the more expressive retreat of the Serra Geral *front* in this region—many of the small drainage headwaters that before flowed westward were captured by the Araranguá river tributaries, from the erosive retreat of the escarpment.

The clear differentiation between the relief compartments in the study area is closely related to the occurrence of processes of great magnitude. An extreme contributing factor to the high rainfall in the region is the orographic barrier represented by the Serra Geral escarpment. Valdati (2000) registered an average annual precipitation of 2,519 mm on the slopes of Serra Geral, at 220 m altitudes, and 1,766 mm at 70 m altitudes, representing a 30% increase in precipitation in only 160 m of difference. The mean annual precipitation in the study area ranges from 1507 mm in the municipality of Torres, at 8.47 m altitude, and 1823 mm in Cambará do Sul, at 1015 m (INMET, 2020). Thus, the rainfall resulting from the collision of frontal systems added to the orographic effect on the escarpment, conditions the great recurrence of mass movements in the GCCS area, so that in the period from 1974 to 2017 about 16 occurrences of "landslides" were recorded, however, it was not possible to differentiate the types of mass movements in the records of occurrences (Pimenta et al. 2018).

3 Methodology

The procedures adopted for this work are divided into four main steps: (i) delimitation of the watersheds; (ii) generation of morphometric parameters; (iii) evaluation of the susceptibility of the watersheds to the development of debris flows and; (iv) joint analysis of the susceptibility with the records of occurrence of runs and the location of geosites. The ArcGis 10.5 *software* provided by ESRI was used for data processing and analysis.

The data processing was performed from the Digital Elevation Model (DEM) of the TOPODATA project of the National Institute for Space Research (INPE), prepared based on SRTM data with 30 m resolution. For the refinement of the drainage generated from the MDE, the hydrographic base of the State of Santa Catarina of the Secretariat of Sustainable Development (SDS) was used, elaborated on a scale 1:10,000, and produced from data with 1 m resolution.

The spatial cut-off had as a criterion the watersheds of the GCCS that reach the coastal plain compartment, thus excluding the watersheds located exclusively on the escarpment and the plateau. Furthermore, due to the extension of the study area, only basins from the 3rd hierarchical order onwards were analyzed. After the delimitation of the watersheds, 12 morphometric parameters were generated based on the recurrent use in the literature and that presented reliable results, considering both the parameters related to the hydrographic network and the relief of the basin (Table 1).

Once the results of the morphometric parameters for the drainage basins were obtained, Pearson's correlation coefficients were generated in order to evaluate the relationships between the morphometric parameters (referring to the drainage network and the surface of the hydrographic basin) and the lithology/structural constraints.

To evaluate the susceptibility to debris flows, the methodology of Gramani et al. (2005) was used, developed to identify the degree of criticality of the basins to the outbreak of runs. The criticality of the basins is obtained from the relativization of the parameter values for the basins, so that its result points to a higher or lower relative criticality. The method proposes a qualitative analysis based on the results of the parameters considered as conditioning factors of the debris flows, as follows: percentage of the area with slope above 30° (A_{30}); inverse of the basin area ($\frac{1}{A}$); basin width (H); main channel slope (S_{L} and; inverse of the circularity index ($\frac{1}{Ic}$).

According to the methodology, the parameter values were transformed into dimensionless indices by dividing each parameter of the basin by the smallest value of the same parameter obtained for the other basins, so that the smallest value of each parameter is always equal to 1.0. After this step, the values were weighted from weights defined according to their importance in the process of generation of runs, as follows: 2.5 for A_{30}; 0.5 for $\frac{1}{A}$; 1.0 for H; 0.5 for S_{L} and; 0.5 for $\frac{1}{Ic}$. For each basin, the weighted values of each parameter were added, which were normalized again, resulting in a criticality index for the basins.

Table 1 Morphometric parameters used in the research

Morphometric parameter	Unit	References
Stream order (*O*)	Ordinal	Strahler, (1945); Christofoletti (1980)
Basin area (*A*)	Km^2	Keller (1992); Castro (2003); Zavoianu (1985); Morisawa (1962), Jakob (1996)
Percentage of area above 30° (A30)	%	Kanji and Gramani (2001); Gramani et al., (2005)
Basin perimeter (*P*)	km	Souza, (2005); Zavoianu (1985)
Drainage density (Dd)	km/km^2	Horton (1945); Morisawa (1962); Zavoianu (1985)
Ratio relief (*Rr*)	M/km	Horton (1945); Jakob (1996), Gomes (2016)
Main stream length (*L*)	km	Horton (1945), Zavouianu (1985); Souza (2005)
Streams length (*Ls*)	km	Souza (2005); Horton (1945); Morisawa (1962)
Mean slope (*S*)	Degrees (°)	Kanji and Gramani 2001; Jakob 1996); Horton (1945)
Main chanel slope (Sl)	Degrees (°)	Souza (2005); Morisawa (1962)
Relief (*H*)	M	Jakob, (1996); Gomes (2016); Dias (2017)
Circularity index (*Ic*)	Dimensionless	Kanji and Gramani 2001; Gramani et al., (2005), Morisawa (1962)

From the result of the criticality index, homogeneous intervals were defined for the categorization of the basins in susceptibility classes to the development of debris flows, which are: LOW, MEDIUM, HIGH, and VERY HIGH.

Finally, the result of the classification of the basins in relation to the susceptibility to the development of runs was confronted with the records of occurrence of mass movements in the territory of the GCCS, during the period from 1974 to 2017 (Pimenta et al. 2018), and with the location of the geosites, in order to ascertain the location of the basins of greater susceptibility in relation to them.

4 Results and Discussions

The delimitation of the basins for the evaluation of the susceptibility to the development of debris flows resulted in the definition of 25 drainage basins in the territory of the GCCS (Fig. 2). The results of the application of morphometric parameters show that the basins of the study area have distinct configurations, with significant differences in their morphometric characteristics, although inserted in the same geological–geomorphological context.

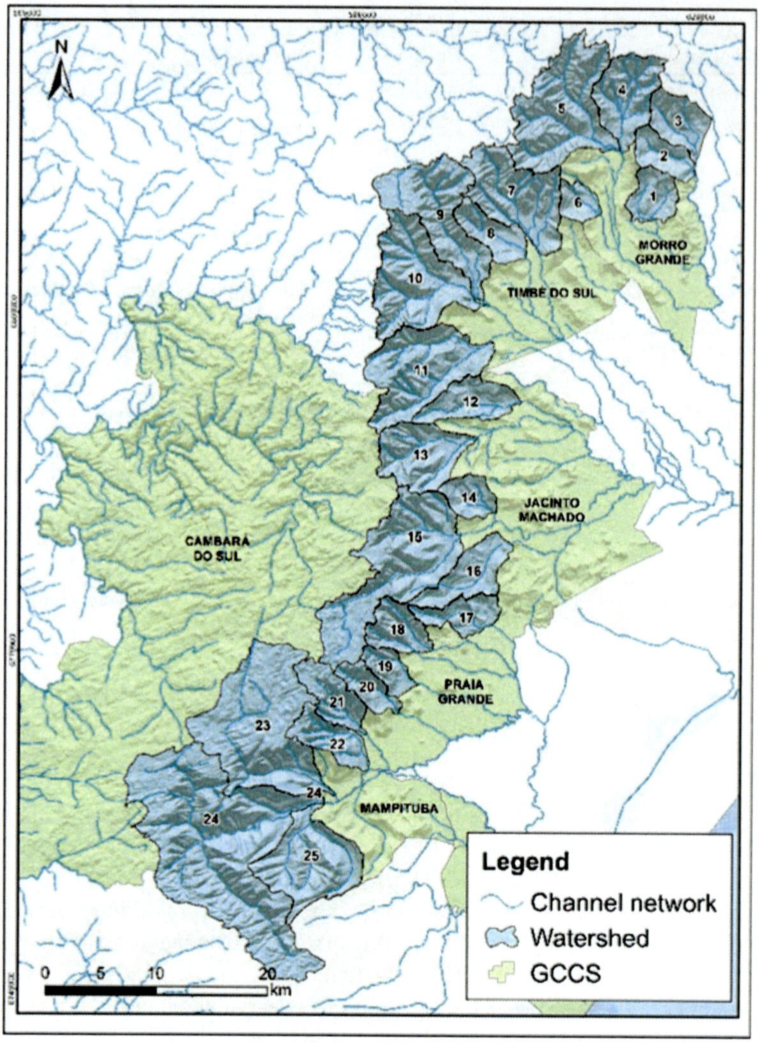

Fig. 2 Result of the delimitation of drainage basins

The range of the parameter results was: basin area (A), between 10.48 and 184.67 km^2; relief ratio (R_r), between 54.82 and 174.41; basin perimeter (P), between 13.45 and 88.89 km; area above 30° (A_{30}), between 1.4 and 48.8%; total channel length (L_S), between 25.7 and 306.5 km; drainage density (D_d), from 1.1 to 3.4; main channel length (L) from 4.35 to 17.93 km; mean basin slope (S) from 10.66° to 27.17°; main channel slope (S_L) from 6.7 to 15.2°; altimetric range (H) from 641 to 1183 m; circularity index (Ic) from 0.29 to 0.74 and; hierarchical order (O) from 3° to 6°. In the GCCS territory, the largest basin is the 4th order $B24$, with an area of 184.67 km^2, 88.89 km of perimeter, 17.93 km of main channel, and average slope of 17.62°. The smallest basin (*B*6, with an area of 10.48 km^2) has the smallest perimeter (13.45 km) and main channel (4.71 km) values, average slope values (14.45°), and hierarchical order equal to the 4th order (Table 2).

Among the 66 correlation coefficients generated (Table 3), only 14 significant correlations were obtained: A/P (0.98); A/R_r (−0.75); A/L (0.89); A/L_S (0.91); A_{30}/S (0.93); A_{30}/H (0.75); P/R_r (−0.80); P/L (0.94); P/I_c (−0.78); P/L_S (0.93); R_r/L (−0.88); L/Ic (−0.81); L/L_S (0.84) and; S/H (0.71). The parameters that presented

Table 2 Results of morphometric parameters

Watershed	Morphometric parameter										
	O	A	A_{30}	P	D_d	R_r	L	S	S_L	H	I_c
*B*1	4	17.1	1.5	19	3	150	4.4	11	9.3	652.66	0.58
*B*2	3	15.9	5.7	18	2	105	6.1	11	10	641.18	0.63
*B*3	4	21.5	16	20	3	152	7.5	18	9.4	1138.6	0.66
*B*4	4	35.1	36	28	3	123	9	17	13	1113.3	0.57
*B*5	6	72.8	48	52	3	105	11	27	15	1183,6	0.34
*B*6	4	10.5	8	13	2	137	4.7	14	9.1	651.3	0.73
*B*7	4	51.3	36	42	3	103	11	23	8.8	1136.5	0.37
*B*8	4	19.1	17	22	3	145	7.6	16	8.2	1101.1	0.51
*B*9	4	55.8	28	47	2	81.5	14	19	8.7	1158	0.32
*B*10	5	61.9	30	43	2	95.2	12	23	8.5	1152.5	0.41
*B*11	5	62	27	42	3	81	13	21	9.2	1060.6	0.44
*B*12	4	24.4	4.4	25	3	83.5	9.8	13	6.7	823.55	0.48
*B*13	4	37.1	24	29	2	114	9.4	21	10	1074.4	0.57
*B*14	3	14.4	1.8	16	2	140	5.2	12	7.4	735.2	0.74
*B*15	5	89.2	21	62	2	63.3	16	18	15	1015.1	0.29
*B*16	3	26.9	1.6	29	1	88.8	10	11	9.6	888.85	0.42
*B*17	3	17.9	5.4	24	2	118	7.8	14	12	917.22	0.40
*B*18	4	19.7	41	21	2	164	6.2	26	14	1020.4	0.57
*B*19	4	12.1	24	17	3	174	5.7	19	10	992.41	0.54
*B*20	3	13.5	32	20	2	149	6.9	23	12	1018.7	0.45
*B*21	3	21.5	26	24	2	136	7.5	19	11	1027.1	0.46
*B*22	3	20.3	18	23	2	130	7.8	18	7.1	1012.5	0.48
*B*23	4	102	9.5	58	2	71.1	15	11	8.3	1038.9	0.38
*B*24	4	185	20	89	2	54.8	18	18	8.4	982.97	0.29
*B*25	3	53.2	15	38	1	89.4	11	19	9.5	977.62	0.47
Minimum	3	10.5	1.5	13	1	54.8	4.4	11	6.7	641.18	0.29
Avarage	4	42.4	20	33	2	114	9.5	18	10	980.57	0.48
Maximum	6	185	48	89	3	174	18	27	15	1183.6	0.74
Standart deviation	1	39.1	13	18	1	32.9	3.6	4.7	2.3	163.08	0.13

Table 3 Result of Pearson's correlation coefficients generated for the 12 morphometric parameters used

	O	A	A_{30}	*P*	D_d	R_r	L	S	S_L	H	I_c	L_s
O	–	–	–	–	–	–	–	–	–	–	–	–
A	***0.41***	–	–	–	–	–	–	–	–	–	–	–
A_{30}	***0.52***	*0.19*	–	–	–	–	–	–	–	–	–	–
P	***0.47***	**0.98**	*0.26*	–	–	–	–	–	–	–	–	–
Dd	***0.54***	*−0.2*	*0.26*	*−0.17*	–	–	–	–	–	–	–	–
Rr	***−0.3***	**−0.75**	*0.07*	**−0.8**	*0.23*	–	–	–	–	–	–	–
L	***0.44***	**0.89**	*0.24*	**0.94**	*−0.181*	**−0.88**	–	–	–	–	–	–
S	***0.47***	*0.16*	**0.93**	*0.22*	*0.15*	*0.1*	*0.2*	–	–	–	–	–
Sl	***0.32***	*0.02*	***0.509***	*0.09*	*0.02*	*0.1*	*0.03*	***0.411***	–	–	–	–
H	***0.43***	***0.321***	*0.75*	0.411	*0.14*	*−0.15*	***0.5***	**0.71**	*0.22*	–	–	–
I_c	***−0.3***	***0.321***	***−0.344***	−0.78	*0.16*	**0.67**	**−0.81**	***−0.31***	*−0.19*	−0.54	–	–
Ls	***0.67***	0.91	***0.396***	0.93	*0.15*	**−0.7**	**0.84**	***0.328***	*0.131*	***0.44***	*−0.67*	–
Legend:	*Weak*		***Mean***		**Strong**							

Legend: hierarchical order (*O*); basin area (*A*); percentage of the area above 30°(A_{30}); perimeter (*P*); drainage density (D_d); relief ratio (R_r); length of the main channel (*L*); average basin slope (*S*); average slope of the main channel (S_L); altimetric amplitude (*H*); circularity index (I_c); total length of the channels (L_S)
Bold value represents mean and strong results of Pearson's correlation coefficient

the highest significant relationship with the others were the area (A) and perimeter (P), which also presented a strong positive relationship with each other and with the parameters main channel length (L) and total length of the channels (L_S), besides a strong and inversely proportional relationship with the relief ratio (R_r). Still, the perimeter (P) showed a strong negative relationship with the circularity index (Ic), which in turn, also has a strong inversely proportional relationship with the length of the main channel (L).

The area above 30° (A_{30}) showed a strong positive relationship with the altitude range (H) and with the average basin slope (S), however, the relationship with the slope of the main channel (S_L) was moderate, demonstrating the importance of using these two parameters (S and S_L) separately. Still, the main channel slope (S_L), as well as the drainage density (D_d), did not present any significant correlation with the other parameters. The highest correlation coefficients obtained were: A/P (0.98), L/P (0.94), A_{30}/S, and L_S/P (0.93).

The results of the morphometric parameters indicate typical characteristics of basins located in steep areas (with high values of $H, A_{30,}$ and S) that have a clear structural control, reflected directly in the results of D_d and Ic. The relatively low/median values of D_d in an area with high rainfall index, and shallow and little permeable soils, demonstrate the strong fluvial incision in the fault and fracture lines since this parameter is strongly conditioned by the structures (Morisawa, 1962). In the same way, the low values of I_c, inversely proportional to P, expose the strong structural control over the drainage in the study area, because, in the basins where the main channel is fitted in the lineaments, the drainage tends to follow a rectilinear trace in downstream direction, resulting in a more elongated format for the basins, contrary to those that present little structural control, and present more circular formats. It is worth noting that the basins with higher I_c values have a larger area in the Serra Geral Plateau, where the sedimentary lithologies of the Rio do Rasto and Botucatu Formation predominate, which, in turn, present a lower concentration of lineaments if compared to the Serra Geral Formation (Fig. 3).

The values obtained for the criticality index vary from 1.0 to 8.7, with $B18$ being the most susceptible to the development of runs and $B16$ the least susceptible (Fig. 4). Following the order of criticality, the basins $B16, B1, B14, B12, B17, B2, B23$, and $B6$ were classified as **Low Susceptibility,** with criticality index ranging from 1.0 to 2.7. In general, these basins present small area above 30° (A_{30}), with the smallest values among the basins, varying from 1.45 to 9.51%, and smaller altimetric amplitude (H), with values between 641.2 and 1.039 m, varying from 641 to 917 m for the majority of these basins, with exception of the values obtained for $B23$. The other parameters have varied results, with area inverse ($1/A$) between 0.01 and 0.095, the highest being obtained for $B6$, the smallest basin in the study area; mean slope of the main channel (S_L) between 6.7 and 12°, with the lowest value obtained in this class and for the other GCCS basins, for $B12$ and; circularity index inverse ($1/I_c$) between 1.35 and 2.63.

The basins with criticality index between 3.4 and 4.5 ($B25, B3, B24, B8, B22$, and $B15$) are the **Medium Susceptibility basins**. These basins have varied characteristics: area above 30° (A_{30}) between 15.28 and 20.73%, with values lower than the average

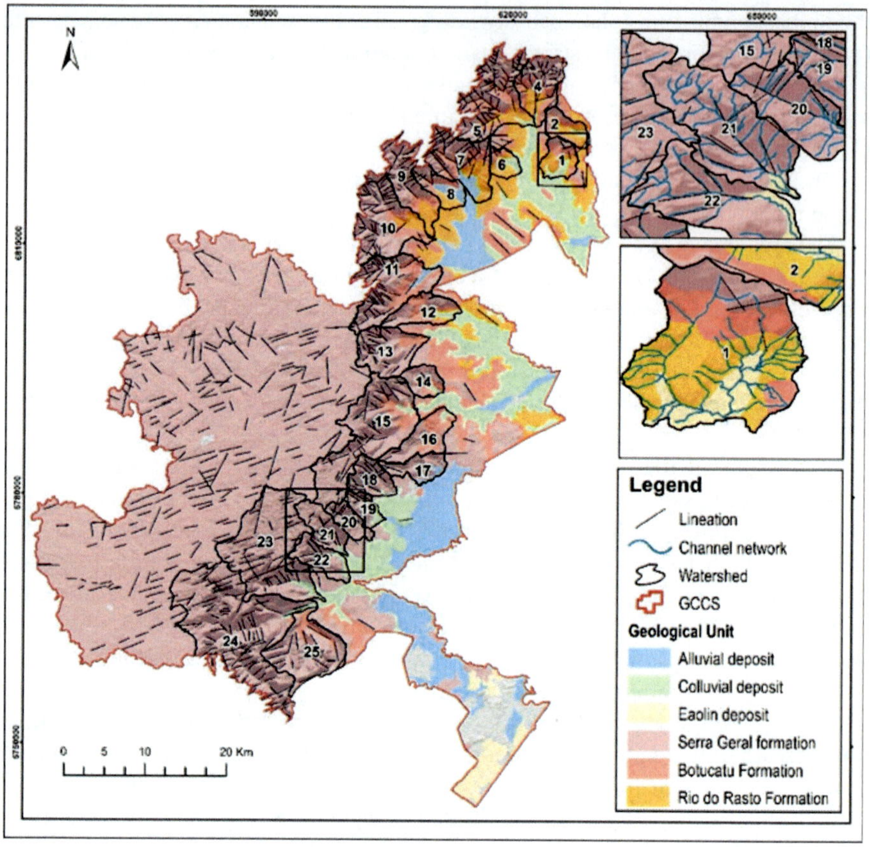

Fig. 3 Geology and structural lineaments of the study area

of the basins, with the exception of the result obtained for *B*15 (20.7%); inverse of the basin area (1/*A*) between 0.005 and 0.052; altimetric amplitude (*H*) between 977.6 and 1,139 m; average slope of the main channel (S_L) between 7.1 and 15°, with the highest value for *B*15, which also presented the highest value in relation to the other basins of the study area and; inverse of the circularity index ($1/I_c$) between 1.52 and 3.45.

The **High Susceptibility** basins obtained a criticality index between 5.1 and 6.3: *B*13, *B*11, *B*19, *B*9, *B*21, and *B*10. These basins have area values above 30° (A_{30}) between 23.61 and 27.58% and altimetric amplitude (*H*) between 992.41 and 1157.95 m, with both parameters for this class with values above average. The parameters mean main channel slope (S_L), inverse of basin area (1/*A*), and inverse of circularity index ($1/I_c$) presented varied values: S_L between 8.67 and 10.67; 1/*A* between 0.016 and 0.047 and; $1/I_c$ between 1.75 and 3.13.

Finally, the basins classified as **very high susceptibility** are the basins B20, B5, *B*4, *B*7, and *B*18, which present a criticality index between 7.1 and 8.7. In general,

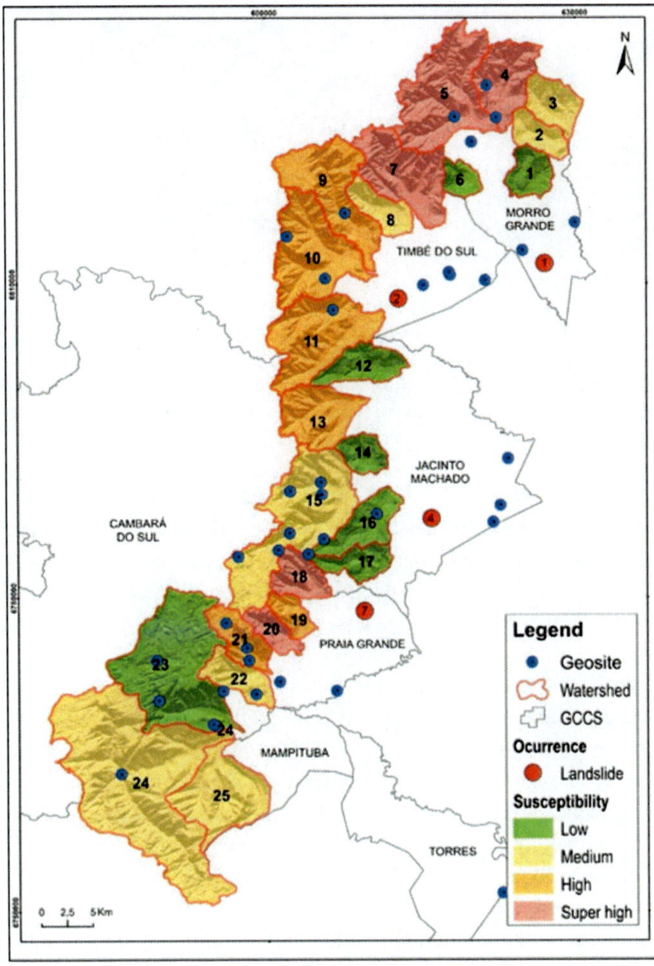

Watershed	Criticality
B18	8.7
B4	7.4
B7	7.4
B5	7.4
B20	7.1
B10	6.3
B21	5.8
B9	5.8
B19	5.7
B11	5.6
B13	5.1
B15	4.5
B22	4.3
B8	4.2
B24	4.2
B3	3.9
B25	3.4
B6	2.7
B23	2.2
B2	2
B17	1.9
B12	1.5
B14	1.3
B1	1.1
B16	1

Fig. 4 Result of the classification of susceptibility to the development of debris flows for the drainage basins. On the right, location of the basins, geosites, and number of landslide occurrences per municipality. On the left, criticality index of the basins and susceptibility classification

these basins present high area above 30° (A_{30}), with the highest values among the basins, varying from 31.82 to 48.06%, and high altimetric amplitude (H), with values between 1018.72 and 1183.62 m. The results of the other morphometric parameters obtained for this class were: average slope of the main channel (S_L) between 8.82 and 14.57°; inverse of the basin area ($1/A$) between 0.014 and 0.074 and; inverse of the circularity index ($1/I_c$) between 1.75 and 2.94.

The parameters that had the greatest weight in determining Very High Susceptibility basins are: A_{30}, H and S_L. Considering that debris races are processes induced by gravity (Costa, 1984), the altimetric amplitude and the slope are closely related to the susceptibility to the development of races, and it is expected that basins of

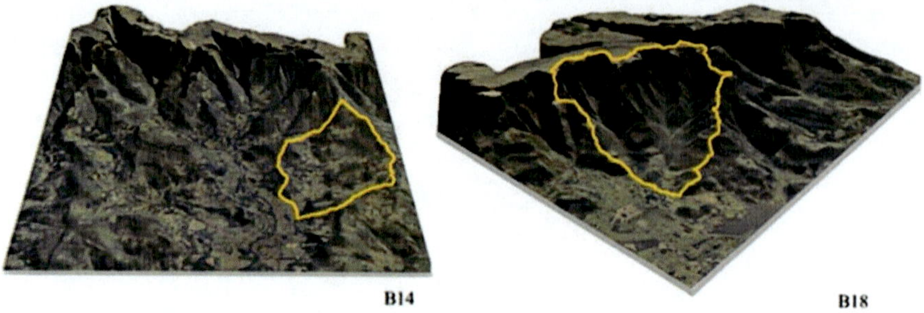

Fig. 5 Three-dimensional representation of basin $B14$, classified as low susceptibility, and basin $B18$, classified as very high susceptibility, representing the basins of these classes. While the basins of low susceptibility tend to present a more circular format and lower slope, the basins of very high susceptibility, in general, present more elongated formats and steep slopes

higher susceptibility present more critical values for these parameters (Fig. 5). Thus, the larger the basin width, the greater the distance traveled by the run and, therefore, the greater the amount of material remobilized by the flow, increasing its destructive power. Gomes (2016) found that basins with occurrence of events of greater magnitude have higher average values of H than basins with evidence of runs of lower magnitude.

The values A_{30}obtained demonstrate characteristics related to the morphology of the relief, indicating the percentage of the steep area. Several authors have attributed critical slope values to the initiation of runs (Costa 1984; VanDine 1996; Kanji and Gramani, 2001), however, according to the methodology proposed by Gramani et al. (2005), the critical slope values are those above 30°. For the basins classified as Very High Susceptibility, the values of A_{30}demonstrate that the basins in this class have approximately 1/3 of the area located above 30°, with almost 50% for $B5$.

When considering the occurrences of mass movements in the municipalities of the GCCS, it is observed a certain tendency that the basins with the highest susceptibility to the development of landslides are located in the municipalities with the highest number of landslides occurrences. The city of Praia Grande has a record of seven landslide occurrences, with two basins classified as Very High Susceptibility ($B18$ and $B20$) and two as High Susceptibility. In the municipality of Jacinto Machado, where four cases of landslides were registered, only one ($B13$) of the six basins of the municipality was classified as High Susceptibility. In Timbé do Sul, which registered only two occurrences, of the six existing basins, three were classified as High Susceptibility ($B11$, $B10$, and $B9$) and one as Very High Susceptibility to the development of runs ($B7$). Finally, in the municipality of Morro Grande, which presents only one record of a landslide, two basins ($B4$ and $B5$) were classified as Very High Susceptibility.

Although the basins with the highest susceptibility are located in areas with a record of landslides, the relationship between the classification of the basins and the record of mass movement occurrences in the GCCS has limitations. The survey of the occurrences of extreme events in the municipalities of the GCCS between 1974 and 2017 (Pimenta et al., 2018) was carried out based on the collection of secondary data in order to elaborate a history of the occurrence of these events. However, the inventory does not present the coordinates or sub-basins affected, with the locality (by neighborhoods) cited only in some records, and the typologies of mass movement process are not differentiated, being referred to generically as landslides. In addition, it is worth noting that the records of these events are made on the basis of the human factor, i.e., for these events to be recorded, the report by the population is necessary, inferring that the places with greater population densification also have a greater number of records, while in protected areas, the records tend to be scarcer. It is noteworthy that the areas of high slope, which, as shown, have greater susceptibility to the development of debris runs, are protected areas or areas with limiting factors to human occupation. Thus, it is estimated that the number of occurrences of mass movements is much higher than the inventory.

Finally, the result of the classification of the basins' susceptibility to the development of debris flows was carried out from the weighting between the parameters, so that the result of the classification does not result in a real morphometric susceptibility to the occurrence of runs, but a hierarchization of the drainage basins considered. In this way, the basins of low susceptibility do not necessarily have a low susceptibility to the occurrence of this process, but they have a lower susceptibility in relation to the other basins of the area, considering that the whole area of the GCCS has evidence of mass movements (Fig. 6) (Duarte, 1995).

In relation to the method used for evaluating the susceptibility to the development of debris flows, based on the criticality index proposed by Gramani et al. (2005), it should be noted that the methodology was developed based on the experience of these authors in the geological–geomorphological compartment of the Serra do Mar and prepared for use in this same context. Thus, the results of the susceptibility to runs should be used with caution. However, despite the results showing a relative susceptibility between basins, it is understood that the basins of the study area are naturally susceptible to these processes since several evidences are found, from landslide scars and colluvial-alluvial deposits. In addition, many of the basins classified as Very High and High Susceptibility have geosites present in their drainage area, demonstrating the importance of identifying the susceptibility of the basins, including for the safe visitation of visitors to the GCCS.

5 Final Considerations

The determination of the susceptibility to the development of debris flows in the GCCS shows that most of the basins have characteristics that indicate a high susceptibility to the occurrence of this process. In the same way, the records of occurrence of extreme events in the territory of the GCCS, as well as the existing evidence, corroborated this result, demonstrating that the occurrence of these processes is not uncommon in the study area, even though the inventory presents limitations.

Fig. 6 Evidence of hydrogeomorphological processes of great magnitude in the territory of the GCCS. **A** Dry valley of a channel abandoned from the occurrence of torrential events. Photo: Sugiyama, 2021; **B** Deposition of pebbles and boulders away from the scarps, demonstrating fluvial reworking in torrential events. Photo: Jairo Valdati, 1998; **C** Subsurface deposits demonstrating the recurrence of hydrogeomorphological processes in the GCCS territory. Photo: Jairo Valdati, 1998

The morphometry of the study area demonstrates the importance of watercourses in the configuration of the relief, from processes of fluvial incision and adjustment of the base level, which in turn condition the greater or lesser energy of the hydrogeomorphological processes. The parameters that presented the greatest weight in the determination of the Very High Susceptibility basins were A_{30}, H, and S_L, and it is expected that the basins of greater susceptibility will present more critical values for these parameters since debris flows are processes induced by gravity and are closely related to the slope and altimetric gradient of the basin.

The method used to evaluate the susceptibility to the development of debris flows has limitations, mainly due to the fact that it was developed in areas with different physiographic characteristics. Future works involving the identification in the field of evidence of the occurrence of debris flows of different magnitudes may contribute to the validation of the susceptibility determined from the morphometric parameters.

References

Cabral VC, Reis FAGV, D'Affonseca FM et al (2021) Characterization of a landslide-triggered debris flow at a rainforest-covered mountain region in Brazil. Natural Hazards 108:3021-3043. https://doi.org/10.1007/s11069-021-04811-9

Castro ALC (2003) Manual de desastres: desastres naturais. Ministério da Integração Nacional, Brasília

Christofoletti A (1980) Geomorfologia. Editora Blucher, São Paulo

Cherem LFS, Faria SD, Zancope MHC et al (2020) Morphometry of basins. In: Barros LFP, Junior APM (ed) Hidrogeomorfologia: Formas, processos e registros sedimentares fluviais. Bertrand Brasil, Rio de Janeiro, p 175 - 217

Corrêa CVS (2018) Morphometric modeling to evaluate the potentiality of watersheds to debris flows: proposal applied in Caraguatatuba (SP) and São Sebastião (SP). Ph.D. thesis, Universidade Estadual Paulista, São Paulo

Costa JE (1984) Physical geomorphology of debris flows. In: Costa JE, Fleisher JP (ed) Developments and applications of geomorphology. Springer-Verlag, Berlin, p 268 - 317

De Scally F, Slaymaker O, Owens I (2001) Morphometric controls and basin response in the cascade mountains. Geografiska annaler 83 (3):117-130. https://doi.org/10.1111/j.0435-3676.2001.00148.x

Dias VC, Gramani MF, Vieira BC (2016) Morphometric and morphological parameters as indicators of the magnitude of debris flow in Serra do Mar Paulista. Confins 29: 1-20. https://doi.org/https://doi.org/10.4000/confins.11444

Duarte GM (1995). Coastal Cenozoic deposits and the morphology of the extreme south of Santa Catarina. Ph.D. thesis, University of São Paulo

Gomes MCV (2016) Debris flows and the long-term denudation rates of Serra do Mar/SP. Ph.D. thesis, University of São Paulo

Gramani MF, Olivito JPR, Filho AO et al (2005) Analysis of the potential generation of mass runs in the highland sections of the OSBAT pipeline. In: XI Brazilian Congress of Engineering and Environmental Geology, Florianópolis, 2005

Horn Filho NO (2003) Sectoring of Coastal Province of Santa Catarina based on geological, geomorphological and geographic aspects. Geosul 18(35):71-98

Horton RE (1945) Erosional development of streams and their drainage basins; hydrophysical approach to quantitative morphology. Geological society of America bulletin, 56(3), 275 https://doi.org/10.1130/0016-7606(1945)56[275:edosat]2.0.co;2

INMET—NATIONAL INSTITUTE OF METEOROLOGY (2020) Meteorological Database for Teaching and Research—BDMEP. Conventional stations/Historical series/Monthly data. Available at: <http://www.inmet.gov.br/projetos/ network/research/inicio. php>

Jakob M (1996) Morphometric and geotechnical controls on debris flow frequency and magnitude in southwestern British Columbia. Ph.D. thesis, University of British Columbia

Kanji MA, Gramani MF (2001) Methodology for determining vulnerability to debris flows in small hydraulic basins. In: III Brazilian conference on slope stability, Rio de Janeiro, 2001.

Keller EA (1992) Introduction to environmental geology. Pearson, London

Morisawa ME (1962) Quantitative geomorphology of some watersheds in the Apallachian Plateau. Geological Society of America Bulletin 73(9): 1025-1046

Pellerin J, Duarte MG, Scheibe LF et al (1997) Timbé do Sul - Jacinto Machado: Preliminary assessment of the extent of the catastrophe of 12/23-24/95. Geosul 12(13):71-87

Pimenta LHF, Freitas MJCC, Lin SC (2018) Integrated and Participatory Plan for Disaster Risk Management in the Geopark Project Territory of Caminhos dos Cânions do Sul

Pissarra TCT, Politano W, Fernando AS (2004) Evaluation of morphometric characteristics in the soil-surface relationship of the córrego rico watershed, jaboticabal (SP). Brazilian Journal of Soil Science 28(2):297–306

Pontelli ME (2005) Pedomorphostratigraphy of alluvial fans deposits: Itoupava River basin, southern Santa Catarina State. Ph.D. thesis, Federal University of Santa Catarina

Santos YRF (2021) Cartografia geomorfológica de detalhe aplicada ao geopatrimônio: geomorfossítios do projeto Geoparque Caminhos dos Cânions do Sul SC/RS. Dissertation, Universidade Federal de Santa Catarina

Slaymaker O (1990) Debris torrent hazard in Eastern Fraser and Coquihalla Valleys. Western Geography 1(1):34-48

Souza CR (2005) Suscetibilidade morfométrica de bacias de drenagem ao desenvolvimento de inundações em áreas costeiras. Revista Brasileira de Geomorfologia 6(1) https://doi.org/10.20502/rbg.v6i1.38

Valdati J (2000) Riscos e desastres naturais: área de risco de inundação na sub-bacia do rio da pedra - Jacinto Machado/SC. Dissertation, Universidade Federal de Santa Catarina

VanDine DF (1996) Debris flow control structures for forest enginnering. British Columbia ministry of forests research program, Victoria

Vieira BC, Vieira ACF, Fernandes NF et al (1996) Estudo comparativo dos movimentos de massa ocorridos em fevereiro de 1996 nas bacias do Quitite e do Papagaio (RJ): uma abordagem geomorfológica. In: II Pan-american symposion on landslides/brazilian conference on slope stability, Rio de Janeiro, 1997

Zavoianu I (1985) Morphometry of Drainage Basins. In: Developments in water science. Elsevier Science, Amsterdam